D0152814

The Spatial Reformation

INTELLECTUAL HISTORY
OF THE MODERN AGE

Series Editors
Angus Burgin
Peter E. Gordon
Joel Isaac
Karuna Mantena
Samuel Moyn
Jennifer Ratner-Rosenhagen
Camille Robcis
Sophia Rosenfeld

The SPATIAL REFORMATION

Euclid Between Man,
Cosmos, and God

Michael J. Sauter

PENN

UNIVERSITY OF PENNSYLVANIA PRESS

PHILADELPHIA

Copyright © 2019 University of Pennsylvania Press

All rights reserved. Except for brief quotations used for purposes of review or scholarly citation, none of this book may be reproduced in any form by any means without written permission from the publisher.

Published by
University of Pennsylvania Press
Philadelphia, Pennsylvania 19104-4112
www.upenn.edu/pennpress

Printed in the United States of America on acid-free paper
1 3 5 7 9 10 8 6 4 2

Library of Congress Cataloging-in-Publication Data

Names: Sauter, Michael J., author.
Title: The spatial reformation : Euclid between man, cosmos, and God / Michael J. Sauter.
Other titles: Intellectual history of the modern age.
Description: 1st edition. | Philadelphia : University of Pennsylvania Press, [2019] | Series: Intellectual history of the modern age | Includes bibliographical references and index.
Identifiers: LCCN 2018020825 | ISBN 9780812250664 (hardcover)
Subjects: LCSH: Europe—Intellectual life—History. | Space—Philosophy—History. | Euclid—Influence. | Geometry—Social aspects—History. | Space—Social aspects—History. | Europe—Civilization—History.
Classification: LCC CB203 .S28 2019 | DDC 940.1—dc23
LC record available at https://lccn.loc.gov/2018020825

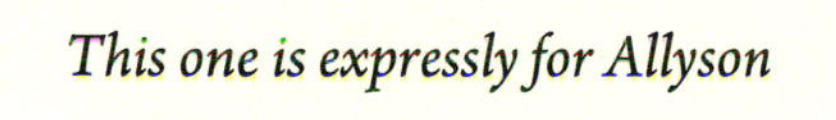

This one is expressly for Allyson

But you must note this: if God exists and if he really did create the world, then, as we all know, he created it according to the geometry of Euclid and the human mind with the conception of only three dimensions in space. Yet there have been and still are geometricians and philosophers, and even some of the most distinguished, who doubt whether the whole universe, or to speak more widely the whole of being, was only created in Euclid's geometry; they even dare to dream that two parallel lines, which according to Euclid can never meet on earth, may meet somewhere in infinity. I have come to the conclusion that, since I can't understand even that, I can't expect to understand about God. I acknowledge humbly that I have no faculty for settling such questions, I have a Euclidean earthly mind, and how could I solve problems that are not of this world? And I advise you never to think about it either, my dear Alyosha, especially about God, whether he exists or not. All such questions are utterly inappropriate for a mind created with an idea of only three dimensions.

—Fyodor Dostoyevsky, *The Brothers Karamazov* (1880), trans. Garnett

Contents

Preface

This book took too long to complete and was supposed to be a study of public clocks and time discipline in the late eighteenth century. These two things are related. Initially, I was interested in how, at the eighteenth century's end, people across Europe yielded the right to define local time to astronomers, an exclusive group whose knowledge and skills were expressly not local, but were tethered to a far-flung community of specialists. My work on this matter appeared in 2007 as a journal article, in which I detailed eighteenth-century Berlin's experience with modern time regimes.[1] At that moment, I intended to write a book on Europe's elaboration of a system of uniform public time. This work would have concentrated on Geneva, London, Berlin, and Paris, since these cities installed the first citywide master clocks.[2] Everything went awry, from there.

In order to pursue the formation of what I understood to be modern time sense, I undertook to study the history of astronomy. (I believe, now, that I was wrong about time discipline's "modernity." This discipline was, in fact, early modern—for reasons that will become clear.) Thus, when I first expanded my inquiry, I was interested in how astronomy's maturation into a continental discipline sustained multiple takeovers of local time regimes. In short, I wanted to explore how time ceased to be a form of "local" knowledge, in the sense that Clifford Geertz has used the term, and became a "scientific" form of knowledge that was incorporated rapidly into the European state's burgeoning disciplinary apparatus.[3]

As I tumbled backward into the history of astronomy, however, I detected similarities in early modern astronomical textbooks with philosophical works that I had previously studied. I recalled, for instance, that the German philosophers Immanuel Kant and Johann Gottfried Herder had deployed in their writings the same astronomical knowledge that I was encountering within eighteenth-century astronomical works. Herder, for instance, dedicated the entire first book of his anthropological work *Ideas on the History of Humanity* (1788–1791) to a description of the cosmos, imagining for his readers terrestrial and extraterrestrial realms that he had never seen, before analyzing a humanity that he could not have fully known. Thus, the great unseen cosmos—to which Herder had access through books, scientific instruments, and the university lectures that he heard in Prussian

Königsberg—was an essential (and essentially imagined) frame of reference for his anthropology. When I noticed that the initial volume of Herder's work appeared one year after Berlin, Prussia's capital city, had installed its first master clock, it appeared that astronomy was behind a much broader change.

Further research in the history of anthropological thought revealed that Herder was not alone in projecting an unseen cosmos as the essential backdrop for a proper study of humanity. The same phenomenon appeared in other fundamental eighteenth-century works of anthropology, including *Natural History* (1749–1804) by the Comte de Buffon, "Essay on Man" (1734) by Alexander Pope, and *Reasonable Thoughts on the Workings of Nature* (1725) by Christian Wolff.[4] Moreover, as I examined seventeenth-century anthropology and, in turn, looked to its sixteenth-century predecessor, I observed that an imagined cosmos usually framed the attending anthropological study. Against this backdrop, it immediately became important that eighteenth-century cosmology was expressly heliocentrist, while its sixteenth-century predecessor remained geocentrist. In this respect, the shift from geocentrism to heliocentrism indicated that changes in cosmology might have wrought changes in anthropology. My pursuit of the potential interconnections between the two disciplines became, in the end, Chapter 6, "Strangers to the World," in which I explore the background to (and significance of) anthropology's embrace of heliocentrism.

This project's conceptual starting point thus forms the resulting monograph's final chapter. I note this fact in order to elucidate how an investigation into the rise of public clocks metastasized into a history of spatial thought. This book was researched and written backward. The various astronomical texts that I read pulled me back *nolens volens* to earlier times and into other disciplinary fields. It turned out, for example, that eighteenth-century and seventeenth-century textbooks in astronomy linked the discipline to geography, often claiming cheekily that geographers would be lost were astronomers not around to tell them where they were. Nor did geographers dispute the point. Subsequent research into the terrestrial discipline, beginning with the eighteenth century and moving back into the seventeenth, then led me into yet more branches of learning, including sixteenth-century cosmography and fifteenth-century globe making. Thus, I began to see how astronomers, geographers, cosmographers, and globe makers *produced* collectively what Europeans could not see. Not coincidentally, I became lost in space.

While pursuing what struck me as interrelated traditions of imagining unseen realms, I noticed that texts from the varied disciplines paid homage to geometry. More significantly, perhaps, the works in question often recommended that readers study early modern geometry's cornerstone, Euclid's *Elements*. That the exhortations were made so easily and so regularly suggested, in turn, that ancient geometry had already become fundamental to early modern thought, before the astronomical issues that I was pursuing rose to prominence. Thus, I began to formulate—fitfully and amid frequent bouts of swearing—a "geometric" reading of European intellectual history that produced (in reverse order) Chapter 5, "Modest Ravings," Chapter 4, "Eden's End," Chapter 3, "Divine Melancholy," and Chapter 2, "The Renaissance and the Round Ball." In each chapter I explore how Euclidean geometric space's reception not only changed the discipline in question, but also wrought a broad consensus: that geometry undergirded all forms of natural knowledge.

The unraveling of my project on public clocks did not, however, end there. Further research on Euclid's early modern career revealed just how widely his *Elements* had diffused, as I discovered that no important early modern figure, from Leonardo da Vinci to G. W. F. Hegel, failed to read ancient geometry's jewel. Nevertheless, although I had a sense for geometry's significance, I had no explanation for why its rise would have wrought such profound changes. In order to address this problem, I initiated the research that culminated in Chapter 1, "From Sacred Text to Secular Space." In this chapter I trace how Euclid's *Elements* diffused and why its cultivation of *nothing* clashed with the anthropological, cosmological, and theological traditions that long suffused medieval thought. I concluded, thus, that geometry's history suggested a profoundly different take on European intellectual history's scope and development.

The final break with my original project came with the realization that Euclid's career identified both the beginning and the end of a deeper transformation. Additional investigations revealed that prior to 1350, Euclid's *Elements* was read quite differently than it was after 1350. Medieval students of geometry absorbed a relatively narrow view of space from the *Elements*, because they usually concentrated on the initial six books, which cover plane geometry alone. Early modern readers, in contrast, delved increasingly into books seven through thirteen, which cover the continuity of number and explicate the foundations of spherical geometry. Concomitantly, as I incorporated this discovery into my narrative, I came to appreciate more fully the significance of post-1850 changes in geometry. Euclid's reign ended when mathematicians invented systems of spatial reckoning that functioned beyond Euclid's vision of three-dimensional space. Thus, it seemed possible to organize a history of European thought with respect to Euclid's rise and decline.

The problem remained, nevertheless, to create an analytical approach that captured both the depth and breadth of Euclid's effects. Additional research in ancient and medieval intellectual history cast light on an intellectual framework that I call the Western triad of anthropology, cosmology, and theology. I cannot trace this triad's long history in this book, but I offer instead an introductory sketch on which I will expound in a subsequent work.[5]

From the beginning of the Christian era until about 1350, anthropology, cosmology, and theology were locked in an alliance that required that a change in any one (say, a new emphasis in the vision of God) be matched by changes in the other two. Although the Western triad was not initially hostile to homogeneous space, it became so as a consequence of Neoplatonism's rise in the late third century AD, which (crucially, for my story) cultivated a hierarchy of being. I cannot properly explicate the philosophical issues here. Instead, I must simply assert that hierarchical being and homogeneous space did not mix well. Thus, when early Christian thought absorbed Neoplatonism's vision of a great chain of being, in the fourth century, this guaranteed Euclidean space's long marginalization within Western thought.

Once I had formulated the triad as an analytical frame, I began to see why Euclid's full return—comprising both the *Elements*' planar and its spherical space—had wrought such profound changes. On the one hand, this space slowly dissolved the triad's conceptual bonds and, in so doing, inspired transformations across multiple areas of thought. On the other hand, space's profusion through the triad altered each of its elements so thoroughly that, by 1850, the three could not be allied again. Indeed, in the end, anthropology was left to stand alone, as the European vision of Man became so expansive that it simply absorbed both the cosmos and God.

The pursuit of Euclidean space across time could not explain, however, why contemporary scholarship has underemphasized this phenomenon's effects. Continuing my study of geometry beyond the year 1850 suggested a potential answer: by about 1900, Continental philosophy had incorporated non-Euclidean geometry so fully that three-dimensional space lost its *philosophical* significance. This transformation had one profoundly important effect. As I detail in the conclusion, postmodern thinkers followed modern philosophers in denying philosophical validity to Euclidean visions of space—and for exactly that reason, postmodernism has greatly underestimated geometry's *historical* significance. In effect, on account of developments in modern mathematics, postmodern thinkers are blind to precisely the thing that distinguished early modernity from modernity.

Having traced the genesis of this book's interpretive outlines, I turn to what I see as its methodological contributions. As an intellectual historian by training, I have long emphasized the study of printed and written texts. While researching this book, however, I borrowed from historians of science who insist (rightly, in my view) on the significance of material culture to all aspects of early modern culture.[6] This scholarship's potential to send intellectual history in new directions crystalized for me in 2012, during a visit to the Museum of Natural Sciences in Raleigh, North Carolina. I was carrying my four-year-old daughter when she extended her arm and spoke one word, "Erde," which is German for Earth. When I turned, I saw a large terrestrial globe. Obviously, in my enthusiasm I had subjected the poor child to the study of globes. Nevertheless, my moment in Raleigh threw into relief two things. First, knowledge of unseen spaces and places is learned, which makes its acquisition a historical phenomenon. Second, illustrations of unseen things are not simply representations of something real, but are also products of the human imagination. Thus, beneath what many scholars have seen as a history of discovery lies, in turn, a history of humanly elaborated spatial regimes. It is along these lines, that I call particular attention to this book's cover image, because it highlights how there is no separating the history of spatial thought from the pedagogical system that taught (and still teaches) everyone how to imagine the unseen.

All this leads me to this work's title. Given that early modern culture infused space through (and into) *everything,* it seemed appropriate to summarize the concomitant intellectual changes in a term that connected space to its multivalent anthropological, cosmological, and theological resonances. Thus, I have chosen the title *The Spatial Reformation* in order to capture the multifarious churning that Euclid brought to early modern thought, before modern mathematics superseded his vision of space. Both this work's title and the analyses within each chapter represent not the discovery of something new, but the recovery of something old. I suggest, in short, that our ostensibly postmodern gaze now trace a line within a fully realized three-dimensional space that (for me) was first demarcated by a little girl's extended arm.

Introduction

The Spatial Reformation

> Thus man is God, but not absolutely, because he is man; humanly, therefore, he is God. Furthermore, man is a world, but not comprising All, because he is man.
>
> —Nicholas of Cusa, "On conjecture" (1443), in *Opera omnia: Iussu et auctoritate Academiae Litterarum Heidelbergensis ad codicum fidem edita*, ed. Hoffman and Klibansky

Nicholas of Cusa's characterization of humanity suggests an unusual starting point for a history of European thought between 1350 and 1850: God's departure from Europe's intellectual center, and humanity's corresponding claim to that spot. Cusa did not initiate the swap itself, but he sensed its onset and also intuited the most significant among its implications, namely that to celebrate humanity was to alter the relationship between Man and God. Writing in 1603, William Shakespeare hit upon a similar theme in Hamlet's exclamation: "What a piece of work is a man! How noble in reason! how infinite in faculty! in form, in moving, how express and admirable! in action how like an angel! in apprehension how like a god!"[1] The prince of Denmark was not entirely taken with the human species, of course. Nevertheless, Shakespeare and Cusa were mightily impressed, as evidenced by how readily they associated human creativity with the divine.[2] And neither Cusa nor Shakespeare was unique in this, but both participated actively in a broader rise in human assertiveness—and, perhaps, in arrogance, as well. If understood with these issues in mind, these writers call attention to a theme that cut from the fourteenth century into the nineteenth, the rise of Man and the decline of God.

Having pinpointed an effect, I turn to what I see as its primary cause: the complete return of idealized, homogeneous geometric space.[3] Beginning in the second half of the fourteenth century, early modern thought slowly reappropriated the three-dimensional space that lurked within Euclid's *Elements*. Medieval thinkers had also studied Euclid's greatest work, but they generally concentrated on the planar geometry in the *Elements'* first six books.[4] Early modern thinkers, in contrast, increasingly studied the entire *Elements*, including especially its final three books, which cover spherical geometry. Armed with a fuller view of space, they applied the new whole jointly to Heaven and Earth. As a result, early modern culture began to *see* a variety of things from a perspective that had once pertained exclusively to God, from "above."[5] The cultural effects were profound. In 1611, John Donne wrote:

> On a round ball
> A workeman that has copies by, can lay
> An Europe, Afrique, and an Asia,
> And quickly make that, which was nothing, All.[6]

The Bible had long taught that God created the cosmos *ex nihilo*. Thus, the new three-dimensional mode of representation sustained a human perspective on Creation that was, as it turned out, frighteningly akin to God's view, insofar as humanity both commanded *nothing* and dared to transform it into *All*.

Donne's discussion of globe making calls attention to a theme that has yet to be integrated into the study of European intellectual history, namely the exceedingly productive relationship between projected geometric space and material culture. As the quotes above suggest, early modern Europeans recalibrated the relationship between humanity and God through the transmutation of imagined space into both printed works and material representations of unseen things. For example, although Nicholas of Cusa never saw a terrestrial globe, we know that he owned a celestial one, with which he subjected Heaven itself to his own (all too human) perspective.[7] Cusa obviously never contemplated the celestial realm from *above*, any more than Donne had gazed onto the earth. Yet both the former and the latter accepted that humanity could study and treat as a source of knowledge an imagined and geometrically infused perspective on Creation.

Cusa and Donne's experience with material objects reveals the cultural depths at which Euclid's reception worked. In addition to being a creative process, the projection of space was a mathematical one, insofar as people needed geometry and arithmetic in order to produce (or to consume) books and objects that dealt with the great unseen. With a panoply of material items on hand, the educated elite learned to imagine the cosmos in greater detail. This process inaugurated, in turn, a progression of continual orientation and reorientation that ultimately transformed European anthropological traditions. As time passed, for example, European thinkers came to believe that geometry's core concepts were ensconced inside the human mind itself, as opposed to the physical cosmos—and this further shifted extant visions of humanity's relationship to both God and the cosmos. Thus, to reflect on any spatial theme, including mere terrestrial globes, became a way to reflect on humanity and its ostensibly marvelous abilities. The rise and diffusion of a fully Euclidean sense of space was, therefore, a first step in what became the thoroughgoing humanization of European thought and culture.

It is important to underscore that I am pursuing a specifically intellectual historical study, because the early modern mathematizing of space has been an important theme within the history of science. The great French scholar Alexandre Koyré explored in multiple works, for example, the philosophical problems that the new space posed for medieval Aristotelian cosmology.[8] As Koyré saw it, the shift from the medieval cosmos to a modern successor—from the closed world to the infinite universe—emerged from geometric space's gnawing marginalization of venerable philosophical assumptions about the cosmos. In essence, Koyré held that geometry's spatial continuity rendered the traditional Aristotelian view of space and its relationship to the cosmos untenable. (I have more to say about this issue below.) Yet, as correct as Koyré was, his argument left an ampler context unexplored, since geometry wrought many and various changes in not only anthropology but also theology—to name only the most prominent areas of thought that were affected. Moreover, if we include these two realms of thought within our purview, we see that changes in anthropology and theology not only preceded changes in cosmology, but also may have produced them.[9] In this sense, the rise of a new spatial regime changed much more than just an ancient perspective on the physical cosmos.

With this fuller intellectual historical context in mind, I concentrate on the tension that homogeneous space injected into the European vision of humanity's relationship with God. Against a specifically medieval backdrop, the cultivation of a mathematical foundation for knowledge of the cosmos elevated humanity at God's expense. Nicholas of Cusa studied both mathematics and astronomy and, not coincidentally, his vision of humanity's creative capacities was quite august. He wrote in *On conjecture*, the work from which this chapter's epigraph is taken:

> For it is not that [Man] reaches beyond himself, while he creates, but where he unfolds his virtue, he extends himself from himself. And he does not make any one thing new, but through the whole, which he clearly brings forth, obtains knowledge of what was in himself. For we have said that everything exists humanly in itself. So just as humanity's virtue is to be able to advance toward the whole, [it is] in this wise a whole in itself, nor is to proceed to the illuminated whole through the wonderful vigor in itself anything other than to fold the universal humanly into itself.[10]

Cusa's profoundly abstract exploration of human creativity's chief virtues constituted a powerful anthropological statement. More significantly, with respect to the arguments that I make in this book, it also justified the production of a material culture that dared to encompass Creation, as the routine reproduction of the *All* sustained ever-grander claims to once-divine forms of knowledge.[11]

Space and Time

Cusa, Shakespeare, and Donne are three examples of a fundamental transformation in European thought that I call the Spatial Reformation. Unlike its famous cousin, the Protestant Reformation, which began in 1517 with Martin Luther's theological rebellion and ended in 1648 with the Peace of Westphalia, the reformation in space started around 1350 and ended around 1850. Although the Spatial Reformation was related to that other reformation, it was, at once, both a broader and profounder intellectual process. Emerging from fourteenth-century debates about God's will, the Spatial Reformation was characterized by the thoroughgoing application of geometry's idealized space to both Heaven and Earth. This process already constituted a revolution in thought, since between 350 and 1350 a fully three-dimensional geometric space had been reserved to the Heavens. The Spatial Reformation extended, in turn, into fifteenth- and sixteenth-century discussions of the cosmos's structures—being crucial to the rise of heliocentrism—before maturing into the physical science and philosophy of the seventeenth and eighteenth centuries.[12] In the nineteenth century, however, this reformation dissolved, when mathematicians reached beyond Euclid and began to think along radically new lines. Euclid's rise and decline denoted, in sum, an epoch in European thought, in the course of which the respective fates of humanity, the cosmos, and God were determined.

To organize an intellectual history against the backdrop of developments in geometric thought challenges contemporary approaches to historical periodization in fundamental ways. I present the specifics further below in this introduction. Here, I sketch a general outline by juxtaposing the Spatial Reformation to what many historians characterize as the early modern period. When understood most broadly, this period runs from 1400 to 1800. (There is also a narrower view to which I return below.) From an intellectual historical perspective, this period includes the ages of the Renaissance, the Protestant Reformation, the Scientific Revolution, the Enlightenment, and, of course, the French Revolution—all of which have been celebrated as emancipatory steps on our collective journey to modernity. If, however, we apply geometric space's history to this expanse, two things become apparent. First, a "long" early modern period is revealed that runs from 1350 to 1850, with fundamental changes in spatial thought demarcating this early modernity's bounds. Second, homogeneous space becomes, following the anthropological logic that I have outlined, the broadest of emancipatory phenomena, since early modern thinkers not

Figure 1. "The Blue Marble," 1972. Courtesy of NASA.

only celebrated geometry but also deployed its idealized space in a way that liberated humanity from both God and His cosmos. In this sense, to concentrate on space opens new vistas onto not only early modernity's emergence, but also its dissolution after 1850.

With the problem of periodization in mind, it is particularly significant that we moderns remain conversant with early modern ways of imagining and producing space. To this day we project much of our cosmos, as contemporary media routinely put before us things that few humans ever truly see. I highlight our connection to this older space through two examples. The first is the "Blue Marble," a contemporary photograph of Earth (see figure 1). Originally taken on 7 December 1972 by the crew of Apollo 17 just as their capsule was leaving Earth's orbit, this image has diffused widely.[13] Indeed, thanks to it (and to other images from NASA), all of us *down here* know what Earth looks like from an *up there* that is beyond our ken. The second example comes from another of Donne's poems, "Good Friday, 1613. Riding Westward." It reads:

> Hence is't that I am carryed toward the West
> This day, when my Soule's form bends to the East.
> There I should see a Sunne, by rising set,
> And by that setting endlesse day beget.[14]

Mixing his era's spatial knowledge with his religious yearnings, Donne juxtaposed his own physical motion with a sacred and immobile place. (One must keep in mind that Donne was a geocentrist.) In this context, the intense juxtaposition of Donne's sense for physical movement with the "rootedness" of his religiosity reveals that his soul bent around the world while his space remained true. And this was merely a first step, as space would develop in a way that ruptured the connections between the sacred and a specific place. As my discussions below of the great French thinker Blaise Pascal show, once heliocentrism *resituated* Earth, the gap between a "rooted" religion and a pervasively mathematical cosmos became impossible to ignore.

By peering back into the early modern period through the "Blue Marble," we can grasp why homogeneous space was so intellectually corrosive. Early modern culture remained deeply Christian, which meant that the Bible's outline in the book of Genesis of a spatially hierarchical, geocentric cosmos, in which God was perched *above* and Man languished *below*, remained a basic frame of reference.[15] The religious ranking of different spaces (sacred *above* and profane *below*) illuminates homogeneous space's theological implications: once the entire cosmos was subject to a single space, the little God was as free to occupy the *up there* and to look back as he was to plunk himself in the *down here* and to look up. Put another way, *above* and *below* no longer had the same divine/mundane resonances, once a human unity enveloped them.[16] Although modern space programs have diffused images that incorporate an extraterrestrial realm, the mental unity that currently envelops both the celestial and the terrestrial dates to the Spatial Reformation's implicit claim to God's mantle.

Against the backdrop of empty space, I now propose to situate the Spatial Reformation against a crucial work on medieval scientific thought. The medievalist Amos Funkenstein has argued persuasively in *Theology and the Scientific Imagination* that late medieval theological speculation about God's nature was essential to the rise of modern scientific protocols.[17] He identified the key change in how a new vision of God sustained new ways of thinking about physical causality and placed this process of rethinking between roughly 1300 and 1700. With Funkenstein's groundbreaking interpretation in mind, I return to the previous quotes from Nicholas of Cusa. It is significant that in this chapter's epigraph Cusa mentioned God at the same moment that he praised humanity's creative power. He had to do these two things together, since the human claim to world-making power comprised an anthropological upgrade—and, in Cusa's time, any shift in humanity's value could only be understood with reference to God's transcendence. We can see why this was the case in the second quote, where Cusa concluded that humanity folded "the universal humanly into itself." Any attempt to retouch the universal mandated a change in the relationship between God and humanity.

Along these lines, one of this book's chief assumptions is that medieval anthropological, cosmological, and theological currents worked together, until homogeneous space resituated the intellectual whole. In order to understand why this was the case, we must circle back to late antiquity. From at least AD 350, when early Christian thinkers, such as Saint Augustine, constructed medieval thought's foundations, the Western intellectual tradition contemplated humanity and God both jointly and hierarchically, with any discussion of the former leading the mind up to the latter (and back down again). Moreover, given God's role as the cosmos's Creator, early examinations of the physical world suggested a triadic relationship between (1) Creator and Creation, (2) Creation and humanity, and (3) humanity and Creator.[18] As a result, in the millennium prior to 1350, humanity, the cosmos, and God became so intertwined that a change in any one mandated revisions to the other two. I call this intellectual frame the Western triad. I cannot trace its early history in this work, but I propose it as a heuristic tool that, when applied to the period after 1350, allows us to understand why homogeneous space became such a prominent intellectual force.[19]

The post-1350 confrontation with homogeneous geometric space drove European thinkers to torque repeatedly the interrelationship among the triad's components. Consider that in the roughly two centuries that followed Cusa's death in 1464, Europe produced in 1543 the heliocentrism of Nicolaus Copernicus, in 1637 the philosophical subjectivity of René Descartes, in 1651 the secular politics of Thomas Hobbes, and in 1687 the physics of Isaac Newton.[20] These achievements comprise a goodly portion of the modern world's intellectual patrimony—and their separate histories have been well written. Rather than retell each story, however, I consider them through their shared effects on the Western triad. Along these lines, I offer the example of Isaac Newton's view of God's relationship to space. In a later edition of the *Mathematical Principles of Natural Philosophy,* Newton reflected on God's physical position:

> [God] is Eternal and Infinite, Omnipotent and Omniscient; that is, his duration reaches from Eternity to Eternity; his presence from Infinity to Infinity; he governs all things, and knows all things that are or can be done.

> He is not Eternity and Infinity, but Eternal and Infinite; he is not Duration and Space, but he endures and is present. He endures forever, and is everywhere present; and, by existing always and everywhere, he constitutes Duration and Space. Since every particle of Space is always, and every indivisible moment of Duration is everywhere, certainly the Maker and Lord of all things cannot be never and nowhere.[21]

Although Newton's words highlighted an important theological issue, they also incorporated anthropological claims that were more than a bit presumptuous. Thus, even as Newton celebrated his God's ubiquity, he also *put* Him into something of an undisclosed location.[22] Humanity now dared to tell God where he was allowed to be.[23]

Newton's almost plaintive attempt to *situate* the divine was a waypoint in a longer anthropological journey. In the eighteenth century, for instance, European thinkers renovated their anthropological traditions with express reference to Newtonian science.[24] Along these lines, no thinker better illustrates homogenous space's effects on the European discussion of humanity's relationship to God than does Immanuel Kant. Kant's greatest work, *Critique of Pure Reason*, which appeared in 1781, transferred to Continental philosophy Isaac Newton's pervasive and uniform sense of space and time. Not coincidentally, Kant recognized that a philosophy that was rooted in homogeneous space imposed limits on reason's ability to contemplate the divine. He wrote, for instance, that when it came to natural knowledge, there was nothing outside of reason: "All our cognition starts from the sense, goes from there to the understanding, and ends with reason, beyond which there is nothing higher to be found in us to work on the matter of intuition and bring it under the highest unity of thinking."[25] Yet, when he turned to God, Kant jettisoned his rational toolkit: "Thus I had to deny knowledge in order to make room for faith."[26] Kant took an important step beyond Newton: although he held that one ought to believe in God, he never suggested that one could *locate* Him, either. Thus, under the guise of a rigorously elaborated epistemological humility, humanity dared to become more independent of the divine.

As heirs to both Newton and Kant, we moderns harbor in our minds the remnants of a spatial imagination that *situated* early modernity's many and varied unseen things. I illustrate the point by returning to figure 1. The reader will likely have noticed that Earth as pictured is "upside down," or disoriented—a word that I use in its original sense of "bereft of an east." This inversion demonstrates, in my view, that we invariably bring our own space to the unseen. I accentuate this point with a question: How do you, the reader, know that this sphere not only looks as it appears in the image, but also that it has an "up"? The latter part of this query is more important than it may seem, since in daily life "up" is constituted by a glance overhead. The lived aspect of the relationship between "above" and "below" means little, however, when we cut our bond with *terra firma*, which is exactly what a terrestrial globe (or an image of Earth) requires that we do. Since most of us do not see Earth from above, when we view either an image or a material object, we cannot but put our planet inside a realm that is beyond our experience—and this collectively imagined spatial realm has a history.

Although we are products of the early modern encounter with space, it is important to recall that we stand beyond this space's most basic assumptions. As this book's epigraph from Dostoevsky's *The Brothers Karamazov* highlights, modern spatial sense is informed by nineteenth-century innovations in mathematics that superseded geometric three-dimensionality.[27] In subsequent chapters I offer more details on both this process of mathematical change and the philosophical clefts that it has produced. The point here, however, is to recognize that the modern embrace of a post-Euclidean space distorts how we see early modern thought. Our modernity is, in short, anchored to a peculiar situation, in which we absorb (and use) the early modern spatial imagination, even as we remain aware that its regularities have been superseded.

The Harvest of Homogeneity

The relationship between early modern and modern spatial senses becomes more obvious when we recognize that *two* ancient works dominated intellectual life between 1350 and 1850, the Bible and Euclid's *Elements*. Both works had been accessible to the medieval culture that preceded the early modern one. Both, however, were read in new ways after 1350—and this had profound implications for European thought up through 1850. The post-1350 rereading of these two very different works flowed from the same source, whose upsurge I discuss just below. Here, however, I emphasize an overlooked problem: the Bible and the *Elements* were profoundly in tension on the matter of space. Consider that the biblical view of the cosmos incorporated a hierarchical arrangement of Heaven and Earth (*above* and *below*), while the *Elements* denied such a hierarchy through its cultivation of an intrinsically uniform idealized space. As a result, when Euclid's thought was applied fully to God's cosmos, a transformation was inevitable.

In order to pinpoint what Euclid's reemergence after 1350 signified, I survey his "initial" return in the twelfth century. The full text of Euclid's *Elements* first returned to Europe in the twelfth century, as complete Latin translations of Arabic versions began to appear around the year 1125.[28] (Versions in the original Greek still existed, but they remained in Constantinople.) The immediate effects of the translations were minor, as Euclidean geometry was read against the Bible's hierarchical cosmos, with the result that three-dimensional geometry was applied to the Heavens—and manifestly not to the Earth.[29] The tendency toward hierarchy had a long history, none of which I can reconstruct, except to emphasize that it became fixed in early Christian thought in the fourth century with the absorption of Neoplatonism's vision of hierarchical being.[30] The cultivation of hierarchy became, in turn, more philosophically rigid, when Aristotle's philosophical corpus arose in the twelfth century, or roughly the same time as Euclid's *Elements* returned. Significantly, Aristotle denied spatial homogeneity and cultivated what I will call a fractured approach to space.[31] On the one hand, he held that space was not diffused in a fully geometric sense, but that *spaces* were defined by what contained them. On the other hand, he transferred this idea to the physical cosmos and argued that concentric crystalline spheres surrounded Earth, with each sphere containing its own space. Thus, for Aristotle, the cosmos did not exist within *a* space, but comprised *spaces*.

The high medieval appropriation of Aristotle yielded a Christian hierarchical cosmovision that, from our perspective, was riven by a fundamental spatial dichotomy.[32] Thus, consistent with the Bible's teachings, Heaven, where God resided, was understood to be both *above* and perfect, while Earth, on which humanity dwelled, was *below* and imperfect. The boundary between the two realms was demarcated by the moon's crystalline sphere, with the higher planets borne by other spheres. In this context, it was critical that Aristotle also argued that the Heavens were qualitatively different from the Earth below, because this paralleled Christianity's view that the God resided *above*. Moreover, since medieval Christians believed their God to be omnipresent in the cosmos, they readily absorbed another Aristotelian doctrine, namely that the cosmos was inherently rational. Thus, for Christians, it seemed wholly justified to say that God's cosmological whole could be understood via the divine's diffusion in reason. As a result, to the extent that the Christian Aristotelian cosmos comprised a physical whole, it was rooted not in spatial homogeneity's pervasiveness, but in divine rationality's ubiquity.

Here, I turn to what prominent scholars in the history of science have seen (rightly, in my view) as a powerful source of change in late medieval thought, the profound reimagining of God's relationship to reason. The medieval philosophical-theological bulwarks against Euclidean space were eroded, when, around 1300, European thinkers began to reject "rationalist" approaches to God and to emphasize His will instead.[33] This theological *volte-face* was a complicated process that comprised many critics across Western Europe. I note here only the most important cluster of thinkers, which appeared at the University of Oxford. Between 1300 and 1350, there arose at Oxford a critical tradition, which is called (not altogether correctly) Nominalism. This school *resituated* reason via its emphasis on the divine will and, concomitantly, undermined Christian Aristotelian views of not only the divine's relationship to the cosmos but also humanity's ability to comprehend the divine in rational terms.[34]

Against the Christian Aristotelian backdrop, Nominalism's rise could only have wrought profound changes. Late medieval thinkers, such as Oxford's William of Ockham, began to emphasize the power of the divine will over and above divine reason's ubiquity—and this shift set other important changes in motion.[35] I underscore that the resulting conceptual alterations were theological (and not mathematical), because Ockham expressly limited geometry to substantive things. Thus, when it came to the question of whether points were indivisible, Ockham concluded in his *Quodlibetal Questions* (ca. 1327):

> As regards the arguments for the contrary position, my reply to the first argument for the conclusion that there are indivisibles is that it is not impossible that there should be indivisibles, since an intelligence is indivisible, and likewise an intellective soul. But it does involve a contradiction for an indivisible to exist in a quantum, since it is impossible for either a part of a quantum or an accident of a quantum to be indivisible. Now it has just been demonstrated that no part of a quantum is indivisible. That no accident of a quantum is indivisible is proved by the arguments of certain thinkers, arguments that prove that neither the beginning nor the end of a line is an indivisible accident. And it seems to me that these arguments are conclusive.[36]

Whatever else one may say about Ockham's contributions to science—and there is no doubt that they were significant—his thought was not intrinsically Euclidean.[37] Indeed, later Nominalists had no idea what to do with geometry's nonsubstantial points either.[38] As a result, Euclidean space initially affected only the intellectual margins. With respect to the Heavens, for instance, new approaches to space corroded only traditional notions of heavenly motion's ostensible perfection, while, with reference to the Earth, space highlighted only weaknesses in Aristotle's belief that the Earth was made up of four elements. (I discuss this more below.) Ockham's influence on space's initial rise must therefore be understood as indirect, with his theological critiques opening the door through which homogeneity rushed into the Christian Aristotelian world.

Ockham's theological insights marginalized divine reason in a way that enabled homogeneous space's broader diffusion through the medieval mental cosmos. Keeping Isaac Newton's ideas in mind, let us examine Nominalism's effect on God. The theological emphasis on God's will loosened reason's connection to both God and the cosmos and, as Cusa's musings already suggested, liberated Him from Creation (not to mention vice versa).[39] Consider the following, again from the *Quodlibetal Questions*:

> If God were to make a quantity without any place, then the sort of position that is a differentia of quantity would exist in the absence of any position

> of the sort that is a category—and this is because the quantum would not in that case be present to a place. But if one assumes that a quantum is present to a place, then this is not valid: "God is able to make a quantum without this position, and without that position, and so on for each of the singulars; therefore, he is able to make a quantum without *any* position." An example: If the sun did not exist, then God would be able to make a body without any [of the possible] distances from the sun, since the sun would not exist; but once the sun is posited, it is impossible for there to exist within the world a body that lacks all [of the possible] distances from the sun, and yet a body like this is related contingently to any given distance from the sun.[40]

Ockham's position was traditional, insofar as space became possible, only *after* God had completed his work. (He did have to make *something*, first.) Nevertheless, since the physical whole remained a contingent product of the will, as opposed to an expression of reason, God's *place* within Creation could no longer be justified in wholly "rational" terms. With Christian Aristotelianism's admixture of God and cosmos beginning to separate, late fourteenth-century thinkers were able to pursue different (and increasingly different kinds of) inquiries into the natural world.[41]

Nominalism's Voluntarist Turn opened the European mind to space. As historians of science have detailed, the late fourteenth century hosted a golden age of cosmological speculation that, in turn, hinted at the first stirrings of modern science.[42] Although this radically speculative tradition was significant for science's emergence, it was also part and parcel of the broader reformation in space. For example, in the fourteenth century, two speculative schools emerged that were heavily influenced by Nominalism and that produced important departures in mathematical approaches to space. They are known as the Oxford Calculators and the Parisian Terminists.[43] The first coalesced around Merton College in Oxford; its most famous members are Thomas Bradwardine (1290–1349), William Heytesbury (1313–1373), and Richard Swineshead (fl. 1340–1354).[44] The other was centered on the University of Paris and boasted the likes of Jean Buridan (1295–1363), Albert of Saxony (1316–1390), and Nicole Oresme (1320–1382).[45] All the people whose names I have listed are rightly considered giants of medieval science—and this is because they directed European thought to apply space ever more fully to the natural world.

I cannot reconstruct the significance of the many innovations that resulted, noting only that each of the thinkers combined Nominalism's theological stirrings with new emphases in mathematics. With respect to the Heavens, both Thomas Bradwardine and Nicole Oresme made advances in the study of impetus, as they sought ways to understand motion as being rooted in individual objects, rather than in a rationally infused cosmos, in which things that lay *above* moved naturally in circles.[46] With respect to the Earth, meanwhile, Jean Buridan and Oresme began to apply geometry to the terrestrial realm in a way that suggested that Aristotle's four elements could not be arranged symmetrically—with the heavier soil always beneath the water and the air—but had to be distributed unequally, so that some soil reached above the water.[47] (Otherwise, humanity would have no place to live.) The increasingly geometric approach to natural knowledge then extended up through Pierre d'Ailly (1351–1420) and, significantly, reached into the work of Renaissance humanist Gregor Reisch (1467–1525), who also happened to teach Sebastian Münster, the cosmographer whom I discuss in Chapter 4. None of the thinkers that I have mentioned folded heavenly and terrestrial space into a uniform physical system, as later thinkers would do. Nevertheless, their separate contributions open to scrutiny the early stages of a process in which the command of homogeneity allowed humanity to rethink the *All* that it could not see but that it could put into space.

The Material Culture of Nothing

Against this theological-mathematical backdrop I make two methodological points. First, I suggest that intellectual historians should begin to concentrate more narrowly on how early modern thinkers appropriated their Euclidean spatial toolkit. Second, with this recommendation in mind, I add that the contemporary discipline also needs to confront more directly the material culture of space, since the production and manipulation of objects played an important role in how people imagined unseen spaces and places. Along these lines, it is significant that John Donne almost certainly saw a terrestrial globe, and that Nicholas of Cusa definitely owned a celestial one, given that these two objects represented, in each instance, one part of the early modern cosmos. As I explain in subsequent chapters, it was the express immersion of *below* and *above* inside a pool of imagined space that drove the production of new perspectives on humanity, the cosmos, and God.

My determined inclusion of material culture within intellectual history has the additional virtue of situating the Spatial Reformation within historical time. I illustrate this point by presenting three examples, all of which come from Nuremberg, a city in central Germany that played an important role in diffusing the new spatial culture. The first example appears in figure 2 and is the oldest surviving terrestrial globe. Completed in 1492 by a merchant named Martin Behaim, it is known in German as the Erdapfel, or "terrestrial apple."[48] Its cartographic representations comprise only the Old World, since Christopher Columbus had yet to return to Europe with news of the new one. Nevertheless, although the Franconian globe maker probably never traveled further west than Portugal—and

Figure 2. Behaim, "Erdapfel," 1492. Globus, Inv.-Nr. WI 1826. Courtesy of Germanisches Nationalmuseum Nürnberg. Photo by Jürgen Musolf.

certainly never hovered high enough to see the terrestrial whole—his globe was oriented "correctly," that is, with respect to an extraterrestrial perspective. Indeed, in many ways, Behaim's Erdapfel is more correct than our "Blue Marble," because, unlike the latter, it is displayed with its axis tilted.

Aside from the Erdapfel's orientation, two other characteristics underscore the homogeneity of this globe's underlying space. First, given its design features, we can be certain that users manipulated this item as we would. They spun it, walked around it, looked down on it from above or up to it from below, and, as a result, they *knew* that Earth had an "up." Moreover, they also probably used this mode of projecting space to locate on that sphere the peoples of whom they were aware, such as Asians and Africans. (This practice's fossilized remains are apparent in our contemporary habit of overlaying racial characteristics onto geography; for example, we denote people as Asians, Africans, and Caucasians even if they were born and raised in Freehold, New Jersey.) Second, by using this globe, Nurembergers also demonstrated that they accepted that the terrestrial surface could be put onto a sphere, and also that this humanly constructed whole could be treated as a source of knowledge.[49] In itself this is already a significant change, as during the millennium prior to Behaim's *floruit* no one thought to put the terrestrial surface onto a sphere. (Medieval maps were circular, while well into the fifteenth century cosmological drawings cultivated a perspective that moved "out" from a flatly rendered Earth rather than "up" into three-dimensional space.) A merchant from the middle of Europe had thus dared to make space's nothing into *All*—and this act underscores that a profound intellectual change was already underway, before Columbus returned.

The early modern encounter with space, however, comprised more than just the construction and contemplation of terrestrial and celestial globes. As European projections of nothing expanded in scope, the human mind was suddenly free to range amid an ever-vaster unseen.[50] Here, I turn to the second example, from *Short Introduction to the Noble Science of Astronomy*, a textbook published in 1708 in Nuremberg by the astronomer Johann Gabriel Doppelmayr.[51] This work provided a celestial tour that, in contrast to medieval approaches, explicitly let go of Earth as a reference point and began, instead, with the sun: "But we begin in the middle of our system and suppose ourselves to stand in the sun's center, [looking outward] with our own eyes and without any hindrance, whether from the Sun's body or its powerful light, so that in every direction we can see before us, just as we do from the earth, all the stars in their proper order and size, because although our Earth moves around the sun in a great circle over the course of a year, we will not sense the slightest change, on account of the immeasurable distances involved."[52] Boldly going where no mind had gone before, the late seventeenth-century astronomer cultivated a more expansive sense of space than did the late fifteenth-century merchant. (Indeed, the former was a heliocentrist, while the latter could not have been.) Nevertheless, Doppelmayr's vision of the extraterrestrial drew from the same well that had irrigated Behaim's work. The astronomer projected a human perspective onto an unseen realm and fixed the whole inside material culture (in this case, his book's illustrations). Indeed, the text's title trumpeted that the work put the cosmos "clearly before the eyes." A human mind that positioned itself above Earth and looked down could just as well travel to the Sun and look back—all without leaving home.

The final example, which appears in figure 3, underscores how the Spatial Reformation reshaped European anthropology. Depicted is what I call an anthropological globe. It was completed circa 1825 in Nuremberg by the globe maker Carl Bauer and actually comprises two items, a tiny terrestrial globe and a set of

Figure 3. Bauer, "The Earth and Its Inhabitants," ca. 1825. Collection of the Utrecht University Museum, inv. no. UM-373. Courtesy of Utrecht University Museum, photo by Peter Rothengatter, HetFotoAtelier.

thirty-two anthropological cards.[53] I do not analyze this set in any detail but highlight two characteristics. First, the people depicted are peasants or burghers, with the individuals wearing local dress and displaying racial features that suggested their geographic origins. Second, there are subtle differences in attitude toward the distinct populations, as Western European peasants and burghers are unarmed, whereas the people from "barbaric" cultures in Eastern Europe and other exotic places burnish weapons.[54] For example, in the images of indigenous Peruvian and Canadian peoples, each figure bears a bow and arrow, while the Englishman is a stylish fellow who sports a top hat and a cane. Bauer's globe regularized diversity precisely by fixing humanity's differences *within space*. Put another way, the continent that had invented the terrestrial globe (and had relocated Earth) now put the world's peoples into their places.

Homogeneous space's historical significance is further exemplified, moreover, in the sometimes disorienting effects of spatial emptiness. One little known example appears in a book on maps by the astronomer Wilhelm Schickard that was published posthumously in 1669 in the German university town of Tübingen. In the introduction, the work's editor suggested that Schickard's text would "giv[e] a hand not only to the traveler but also the homebody that amuses and improves himself by reading works of . . . world history, [should either be] led astray in the darkness and become ensnared in error—in body or mind—and [find himself] at a loss, lost in the world."[55] The experience of local space became so intertwined with the new *global* space that merely reading about Earth's expanse could be psychologically overwhelming. Anyone who studied unseen spaces and places without proper maps on hand risked getting lost inside his own living room.

Meanwhile, the production of *extraterrestrial* space could be equally disorienting. In 1654, the great French mathematician-theologian Blaise Pascal wrote in his spiritual notebook, *Pensées*: "The eternal silence of these infinite spaces fills me with dread. How many realms are unaware of us!"[56] Pascal's astral angst takes on added significance, when we consider the following fragment from the text:

> The year of grace 1654
> Monday 23 November, feast of Saint Clement,
> Pope and martyr, and others of Roman Martyrology
> Eve of St. Chrysogonus, martyr, and others.
> From about half past ten in the evening until about
> Half past midnight.[57]

As did John Donne a few decades earlier, Pascal mixed abstract space with a religiously circumscribed sense for *location*. More significant, however, is that the theological stakes were higher, because Pascal, unlike Donne, was a heliocentrist. By denying Earth a central physical-philosophical spot in the cosmos, he put his Savior's birthplace into motion, and one result was a vertiginous sense of rootlessness. If we follow Pascal's path, that is, if we hurl ourselves into emptiness, we glimpse the Spatial Reformation's profoundest effect: it subordinated *everything* to a humanly constructed space.[58] Humanity was now the little god who trembled.

The Rise and Decline of Nothing

Against a backdrop defined by Pascal's anguish, I turn to the Spatial Reformation's chronology. I divide it into two phases. The first ran from 1350 to 1650 and comprised both the reception of classical homogenous space and its application to the medieval mental cosmos (Heaven and Earth, as we shall see). I cover this process in Chapters 1–3. In Chapter 1, "From Sacred Texts to Secular Space," I explain how, as Euclid's vision of space diffused from the fifteenth century into the eighteenth, Europeans related to the unseen in ways that challenged the Western triad's structures. In Chapter 2, "The Renaissance and the Round Ball," I pursue this space's development by examining the invention of terrestrial globes against the backdrop of Christian Aristotelian space, before tracing global space's career into the mid-nineteenth century. In Chapter 3, "Divine Melancholy," I explore spatial homogeneity's enervating effects on the medieval divine via Albrecht Dürer's "Melencolia I" and also use this print to explain why homogeneity's rise *demanded* new approaches to the divine. With that, I conclude my discussion of the first phase.

The Spatial Reformation's second phase emerged from a remarkable intellectual development: in the early seventeenth century geometry came to be seen as an artifact of the human mind, rather than a constituent element of reality. In effect, after roughly three centuries' worth of cultivating Euclidean geometry, Man the geometer commanded space so thoroughly that the mind rethought (once again) space's relationship to the cosmos and God. I explicate this transition in Chapter 4, "Eden's End," in which I pursue how changes in spatial thought drove a shift from sixteenth-century cosmography, which included both Heaven and Earth, to seventeenth-century geography, which concentrated solely upon the terrestrial realm—and all of this occurred with a space that was defined by humanity. The result was, I argue, that by 1650 the most mythical parts of biblical history, including Eden, had been expelled from terrestrial space, while a newfound "Natural Man" became ensconced inside a geometrically sustained State of Nature.[59]

This second phase of the Spatial Reformation extended, in turn, to 1850 and was distinguished, above all, by a full-blown anthropological revolution. I cover this trend in Chapters 5 and 6. In the former, "Modest Ravings," I use the example of secular politics' rise in the seventeenth century to illustrate how homogeneous space resituated all aspects of early modern thought. Using the traditional trinity of secular politics—Thomas Hobbes, John Locke, and Jean-Jacques Rousseau—I show how "modern" politics was not purely a political phenomenon but was one result of space's simultaneous marginalization of God and promotion of humanity. In the latter chapter, meanwhile, "Strangers to the World," I detail how post-1650 applications of space to extraterrestrial realms led anthropologists to conclude that humanity may not be the only sentient being in the cosmos. Thus, although distant realms probably were unaware of us, it still seemed plausible to believe that other *beings*, at least, were looking back. (Pascal would have been appalled.) This radical democratization of visual perspective within a projected space illustrates, I add, that the most important change in anthropology emerged not from Columbus's uninvited landing in the New World, but from homogeneity's victory over *All*.[60]

Homogeneous space's reign ended around 1850, with the *Elements*' final demotion from core mathematical text to an introductory preface. In the first five decades of the nineteenth century, Euclidean space was itself fundamentally resituated by the rise of new geometries that, as Dostoevsky noted above, were not fit for a three-dimensional mind. I explicate both the broader reformulation of space and its intellectual significance in this book's conclusion, "Prosaic Reflections," where I trace how, between roughly 1820 and 1850, mathematicians' critiques of Euclid's assumptions led to a groundbreaking realization: a wholly consistent and logically rigorous geometry could be constructed on non-Euclidean foundations.[61] I then show how this mathematical insight was transferred to twentieth-century philosophy before it passed to the postmodern tradition from which much contemporary scholarly work on the early modern period takes its cues.

Before I consider space's significance for the historical field, I must include a few words about my methods and the cases that I have chosen. First, I have written an expressly intellectual historical work that analyzes idealized and homogeneous geometric space's effects on early modern European thought in general. In the context of this work, I do not pretend to offer a history of mathematics, history of science, or history of any other discipline, however much I may draw on the contributions of experts in these (and other) fields. Second, in order to cast light on the depths of the intellectual changes that were involved, I construct a narrative that emphasizes the analysis of specific examples within the chapters at the expense of elaborating a uniform story across the entire text. Thus, I outline my argument in this introduction and summarize its theoretical implications in the conclusion, while I illustrate broader themes through the separate analyses that are executed in each chapter. For that reason, this book's narrative often progresses implicitly, rather than explicitly, and reflects backward as much as it looks forward. Finally, I chose each case on the basis of both chronological spacing and interdisciplinary impact. For example, the history of science excludes Albrecht Dürer, because he belongs to the history of art, while including (quite justifiably) his rough contemporary, Nicolaus Copernicus. My point in choosing the former over the latter for my foray into the sixteenth century was to expose the true vastness of European culture's debt to Euclid—and a similar intention governed every other choice that I made. The history of space that I offer is thus deliberately unorthodox in just about every way.

With these academic caveats in mind, I return to my core narrative. It was only when the geometric homogeneity inside which humanity had been ensconced collapsed that modernity arose. The consequent *decentering* of Euclid's intuitively accessible space had three consequences for modern European thought—and one related consequence for contemporary research on early modernity. First, Euclid was demoted, albeit without being erased. Although Euclid's geometry continued to be studied in schools, it was (and it still is) understood that Euclidean space had been superseded.[62] Second, as Dostoevsky's discussion of space indicates, God was finally expelled from humanity's mental cosmos, as the new vision of geometric space sealed Him off hermetically from the human mind. Third, radically new physical theories and cosmologies emerged from the new geometry, including Albert Einstein's notion that Newtonian time and space were relative constructs.[63] Finally, and relatedly, with respect to historical research itself, non-Euclidean geometry's importation into post-modern theory had the unintended effect of obscuring the profound depth of Euclid's influence on early modern thought. I hold that, in the end, contemporary theory misunderstands early modernity, precisely because it cannot *see* three-dimensional space's intellectual-historical significance. I return to this point in this book's conclusion.

Making the History of Nothing Matter

The three material objects from Nuremberg also provide a framework that allows us to redefine the early modern period's bounds. With Behaim's Erdapfel in mind, for instance, we can see that the reformation in space preceded the shock caused by Christopher Columbus's return from the New World. Concomitantly, this object also predated the religious and political upheaval of Martin Luther's Protestant Reformation, which the testy monk initiated in 1517. More significantly, Behaim's Erdapfel was not the first terrestrial globe but probably the third, as one may have appeared between 1440 and 1444 in Brussels, while a second was definitely completed in 1477 in Rome.[64] To this I must add that these globes' underlying intellectual apparatus was constructed in the course of the late fourteenth century and the early fifteenth. An express concentration on space leads us, therefore, to an early modern period that begins around 1350.

Taken together, Christopher Columbus and Martin Luther also suggest an alternative to a more narrowly defined approach to the early modern that emerged largely (although not exclusively) from the French Annales School. This school saw the early modern period as extending from roughly 1500 to 1800, which naturally excludes the quattrocento, while including everything that lies between the immediate effects of Columbus's return and the French Revolution's demise in 1799. The resulting "short" view owes much to the idea that the break with the medieval came with the rise of sixteenth-century phenomena, including Europe's exit from its peninsula of Asia, the Protestant cataclysm within Christendom, and the formation of an Atlantic economy. The most famous example of this approach is Fernand Braudel's magnificent *The Mediterranean and the Mediterranean World in the Age of Phillip II*.[65] Braudel's masterpiece contains two volumes, with the first covering the geographic foundations of medieval social and economic structures and the second boasting a history of Spain under Phillip II (1527–1598). This second volume is especially illustrative of the underlying conceptual issues, because it characterizes Spain's decision to sail west (as opposed to invading North Africa) as a *political* one.[66] For Braudel the break into early modernity constituted, to no small degree, an intellectual rebellion against rhythms that were rooted in the Old World's geography itself.

Braudel's manner of overlaying politics onto a broad geographic outlook sustains the "short" early modern period. Thus, Europe's exit westward could only have been political, since the Mediterranean itself mandated the easier trip across the Straits of Gibraltar. Many others had, of course, made that voyage. The ancient Romans dominated both the Straits and the North African littoral before the Vandals crossed southward in the fifth century AD to found a kingdom that was, in turn, destroyed by Byzantium in the sixth century. Military might and imperial fanaticism moved in the other direction, too. In the eighth century, North African Berbers entered the Iberian Peninsula, where they founded a kingdom whose last gasp came in 1492, just before the Spanish crown dispatched Columbus on his journey to the West and also expelled Iberian Muslims and Jews in the opposite direction. For Braudel, Europe became early modern when it defied the Pillars of Hercules and sailed into the Atlantic.

Spain's westward turn was made possible, however, by the previous development of a "global" spatial sense. The Spanish broke with the Mediterranean, precisely because they had access to mental tools that revealed a path out of the basin that had defined their world.[67] It is no accident, for example, that Martin Behaim lived and worked in Portugal, where the most advanced geographic knowledge was being collected. In addition, it is no less an accident that Behaim's globe was made in Nuremberg, as Central Europe had played an important role in joining transalpine artisanal traditions to ancient methods of spatial projection, most of which had come to Italy from Byzantium. The entire continent participated, thus, in this extremely broad and intensely varied intellectual transformation.

A collectively imagined web of space extended across European culture long before the continent's penchant for war and larceny seized the planet. Consider, in this respect, the cartographer Martin Waldseemüller, who in 1507 gave to the New World the name "America." He lived and worked in St. Dié, which lies in the heart of Lorraine, quite a distance from the Iberian coast. Nevertheless, he studied ancient texts, such as Claudius Ptolemy's *Geography*, a fundamental geographic work from the second century AD that had first "returned" to Italy in the early fifteenth century. One direct result was his production of a cosmographical work that would have a profound influence on a fellow named Peter Apian, whom I discuss in Chapter 2. Along similar lines, Christopher Columbus was not simply an intrepid Italian sailor in Spanish employ, but also a Renaissance humanist who consulted many ancient spatial works, including especially Ptolemy's *Geography*.[68] It was, therefore, the intoxicating mixture of old and new knowledge that allowed Behaim to paint the earth onto a sphere, and that also prompted Columbus to plot a path around the *All*.

Building on Braudel's perspective, I suggest that Spain's westward turn should be dissolved into the space that succeeded the medieval Christian Aristotelian frame of reference. For that reason, it is significant that Behaim's Erdapfel was completed well before Phillip II's reign began in 1556. Moreover, Behaim's globe itself was a product of at least a century's worth of reflection on space's relationship to the cosmos and God. As I noted earlier, the first verifiable terrestrial globe was made in 1477, while the theological debates that succored this globe's underlying mathematics became a force in the second half of the fourteenth century. Before Spain and the rest of Europe's thieving cabal broke free of their geographic constraints, the terrestrial globe and its attending space had to become thinkable.

With space's fourteenth-century roots in mind, I turn now to 1850, my proposed endpoint for the Spatial Reformation and, concomitantly, for the "long" early modern period. Taken together, Bauer and Behaim's globes highlight continuities that extended into the nineteenth century. It is important to note, in this context, that Bauer's globe was a direct successor to not only Martin Behaim's spatial sense, but also that of Isaac Newton. The connection to the former is obvious. With respect to the latter, however, I add that the great physicist republished in 1672 an important geographic work, *General Geography,* by the German-Dutch geographer Bernhard Varenius.[69] Varenius is considered by many to be modern geography's founder, because he anchored the discipline expressly to mathematics. (I discuss Varenius in Chapter 4.) Keeping in mind this very terrestrial aspect of Newton's contributions, Bauer's anthropological globe illustrates why the Spatial Reformation's transformation of the relationship between Heaven and Earth had such profound effects on Europe's vision of humanity: the human mind had produced the space inside which the earth and its denizens were *located.*

Here, I return to the Western triad and consider more narrowly how material culture resituated anthropology. Carl Bauer's globe is a crucial example, in this respect, because it defined humanity through its many geographic locations. Homogeneous space, in short, sustained the production of expressly *European* ways of seeing humanity, all of which appeared to be universal, precisely because an idealized space undergirded the resulting whole.[70] Nor was Bauer alone. Consider this report from 1791 of a British expedition to New South Wales. While recounting an excursion inland, the group's leader, Watkin Tench, noted that "at a very short distance from Rose Hill we found that [the aboriginal guides] were in a country unknown to them; so that the farther they went, the more dependent on us they became, being absolute strangers inland."[71] The British were better off, according to Tench, because among them was an astronomer, with the result that "we always knew exactly where we were and how far from home." The expedition's members brought with them a spatial sense that was so deeply entrenched in their minds that it remained operative, even when they were completely lost.

The differences that Tench highlighted between aboriginal and European ways of knowing underscore how the ability to project space shaped Europeans' sense of both themselves and others. In this respect, a useful example is a deservedly obscure poem by Friedrich Schiller, "[A] Saying of Confucius." Published in 1799, it associated Confucius (falsely) with geometry's conceptual rigor.[72] Schiller wrote:

> Threefold is space's measure:
> On and on unremittingly, disquietingly
> Strives *Length*: into the limitless
> Expanse *Breadth* pours itself;
> Fathomless *Depth* sinks itself.
>
> For you a picture these [three] endow:
> You must strive forward restlessly,
> Never stand still, wearied,
> If you wish to see perfection;
> You must unfurl yourself into breadth,
> If the world is to fashion itself for you;
> Into the depths must you descend,
> If Nature is to reveal herself to you.
>
> Only persistence achieves the goal,
> Only fullness leads to clarity,
> And in the abyss lives Truth.[73]

Schiller's poem highlights two issues. First, by the eighteenth century's end it was assumed that geometry's presence indicated that a civilization was advanced.[74] For early modern Europeans, the Chinese were famously civilized, so Schiller naturally ascribed a geometric culture to them—even though this was wrong.[75] Second, the poem's final lines return our attention to space's theological implications. In the book of Genesis, Creation is described thus: "In the beginning, God created Heaven and Earth. But the Earth was empty and unoccupied, and darknesses were over the face of the abyss; and so the Spirit of God was brought over the waters."[76] Within the Christian tradition, it was God who confronted an abyss and responded by creating *something*. Now, following Schiller, it was Euclidean Man who peered into darkness and even dared to seek his own Truth.

Schiller underscores how homogeneity reduced every value and judgment to three-dimensional space. And the Swabian poet-philosopher was hardly alone in confronting this kind of space, as from 1350 to 1850 no major thinker was uninitiated in the geometric discipline as it is presented in Euclid's *Elements*. The list of Euclid's students is impressive, as it includes Nicole Oresme, Nicholas of Cusa,

Leonardo da Vinci, Peter Ramus, Albrecht Dürer, René Descartes, Gottfried Leibniz, Thomas Hobbes, Isaac Newton, John Locke, Denis Diderot, Voltaire, Jeremy Bentham, Immanuel Kant, and G. W. F. Hegel. Every single one of these thinkers played a role in propelling early modern thought forward, and every one of them studied Euclid in detail. Since I am concerned with relating the Spatial Reformation's closing stages to modernity's rise, I concentrate on the last of the thinkers, Hegel.

Although Hegel occupies a prominent position in the historiography of European thought, his relationship to space has yet to be explored fully. Before he distilled Napoleon's exploits into a spirit that suffused nineteenth-century philosophy, Hegel wrote two spatial texts.[77] In 1801, he submitted to the University of Jena his obligatory second doctoral dissertation, the dreaded *Habilitationsschrift*. This work, which bore the characteristically scintillating title *Philosophical Dissertation on the Orbits of the Planets*, evaluated the philosophical significance of a recently proposed law that explained the spatial gap between planetary orbits.[78] The law turned out to be wrong. Nevertheless, for my purposes, it is significant that Hegel derived *philosophical* significance from a humanly constructed perspective on an unseen whole. Moreover, this audacious maneuver had a geometric foundation. Not only did Hegel study geometry as a ten-year-old, but he also wrote, toward the end of his secondary education, *Geometrical Studies*, an unpublished work in which he evaluated the *Elements*' logic.[79] All of Hegel's speculations—whether cosmological or historical—were thus built on *the* work in idealized geometric space.

I sketch out Euclid's relationship to the Spatial Reformation in more detail in the next chapter. For now, however, it suffices to emphasize that Hegel, like all great thinkers of the day, cut his teeth on Euclid before he did anything else of significance. To note geometry's role in Hegelianism's rise is important, moreover, precisely because the contemporary literature views Hegel as a quintessentially "modern" philosopher. He was born in 1770 and died in 1831, which meant that he witnessed the ravages of the "modernizing" French Revolution and the still-worse transgressions of its heir, Napoleon Bonaparte. Yet however "modern" Hegel's views were, they remained tethered to the space that pervaded both Behaim and Bauer's visions of the terrestrial globe. Consider Hegel's take on world history: "These general remarks on the different degrees of knowledge of freedom—firstly, that of the Orientals, who knew only that One is free, then that of the Greek and Roman World, which knew that Some are free, and finally, our own knowledge that all men as such are free, and that man is by nature free—supply us with the divisions we shall observe in our survey of world history and which will help us to organize our discussion of it."[80] Hegel put freedom's ostensible march into global space, placing its origins in the east and associating its development with a chaotic glide westward. It is thus particularly significant that

Hegel began as a Euclidean (and never stopped being one), because his historical speculations functioned inside an established spatial whole. Far from identifying modernity's arrival, Hegel reveals early modern space's perdurance.

Conclusion

Homogeneous space's reign dissolved when contemporary mathematics moved decisively beyond Euclid's *Elements* in the decades after 1850. Thus, the rupture that crystallized within European spatial thought identifies the moment when the early modern period began to dissolve.[81] Along these lines, I suggest two principles that can serve as guides to future research. First, given Euclidean geometry's indisputable intellectual and cultural significance, it would seem justified to affix the early modern period to idealized space's rise and decline. I explore the implications that such an approach has for contemporary discussions of both modernity and its relationship to postmodernity in this book's conclusion. For now, I propose that the early modern period is both "early" and "modern," precisely because its geometric space endures but only in a secondary role. Second, along more methodological lines, early modern intellectual history cannot be reduced solely to the production and consumption of printed ideas, because homogeneous space permeated every corner of early modern European thought and culture, including its longstanding tradition of producing and consuming images and material objects. Put more directly, we cannot understand the five centuries that this book covers if we do not grasp that print and material culture long worked together inside an expressly Euclidean matrix.

With these broad brushstrokes in mind, I look back to the late fourteenth century and specifically to the Parisian Terminist Nicole Oresme. As I have noted, Oresme came under the influence of Nominalist theological currents that diffused from Oxford to the Continent—and his Nominalism drove his contributions to science. Here, I call attention to figure 4, which contains a posthumously produced image of Oresme studying an armillary sphere. The drawing appeared originally in a 1410 manuscript edition of Oresme's *Treatise on the Sphere*, a French translation of Aristotle's *On the Heavens*.[82] This sort of translation/commentary was standard fare, as such treatises had been appearing since the early thirteenth century. I return both to this backdrop and to the use of armillary spheres in Chapter 2. For now, I concentrate on what the image reveals about the material culture of space's intellectual significance.

The fifteenth-century rendition of Oresme and an armillary sphere highlights another triad that the early modern world had inherited from the medieval one. This triad was essentially practical and was composed of human beings, texts (both written and printed), and material objects. Emerging from venerable medieval practices, this triad cut across the entire early modern period and is

Figure 4. Illustration from Oresme, *Treatise on the Sphere*, ca. 1410. Courtesy of the Bibliothèque nationale de France.

probably still around, albeit in an etiolated form. In the context of what I have argued, this triad has two uses. First, it allows us to understand how knowledge of unseen spaces and places was acquired in the early modern period, that is, through the study of texts and the manipulation of material objects. Second, it allows us to isolate within the triad's confines not only the books that were being read but also their individual significance. Thus, when Euclid's *Elements* began its post-1350 rise, its vision of homogeneous space clashed with other works that dominated medieval thought, such as Aristotle's *On the Heavens* or the Bible. (We do not know what Oresme was reading in the image, although it is reasonable to assume that it was *On the Heavens* or, perhaps, Claudius Ptolemy's astronomical work, *Almagest.*) The practical triad puts us into a position, therefore, to understand how the relationship between texts, material objects, and readers challenged older mental worlds, while also producing new ones, as changes in the underlying textual literature and the production of instruments continually wrought new perspectives on the *All*. I return to this theme throughout this book.

Taking off from this point, I emphasize that one way in which we are expressly modern, today, is that we understand and respect Euclid's *Elements*, even as we recognize that its geometry comprises a limited view of space. Our modernity rests, therefore, on our ability to accept alternative notions of space, most of which, unlike Oresme's armillary sphere, cannot be apprehended visually. In this sense, one of modernity's more important (if also greatly overlooked) pillars is its expressly post-Euclidean status.[83] Only when we have come to terms with the fundamentally Euclidean nature of early modern thought can we take full account of what separates its way of envisioning the world from modern approaches. Whether we should understand ourselves as moderns, or as beings who exist beyond modernity, is a question that I leave for the conclusion. For now, I merely suggest that we concentrate on Euclid's career, when examining early modern thought's rise and decline.

Chapter 1

From Sacred Texts to Secular Space

> Arithmetic and geometry are, therefore, the wings of the human mind.
>
> —Philipp Melanchthon, "Praefatio in arithmeticen" (1536), in *Corpus Reformatorum Philippi Melanthonis operae quae supersunt omnia*, ed. Bretschneider and Bindseil

The Protestant Reformation and the Spatial Reformation were interconnected in a fundamental way: both movements cultivated a sacred text that was assiduously translated into the vernacular and evangelized by true believers. The texts in question were the Bible (naturally) and Euclid's *Elements,* which may seem less obviously sacred. Although intellectual historians do not usually juxtapose these works (and not in this way), there are good reasons to do so. First, the Bible and the *Elements* entered print roughly concurrently, as the former appeared in 1454 and the latter in 1482, with flurries of editions blanketing the continent thereafter.[1] Second, their respective narratives intersect in an overlooked way, as both begin with and build on *nothing.* The Bible's first book, Genesis, recounts God's creation of the cosmos *ex nihilo* and affixes to that moment a providential history that assures humanity's destiny in the next world. The *Elements,* meanwhile, begins with substanceless points and perches on top of them a system of projection that makes idealized space applicable to real space, that is, to the physical realm.[2]

The express juxtaposition of God with Euclid reveals that beneath early modern thought's main currents swirl profound changes in space's intellectual context. Both the Bible and the *Elements* offer coherent ways to think about humanity's position not only within the physical cosmos, but also in reference to an ideal realm. Yet their respective *spaces* were incompatible, since Christianity emphasized hierarchy and discontinuity, whereas Euclidean geometry insisted on uniformity and homogeneity. Before continuing in this vein, however, I summarize Christianity's reception of hierarchical space. Between roughly 350 and 1350,

Christian thinkers applied biblical cosmology to antiquity's remnants. In doing so, they overlaid onto pagan philosophical constructions of space their own views on the hierarchy that suffused Creation. According to the Bible, God placed the Heavens "above" and fashioned humanity from the elements that rested "below." The classical heritage, in turn—including especially Neoplatonic and Aristotelian traditions—justified the same physical arrangement, if along more philosophical lines. The ensuing synthesis yielded, in turn, a geocentric cosmos inside which humanity was *situated* both religiously and philosophically. Any imagined journey "upward"—whether from profane to sacred, or from Earth to Heaven—was defined, thus, by hierarchy and discontinuity.

The reappropriation of Euclid's *Elements* after 1350 (and in Nominalism's wake) upset this exquisitely balanced applecart. First, by moving from nothing to perfect figures that were suspended within idealized space, the *Elements* suggested that perfection was not necessarily limited to realms *above*. Second, as human beings applied this idealized space to the real world, including in the production of material culture, it seemed to them that humanity might not have been so utterly inferior to God, after all. The Renaissance Cardinal Nicholas of Cusa, whom I have already discussed, was a student of geometry and arithmetic and, not coincidentally, groped his way to a strikingly positive anthropology. In the mid-fifteenth century, he wrote: "Therefore, as the human world emerges, all things are in fact explained humanly (and with respect to the universe itself, universally). Indeed, all things are folded up humanly in themselves, since the human is [also] God."[3] And this is to say nothing of homogenous space's implications for medieval cosmology and theology, both of which had previously absorbed fractured space. We can thus begin to understand why the Western triad came under such stress, when European spatial sense shifted toward geometric homogeneity.

In spite of Euclidean space's prominence within early modern thought, there is currently no broadly conceived analysis of its effects on European intellectual history. Classic intellectual-historical works on the late Medieval, Renaissance, and Reformation eras do not concentrate on Euclid's *Elements* and its attending spatial sense.[4] The situation is similar for work on the seventeenth and eighteenth centuries, which emphasize early modern thought as a radicalizing prequel to the French Revolution.[5] As for the nineteenth century, homogeneous space has hardly been an issue. University-level surveys, for example, still serve traditional historical-philosophical cocktails of Hegelianism and Marxism, without covering geometry's influence on Hegel himself.[6] Concomitantly, classic works on Hegel and Hegelianism foreground neither Euclid nor ancient geometry.[7] Yet, as I noted in the introduction, Hegel studied geometry as a youth and, later, wrote the minor work *Geometrical Studies*, in which he analyzed Euclid's arguments.[8] Even in the simplest biographical terms, it is clear that Hegel's mature thought arose within the context of a billowing sense of homogenous space. The scholarship has, however, largely overlooked this point.[9] In addition, those intellectual-historical works that do confront space in the nineteenth century concentrate on the fin de siècle, by which point the Spatial Reformation was dissolving.[10] Thus, on the whole, it seems safe to say that the contemporary scholarship has elided Euclidean space from intellectual history.

With space's absence in mind, I turn to the quote from Philipp Melanchthon in this chapter's epigraph. Coming from one of the Protestant Reformation's driving forces, the idea that arithmetic and geometry form the wings of the human mind casts light on how the *Elements* came to serve as not only an ersatz sacred text but also a corrosive philosophical agent. In natural philosophical matters Melanchthon was a thoroughgoing Aristotelian.[11] Yet, in spite of the spatial commitments that went with that, the *praeceptor Germaniae* also dabbled in geometry in a way that undermined Christian Aristotelianism's hierarchical space. A particularly good example comes from a preface that Melanchthon wrote for a 1536 Latin edition of the *Elements*: "No one without some knowledge sees enough of this art, which is life demonstrated. No one without it will be a maker of method. . . . There is here great praise of geometry, which did not cling to inadequate and inferior [human] constructions, but flew into Heaven and transported human minds, which were stuck in the mud, back up to the heavenly throne."[12] If we interweave this quote with the epigraph above, we catch another glimpse of geometry's implications for European thought: any journey "upward" strained the theological and philosophical bonds that had long tethered God's children to His cosmos's center—and Euclid's *Elements* gave European thought the means to cut them.

Homogeneity's Sacred Vessel

In the wake of medieval Nominalism's spread to the Continent, European thinkers became increasingly interested in incorporating mathematics into their knowledge of both the cosmos and God. By the end of the fourteenth century, the most important center of mathematical study was the Papal curia in Rome, which was slowly confronting the need to recalculate the calendar.[13] Rome was also becoming Europe's chief repository of ancient mathematical texts and, not coincidentally, was attracting the continent's best mathematicians, as in the example of Johannes Regiomontanus, who was recruited from the University of Vienna. The continual enlistment of foreigners produced a powerful mixture of intellectual currents, as the northern universities from which these people hailed often cultivated the Nominalism that had spread from Oxford to Paris. Of equal significance, however, was the continual arrival in Italy of Greek manuscripts from the city of Constantinople, which was then being threatened by Ottoman forces.

Among the most important among the returning works were Greek-language copies of Euclid's *Elements*. None of these manuscripts had ever been translated out of the original tongue (medieval Latin versions of the *Elements* usually came from Arabic translations), and this made it seem that "uncorrupted" copies of the original had finally arrived. Of course, more than a millennium of copying had corrupted them, too. This aspect of Euclid's second "return," however, is another story.

After coming under intense study in the fifteenth-century Papal court scene, the recently returned *Elements* took the first step toward cultural ubiquity, when a Latin translation entered into print in 1482 in Venice.[14] Although medieval thinkers had enjoyed access to Latin versions of Euclid's work since the twelfth century, on the whole the medieval encounter with geometry did not compare to the early modern experience in terms of either its breadth or its intensity. From the late fifteenth century until the early nineteenth century, waves of editions appeared in multiple countries and in multiple languages, producing a web that not only reached across Europe but also extended to its colonial settlements. By 1650, at the very latest, there was no way that an educated European could have been ignorant of the *Elements*' lessons, any more than he or she could have been ignorant of the Bible. Indeed, in the history of the book, the *Elements* may be second only to the Bible in the number of editions, with some scholars estimating the total to be over a thousand.[15]

The *Elements*' print diffusion was the precondition for a development that I will call spatial secularization. Spatial secularization constituted the continual distancing of the divine from both humanity and the cosmos, with homogeneous space filling the resulting gap. My approach breaks sharply with regnant views. Whereas I emphasize humanity's imagined relationship to the divine, other scholars generally concentrate on the retreat of religious institutions from society. Along these lines, traditional approaches chronicle how the church receded from public life, only to be replaced by more secular institutions—whether economic, political, or social.[16] I read secularization, however, with respect to changes in the divine that emerged from space's advance.[17] In this respect, spatial secularization is a successor to the fourteenth-century Nominalist reappraisal of God's will, which separated God from the cosmos in a way that allowed humanity to apply mathematics to the whole. Not coincidentally, the projection of a regularized space onto the cosmos became the dominant mode of knowing things.

To contextualize secularization's history via homogeneous space is important precisely because we moderns stand beyond weighty changes in spatial thought. We have a peculiar perspective on space, because we understand Euclid's geometry, even as we have learned to see it in post-Euclidean terms. This situation has profound implications for the history of secularization. First, the *Elements* remains present in contemporary education, in a way that the Bible simply no longer is—which means that we see no inherent conflict between space and God, although there long was one. Second, we do not take the *Elements* seriously as an historical force, precisely because its geometry and, above all, its sense of space have become so innocuous. As Dostoevsky noted, there are other geometries out there, whose implications are more radical and profound than anything that one finds in Euclid; and as Dostoevsky showed us, in this context Euclidean thought can become a refuge for the religious mind.

I illuminate further the differences between my view of secularization and contemporary approaches by looking at our own experience with geometry. The mathematical sequence in secondary education expressly includes a year's study of Euclidean geometry. Here, I call attention to the year 2000 edition of the high school textbook *Geometry* by Ray C. Jurgensen. The text begins by emphasizing geometry's utility and notes that we can use the discipline to find the distance between things on earth, such as a pole, a fountain, and a tree, before also appending an illustration and a proof-like discussion.[18] It then explains how the position of a real thing can be reduced to an imagined point: "Each dot on a television screen suggests the simplest figure studied in geometry—a *point*. Although a point doesn't have any size, it is often represented by a dot that does have some size."[19] Now, compare this rendition to the *Elements'* opening line: "A *point* is that which has no part."[20] Although Jurgensen included contemporary referents in his explication, the sense of the original remains, given that the modern version begins with *nothing* and heads toward *something*. Consistent with what I said about the Bible's space, it is important to underline that Jurgensen's attitudes toward space are predicated on God's absence. Thus, where the sixteenth-century pedagogue Melanchthon saw in geometry a potential path to God, his twentieth-century counterpart, Jurgensen, associated Euclid with his own television's pixels.

The *Elements'* diffusion in schools makes it a unique intellectual historical phenomenon. If we take the initial 1482 print edition as a starting point, this text has been in print (and in use) for over five hundred years. The same can hardly be said of any other Western work. Historians of mathematics may object that ancient discussions of conic sections are still present in mathematical education, which is true. Nevertheless, three things mitigate this objection's force. First, today's students arrive at conic sections only after having studied Euclidean geometry's basics. Second, in the course of the early modern period, ancient writings on conic sections were never printed as often as the *Elements*. Euclid is, in this respect, the intellectual tie that binds, with his greatest work serving as the *alpha* and the *omega* of the passage from early modernity to modernity.

Euclid's contemporary ubiquity, however, has contorted our view of geometry's history. We moderns find nothing problematic either in Euclid's space or in his imagined points, because neither has larger implications—whether anthropological, cosmological, or theological. Given how distant we are from the Spatial

Reformation's end (and the Western triad's concomitant dissolution), geometric points are about as significant and as objectionable for us as our television's pixels. We must, therefore, historicize Euclid's contemporary innocuity if we are to understand how *space* has skewed the scholarship on the early modern period. Along these lines, we must recognize two things. First, geometry *belongs* mostly to fields other than intellectual history, such as the history of mathematics, art, architecture, and (to a small degree) philosophy, which means that the discipline's space has remained marginal to intellectual history.[21] Second, within these established fields, geometry's history is pitched as a progressive tale. There is nothing wrong with such a view, if the given progression's significance is limited to a specific discipline. The problem, I would suggest, lies in the academic extension of this progressivism to broader periods, because this has occluded how homogeneous space rose and declined. (I deal with this issue in this book's conclusion.)

Overall, geometry's historical significance has been determined by the emphasis on chronicling mathematics' development toward its modern incarnation. By privileging this endpoint, however, the contemporary scholarship has rendered unproblematic the early modern reception of homogeneous space. As I noted in the introduction, the new space rattled European thought, as in the example of Blaise Pascal, peerless mathematician and white-knuckled theologian, who extended his mind outward and blanched. For a thinker with such intense religious inclinations, homogeneity opened a terrifying perspective onto things that should not have been called into question, including humanity's position within the physical cosmos.[22] It is, in this respect, not surprising that after Pascal had reached mathematics' pinnacle, he stopped pursuing the discipline and spent his remaining years contemplating God.

I bring spatial secularization's depths into focus, through a reading of two Renaissance encounters with space. First, I call attention to figure 5, which presents the reverse side of a memorial medallion that Renaissance artist Matteo de'Pasti struck somewhere between 1446 and 1450 in honor of the great polymath Leon Battista Alberti. (Alberti's profile is on the other side.) Alberti knew Euclid's work well and used ancient geometry extensively in his published works on art, architecture, and mathematics.[23] It is, therefore, significant that the medal presents Alberti's personal sign, a winged eye, and his motto, the Latin phrase "Quid tum." These two features direct our attention to homogeneous space's most important effects. First, Alberti's eye resituates humanity in relation to the physical cosmos, because it liberates the human gaze from the terrestrial realm. Whereas humans had once looked "up," they could now look "back," too. Second, the Latin motto evokes exactly the renewal of humanity's relationship to the divine that space demanded, insofar as "Quid tum" translates, roughly, as "What now?" Liberated from the medieval world's fractured and hierarchical space, the mind was free to proceed (humanly) through *everything*.

Figure 5. De'Pasti, bronze memorial medal for Leon Battista Alberti (reverse side), 1446/1450. Courtesy of the British Museum.

With homogeneous space's secularizing effects in mind, I turn to the other example, which stems from that other reformation. Figure 6 depicts Creation and comes from Martin Luther's German translation of the Bible, the first edition of which appeared in 1534.[24] Significantly, the page on which the image appears is situated opposite the initial lines of Genesis, printed on the facing page. Thus, within the text itself the image is connected directly to the printed word, that is, the reader *sees* the things and events that the words describe. Merely a glance reveals the conflict's nature. In the biblical version, the relationship between Heaven and Earth is hierarchical and determined by God's perspective rather than humanity's. Thus God looks down, whereas humanity's perspective demands, on the one hand, that humans look up and, on the other, that this relationship to the Heavens be represented as if the eye were moving straight *back*

Figure 6. Illustration from Luther, *The Bible,* 1545. Courtesy of the Staatliche Bibliothek Regensburg.

from the image's (flat) depiction. Furthermore, anticipating the arguments that I make in Chapter 4, it is no coincidence that Luther's Eden has no distinctive cartographic features, as humanity's garden of innocence cannot exist within the new space. (I have more on this issue in Chapter 4.)

The comparison of de'Pasti's medallion with Luther's Bible illustrates why homogeneous space raised such uncomfortable questions. First, by promoting a human perspective on the cosmos, homogeneity altered humanity's relationship to both Creation and the Creator. Second, in a post-Gutenberg world it was much easier for texts, images, and objects to work together, in order to *make* space through the cultivation of a new mode of spatial representation. (I discuss this mode in the next chapter.) Put another way, the continual production of printed works and material objects sustained a culture of spatial projection that altered humanity's relationship to spaces both seen and unseen. Thanks to the changes within geometry that brought the Spatial Reformation to its end, however, we moderns have lost contact precisely with a mental world in which homogeneous space could have broader implications.

Resituating the Spatial

Euclid's early fifteenth-century "return" from Byzantium inspired ever-greater demand for the text itself. (The medieval Euclid had been based on Arabic versions.) Already in 1460, Johannes Regiomontanus, the German astronomer who rotated between Nuremberg, Rome, and Vienna did a new (and incomplete) Latin translation of a Greek manuscript that circulated among the cognoscenti in Rome. Regiomontanus was, as I noted above, part of the mathematical circle that centered on the Papal curia.[25] Thanks largely to this circle, which also included Nicholas of Cusa, geometry became the subject of intense study and, concomitantly, gained enormous intellectual prestige. I return to this group below; for now, however, I concentrate on Euclid's next great journey, from manuscript into printed work.

The arrival in Rome of Byzantium's cultural wealth highlights two Renaissance trends that were essential to the Spatial Reformation's initial phase. The first was a change in attitude toward medieval textual traditions. As elites gained access to Greek versions of ancient works—the *Elements* included—they began to ascribe greater value to the so-called originals. For thinkers of the period, the Greek versions were closer to classical culture than the "corruptions" that Europe had possessed since the twelfth century. Although the new Latin translations were not necessarily better than the medieval ones, they were significant for expanding the number of textual lineages, with the result that "new" and the "old" came under direct comparison. The resulting process of revision and emendation was critical to the Spatial Reformation's development, because it drove a continental standardization of geometric language.

The second trend was the return of Plato's philosophical corpus. Within Renaissance thought this jewel served as a counterbalance to the established Aristotelian corpus. I cannot trace this history but refer readers to standard works in the field.[26] With reference to Euclid's rise, the Platonic return is significant, because it reveals how Euclidean thought always interacted with (and worked between) multiple intellectual traditions. Thus, it is not merely a matter of understanding the effects of this single text's return, but of noting expressly how Euclid's *Elements* was read in conjunction with other philosophical works and, moreover, that this occurred in a way that allowed early modern thinkers from many different schools to reformulate their intellectual worlds.[27]

Against the backdrop of Plato's second return, it is important to recall that medieval thinkers enjoyed only limited access to Platonic thought. Since the fourth century AD, only one of Plato's works had been available in Latin translation, the cosmological dialogue *Timaeus*—and this version was incomplete.[28] Moreover, the vision of space that Plato cultivated in this text, although geometric in spirit, was by no means Euclidean. Rather than build a complete theory of space that led, in turn, to the construction of geometric forms, in the *Timaeus,* Plato began with forms and used them to draw parallels between the real and the ideal worlds. In no way did this construction of an ideal world constitute, however, a full system of idealized space. There is no doubt that Plato respected ancient Greek geometry.[29] Nevertheless, he did not cultivate in the *Timaeus,* nor in any of his other works, a rigorously argued system of projection the sort that Euclid would collate in the *Elements.*

The scholarship has, thus, missed an important wrinkle in the relationship of Platonic thought to ancient geometry. Although Plato supported geometric research, on account of its utility for astronomy, his primary interest was in using projected space to illustrate philosophical transcendence rather than to cultivate geometry as a discipline. Concomitantly, he used geometric forms to elucidate the philosophical world's meaning for the real world rather than to construct a system of spatial projection that could be applied to the physical realm. Euclid, in contrast, while building on older Pythagorean traditions to which Plato had access, articulated a system of spatial reckoning that enveloped Plato's forms entirely *within idealized, homogeneous space.* Put more narrowly, Euclid began with substanceless points and constructed his system on top of them, whereas Plato began with forms and applied them to the physical world's relationship to an ideal one.

Against this backdrop, it is significant that Plato's initial return coincided, roughly, with Euclid's second reception. (The first reception was, as I noted in the introduction, obscured by Aristotle's twelfth-century return.) In 1438, a complete collection of Plato's dialogues "returned," with the arrival in Florence of George Gemistos Plethon, a Byzantine scholar who possessed an unmatched collection of ancient manuscripts.[30] Plethon gave his copy of the dialogues to Cosimo de Medici, Florence's own *princeps,* who then tasked the great Marsilio Ficino with translating them into Latin. After Ficino completed that task, Western Europe finally enjoyed access to the entire corpus against which Aristotle had formulated his own system.

Although historians have analyzed the exchanges between Renaissance Aristotelians and Platonists, they have not fully considered Euclid's relationship to both of these groups. Euclid and Plato's roughly simultaneous returns produced a temporary alliance that—initially, at least—worked to the geometer's advantage. Platonism proved supportive to Euclid's diffusion in two ways. First, instead of bifurcating the cosmos's space along physical lines, as did Aristotle, Plato posited a philosophical distinction between the physical and the ideal. There was an imperfect physical whole, in which things existed, in addition to the ideal whole, to which philosophers had access. Second, he characterized geometry as a mediator between the two worlds, which suggested, in the context of early modern thought, that this discipline may be able to serve a similar role in Christianity's dialogue between divine and mundane. (I underscore that Plato was referring to ancient geometry, and not to Euclid, since the latter's *floruit* came after Plato's death.) Thus, geometry appeared prominently in the *Timaeus*, where Plato used it to explain the cosmos's physical structures, which was again a convenient alternative to Aristotle.[31] It also appeared in the *Phaedo*, where Plato suggested that geometry's rules were accessible to all human minds.[32] Finally, it appeared in the *Republic*, where Plato used geometry to justify undemocratic theories of governance.[33]

As Plato's stock rose in the Renaissance, Euclid's did, too, especially given how accessible print copies of the *Elements* were becoming. Consistent with my discussion of ancient philosophy above, however, it is important to recall that the Platonic reading of geometry differed in crucial ways from the Euclidean one. Plato was interested primarily in using the ethereal perfection of forms to illuminate the ideal. Thus, he used geometric metaphors when he wished to direct the mind to Truth, while studiously avoiding discussions of space's conceptual foundations. I would suggest, therefore, that rather than read the Renaissance in terms largely of a battle between Aristotelianism and Platonism, we take into account how the resulting skirmishes occurred in conjunction with Euclidean geometry's rise and diffusion. Both Aristotelians and Platonists laid claim to Euclid's space, although for different reasons in each case.

Along these lines, it is significant that Plato's return accelerated homogeneous space's passage into the realms from which medieval Aristotelian thinkers had excluded it. From a Platonic perspective, to apply Euclid's space to the physical world did no harm to the ideal one, since perfection remained *elsewhere*. For that reason, unlike a Christian Aristotelian, a Christian Platonist would have perceived little danger to God's majesty in the application of homogeneous space to the physical world. Equally significant, however, was a concomitant change in attitudes toward real space. Since Plato never cultivated fractured space, as did Aristotle, Platonists could apply Euclidean homogeneity to the entire physical realm, that is, to *above* and *below*. It is thus no accident that historians have underscored Renaissance Platonism's contributions to the rise of physics in seventeenth-century England.[34] As Alberti's Eye illustrates, Platonists were able to *see* the unseen differently than Aristotelians had been able to do.

In this context, it is also important to underscore that Euclidean space's rise altered the philosophical terrain for every returning ancient philosophical system. Consistent with this point, I conclude this section by looking briefly to the broader philosophical context. The fifteenth and sixteenth centuries brought the return of other ancient systems, including Stoicism and Epicureanism.[35] I cannot reconstruct the resulting debates among adherents of the various schools except to note that although much arguing between partisans occurred, there was also one significant point of agreement: everybody prized Euclid. Early modern adherents to Stoicism and Epicureanism read the *Elements*, not only applying its space to their own philosophical systems but also doing their own translations into Latin. And some thinkers did the latter, even though Euclid's emphasis on homogeneity clashed with their core philosophical assumptions. This was especially the case for the Epicureans, who held that reality was based on discontinuity. (I have more in Chapter 5 on the clash between continuity and discontinuity.) As I detail below, with the sole exception of the Skeptics, all the major Renaissance schools had at least one adherent who produced a translation of the *Elements*.[36]

Sacred Space in a World of Print

One cannot understand the full scope of Euclid's influence without accounting for print. Here, I return to my earlier juxtaposition of the Bible with the *Elements*. Like the *Elements*, the Bible had a history of appearing in other languages prior to its being translated into Latin, from the Old Testament's Hebrew and the New Testament's Greek. The first Latin translation was done by St. Jerome at the end of the fourth century AD, which made it slightly older than the first (and lost) Latin translation of the *Elements* by the sixth-century figure Anicius Boethius.[37] Known as the Vulgate, this translation became the Catholic Church's official version of the text and was also the version that Johannes Gutenberg first transferred to typeset.

The growth of linguistic diversity was a key aspect of the Spatial Reformation. Like the *Elements*, the *Bible* was not simply published in Latin, but was soon subjected to an extensive program of translation into the European vernaculars, of which Luther's Bible is the sterling example. Moreover, this progression within the Bible's print diffusion ran parallel, in many ways, to Euclid's experience. When Martin Luther did his translation, he did not simply work from the Vulgate but used copies in the original languages.[38] Thus, regardless of the language involved, both Luther and all other translators had to confront the Bible's tradition of beginning with *nothing*. As a result, Luther's translation of Genesis hardly differed from the Vulgate. Here is the Gutenberg Bible's description of Creation (in this instance, all the quotes are given in the original language): "A principio creavit deus celum et terram. Terra autem erat inanis et vacua: et tenebre erant super faciem abissi: et spiritus dei ferebatur super aquas."[39] Luther produced this: "Am Anfang schuf Gott Himmel und Erden. Und die Erde war wüst und leer / und es war finster auff der Tiefe / Und der Geist Gottes schwebet auff dem Wasser."[40]

Luther's German is not markedly different in structure or meaning from the Vulgate's Latin. And the same phenomenon is evident in the most famous English translation of the Bible, the King James Version, published in 1611.[41] Here are its first lines: "In the beginning God created the Heaven, and the Earth. And the earth was without forme, and voyd; and darknesse *was* upon the face of the deepe: and the Spirit of God mooved upon the face of the Waters."[42] It should not be surprising that the translations are so similar. The narrative is not complicated, and the Vulgate had also formed translators' expectations over the course of a millennium. For early modern culture, the Bible was thus one well-known well-head for a common approach to *nothing*.

The *Elements*' print history was quite similar. The first version appeared in 1482 in Venice and was based on a Latin translation from the thirteenth century. It was reissued again in 1491. Additional versions came slowly, at first, but these were now tied to the Greek language source materials that had been flowing westward from Byzantium. In 1505, Bartolomeo Zamberti published a Latin translation of the *Elements* that was based on Greek originals and, in turn, competed with the thirteenth-century version for the title of most faithful to the original. Indeed, adherents of each side criticized the other side openly, while one reprint of 1516 dealt with the tension by publishing both versions together.[43] In 1533, however, the pivotal edition appeared.[44] Printed in Basel, it was based on Greek sources and, more important, was published in Greek, thus finally making the "original" available.

Recognized today as the *editio princeps*, the Basel version of the *Elements* became essential to Euclid's continued diffusion. Multiple new mixed editions (Greek with Latin) followed, appearing in 1536, 1545, 1549, 1550, 1557, and 1564.[45] Even more significantly, these editions appeared in multiple cities: Paris (twice), as well as Rome, Leipzig, Basel, and Strasbourg. Thus, within thirty years, new editions of the *Elements* peppered Western and Central Europe. Moreover, among the cities mentioned, perhaps the most important was Leipzig (the origin of the 1549 edition), because it had hosted, since 1478, a well-attended annual book fair.[46] Given Leipzig's position on the confluence of three rivers, a book sold there could wind up anywhere in Europe.

The *Elements*' print diffusion set the stage for Euclid's final leap into ubiquity, as the work entered multiple vernaculars. In the four decades that followed the *editio princeps*, new translations appeared in Italian (1543), German (1555), French (1564), English (1570), and Spanish (1576), although not all were based on Greek sources.[47] And with the inception of vernacular traditions the print environment became ever more complicated, as new Latin versions appeared in 1536, 1545, 1557, 1559, and 1566.[48] This series was, in turn, punctuated in 1572, with the publication of the definitive Latin translation by the Italian scholar Federico Commandino. Commandino was both a competent translator and a legitimate mathematician, which made his version a favored source for retranslations into the eighteenth century.[49] And these kept coming, as another German edition appeared in 1562 in Basel, while a French one came in 1598 in Paris.[50]

Remarkably, the market for vernacular translations never reached saturation. Beyond the new Latin versions that never ceased to appear, vernacular ones were published in Dutch (1606, 1617, 1695), English (1651, 1660, 1661, 1685), French (1604, 1615, 1639, 1672), German (1610, 1618, 1634, 1651, 1694, 1697), Italian (1613, 1663, 1680, 1690), and Spanish (1637, 1689).[51] By 1650, when the Spatial Reformation's initial phase came to an end, it was extremely unlikely that any educated person in Europe lacked access to a copy of the *Elements*; and this included those who lived in regions, where vernacular versions had yet to be published, such as the Czech and Polish lands. Indeed, with respect to the latter realm we have the specific example of the Polish-German astronomer Nicolaus Copernicus, who owned a copy of the 1482 edition. Thanks to the mushrooming of printed texts, Euclid's practical reach was becoming virtually unlimited.

The sheer extent of Euclid's diffusion recalls the point with which I ended the previous section. Regardless of their differences, most Renaissance schools of classical thought embraced the *Elements*. (Skepticism was the only major exception.) The *Elements* routinely crossed intellectual boundaries, as Aristotelians, Epicureans, Platonists, and Stoics all read the work, if they were not also involved in translating and publishing it. The Aristotelian Protestant Melanchthon contributed the introduction to a 1536 edition of the *Elements*.[52] Marin Mersenne, the French Friar who was heavily influenced by Epicureanism, published his own version of the *Elements* in 1626.[53] With respect to Platonism and Stoicism, the Italian mathematician Niccolò Tartaglia, who came under the former's sway, edited and published the Italian version of 1543, while French logician Peter Ramus, who was influenced by the latter, did his own version in 1549.[54] And religious boundaries were equally irrelevant, as Catholics and Protestants read Euclid, too. Well-respected editions by the Jesuit (and Aristotelian) astronomer Christopher Clavius and the Protestant (and Platonist) mathematician Isaac Barrow appeared in 1574 and 1655, respectively.[55] It is, therefore, no accident that when seventeenth-century Protestants and Catholics launched counter-attacks against Skepticism (and in defense of their Christian faith), both parties wielded a geometric club.

We now find ourselves in a better position to understand the specific role that language played in geometry's rise. The uniformity and ubiquity of the early modern Euclid emerged, in part, from the qualities of the source itself. Regardless of when a translation was done, all versions diffused the same basic concepts, including points, lines, planes, and spheres. For example, the first line of the 1482 version of Campanus's Latin translation is "Punctum est cuius non pars est"[56] ("A point is that which has no part"). These exact words appear on the first page of Christopher Clavius's Latin translation—not to mention Isaac Barrow's subsequent one.[57] The vernaculars were no different. Niccolò Tartaglia's Italian translation of 1543 begins, "Il ponto è quello che non ha parte."[58] Wilhelm Holtzman's German translation of 1562 hardly differs: "Ain Punct oder tipfflin / wirtt das genant / so khain thail hatt."[59] Nor is the 1564 French translation by Pierre Forçadel de Bézier exceptional, as it begins, "Cela, qui n'a partie aucune se nomme: poinct."[60] And Rodrigo Zamorano's Spanish version of 1576 begins, "Punto es, cuya parte es ninguna."[61] Regardless of the language, the language of space was becoming universal.

I conclude this section by extending the parallel that I drew between the Bible and the *Elements*' entries into the vernacular, via a 1694 German edition of the *Elements*. Translated by Anton Ernst Burckhard von Birckenstein, this version was published out of Vienna under the revealing title *German-Speaking Euclid*.[62] The title reflects the linguistic diversity that characterized the *Elements*' reception, as it recognizes implicitly that the text was being published in other languages. In addition, its first line begins with the same old words (again, I leave the quote in the original): "Der Punct ist ein Tüpfflein / das in seiner Grösse unzertheilig ist / oder in welchem wir allhier keine Theile zu betrachten haben."[63] Up through 1650, regardless of the language that the given translator used, an increasingly *European* context for published versions of the *Elements* lurked behind (and sustained) a common vision of space.

Between Text and Pedagogy

Euclid's second return had widespread intellectual and cultural effects. In mid-fifteenth-century Italy the architect and artist Filippo Brunelleschi, after already having invented perspective, took instruction in geometry from the mathematician Paolo Toscanelli.[64] The latter, in turn, was a sometime member of the mathematical circle in Rome that included, at different points, the Cardinal Nicholas of Cusa, the astronomer Johannes Regiomontanus, and the Italian Renaissance's most enthusiastic student of geometry, Luca Pacioli.[65] For his part, Pacioli was geometry's answer to Saint Paul, writing extensively on mathematical topics and also advocating tirelessly for the study of Euclid—and with impressive results. Pacioli's friend Leonardo da Vinci, to take an example from the world of art, filled his notebooks with doctrines from the *Elements*, including this particular line: "The point, being invisible occupies no space. That which occupies no space is nothing."[66] In many ways, the quattrocento began and ended with Euclid's vision of *nothing*.

As the fifteenth century yielded to the sixteenth, Euclid's thought spread wider, spilling northward over the Alps and diffusing into every intellectual corner. After returning to Poland from Italy, Nicolaus Copernicus repeatedly consulted and copiously annotated his personal copy of the *Elements* before writing a work that extended geometry into the study of spherical triangles, which (significantly) was not among the *Elements'* topics.[67] Copernicus's contemporary, Albrecht Dürer, who probably studied perspective with Pacioli during an Italian sojourn of 1505–1507, also owned a copy, the reading of which prepared him to write *Instruction in Measurement,* a text that made Euclid accessible to artisans.[68] Of Euclid Dürer wrote: "The wholly astute Euclid constructed geometry's foundations, and who understands these well has no need of the following written work, but it is [written] exclusively for the young and for those who have nobody to teach them faithfully."[69] In many ways, what made the Renaissance truly *European* was how its major figures continually engaged with Euclid's vision of space.

Nor did the *Elements'* echoes die in the sixteenth century, but they could be heard through the seventeenth and into the eighteenth centuries. In 1596, the astronomer and physicist Johannes Kepler wrote that "Geometry has two great treasures: one is the theorem of Pythagoras; the other, the division of a line into extreme and mean ratio [the golden section]. The first we may compare to a measure of gold; the second we may name a precious jewel."[70] Kepler's contemporary, the political theorist Thomas Hobbes was smitten, too—at least according to the biographer John Aubrey, who in 1696 reported: "[Thomas Hobbes] was forty years old before he looked on geometry; which happened accidentally. Being in a gentleman's library Euclid's Elements lay open, and 'twas the forty-seventh proposition in the first book. He read the proposition. 'By G——,' said he, 'this is impossible!' So he reads the demonstration of it, which referred him back to such a proof; which referred him back to another, which he also read. And so forth, that at last he was demonstratively convinced of the truth. This made him in love with geometry."[71] Aubrey's tale reflects a broader enthusiasm for geometry. In 1686, John Locke wrote that the study of Euclid was essential to a young person's education, a sentiment that both Jeremy Bentham and Adam Smith echoed in the succeeding century.[72] And in 1786, the philosopher Thomas Reid wrote: "Hence, a man of ordinary capacity finds no difficulty in understanding the definitions of *Euclid.* All the difficulty lies in forming the habit by which the name, and an accurate conception of its meaning, are so associated, that the one readily suggests the other. To form this habit requires time, and in some persons much more than in others."[73] Reid was obviously not alone, both in terms of his own enthusiasm for geometry and the sense of intellectual and cultural hierarchy that the mastery of Euclid's doctrines engendered.

One overlooked implication of Euclid's rise is that it offered new possibilities for ranking both peoples and cultures. This darker side of geometry's history is particularly observable in ostensibly enlightened North America.[74] Thomas Jefferson, who authored the Declaration of Independence, studied geometry intensely, even going so far as to acquire a copy of the *Elements* in Arabic so that he could pursue the study of that language.[75] Euclid's textual career had seemingly come full circle, as the *Elements* now led a prominent member of the Enlightenment back to the Arabic from which the work had once been translated. Significantly, Jefferson filtered geometry through the categories of his own era, as he held that Africans were incapable of understanding the subject's subtleties, while still holding out hope for indigenous Americans.[76] Whatever positive influences Euclid's *Elements* may have had in the New World, the anthropological rankings that emerged from geometry's cultivation became an intellectual backstop for the Old World tradition of slavery that Europeans had exported to the new one.

The belief in the efficacy of studying Euclid characterized the Spatial Reformation from its beginning to its end. I now examine the nature of the broad consensus by considering the sentiments that people expressed when discussing Euclid and the *Elements*. I begin with Isaac Barrow, who preceded Isaac Newton in the Lucasian Professorship of Mathematics at Cambridge. In an oration delivered upon his election to that position, Barrow said:

> For to pass by those Ancients, the wonderful *Pythagoras*, the sagacious *Democritus*, the divine *Plato*, the most subtle and learned *Aristotle*, Men whom every Age has hitherto acknowledge and deservedly honoured, as the greatest Philosophers, the Ring-leaders of Arts; in whose Judgments how much these Studies were esteemed, is abundantly proclaimed in History and confirmed by their famous Monuments, which are every where interspersed and bespangled with Mathematical Reasonings and Examples, as with so many Stars; and consequently any one not in some Degree conversant in these Studies will in vain expect to understand, or unlock their hidden Meanings, without the help of a Mathematical Key: For who can play well on *Aristotle*'s Instrument but with a Mathematical Quill; or not be altogether deaf to the Lessons of natural *Philosophy*, while ignorant of *Geometry*? Who void of (*Geometry* shall I say, or) *Arithmetic* can apprehend *Plato*'s *Socrates* lisping with Children concerning Square Numbers; or can conceive *Plato* himself treating not only of the Universe, but the Polity of Commonwealths regulated by the Laws of Geometry, and formed according to a Mathematical Plan?[77]

Barrow was just clearing his throat, as his comments extended for pages, in the course of which he added that without training in geometry it would be impossible to understand the work of scientists such as Galileo Galilei, William Gilbert, Pierre Gassendi, Marin Mersenne, and René Descartes.[78] Euclid was the key to all scientific knowledge.

With this parade of greats in mind, we can understand how Euclidean geometry came to form early modern thought's very sinew. Barrow subordinated every intellectual tradition, whether ancient or early modern, to Euclidean geometry. This tactic was common, as editors and translators of the *Elements* had long been saying similar things. No one ever expressed doubts about geometry's utility, although the justifications that were offered also varied with the time and place. For the most part, early modern editors advanced two justifications for studying Euclid. First, some held that reading the *Elements* built character, because it taught people to separate truth from falsehood. Second, others lauded its practical benefits, in terms of both the ability to reason and the ability to measure. The former justification was most prominent in the sixteenth century and was used heavily in Protestant regions, while Catholics tended to favor the latter—although the rules were never rigid. Nevertheless, even if there was some divergence among the groups, the editors were generally united in holding that the study of Euclid brought significant intellectual benefits.

The Protestant context, in particular, illustrates the complexity that could lurk behind Euclid's rise. The ethical dimension that Protestants discovered within Euclid's thought reflects, in part, ancient Stoicism's influence on sixteenth-century Christian humanism. Christian humanists were enamored especially of the Roman rhetorician Cicero, whose Stoic view of duty they applied to their own worldviews.[79] Thus, works published where Christian humanism prospered—in Protestant cities, such as Basel, Leipzig, Strasbourg, and Wittenberg—naturally praised Euclid as a teacher of virtue. The same theme appeared in the first English edition of the *Elements* by Henry Billingsley, which was published in 1570 in London. In the preface, Billingsley began with the proposition that knowledge of the arts and science: "[T]eacheth us rules and precepts of vertue, how, in common life amongst men, we ought to walke uprightly: what dueties pertaine to ourselves, what pertaine to the government or good order both of an houshold, and also of a citie or common wealth."[80] For Billingsley, the study of geometry was an essential aspect of a life well lived.

In Catholic regions, the situation differed slightly, because Catholic thinkers still cultivated Aristotle, which made Euclid's reception trickier. The most famous example of a Catholic appropriator is Christopher Clavius, the Jesuit astronomer who taught at the *Colegio Romano* in Rome and vigorously defended Aristotle. In these latter efforts Clavius failed. He succeeded, however, in cultivating a broad respect for Euclid within the Catholic world, as he produced one of the most important of all sixteenth-century translation/commentaries of the *Elements*.[81] When this work appeared in 1574, it was hailed as a triumph and was republished, or cited in many other editions as a source, including ostensibly Protestant ones.[82] Clavius, however, did not pursue ethics in his introduction. Instead, he emphasized the text's pedagogical utility: "Namely in this manner, and in our opinion,

Euclid [is presented] without difficulty for students, especially for those, who as beginners, now start the first study of mathematics; [they] will perceive greater enjoyment and utility."[83] The emphasis on mathematics was not accidental, as the Jesuits devoted much energy to this discipline's improvement, because they saw it as a means for sharpening young minds for the conflict with Protestantism.[84] Thus did Euclid become a central weapon in the Counter-Reformation's intellectual arsenal.

Clavius's mention of utility calls our attention to what was universal in Euclid's appeal, that understanding how to project space was useful. In 1551, the English mathematician Robert Recorde explained geometry's significance in a workbook, *The Pathway to Knowledge,* in which he underscored its utility for artisans: "Carpenters, Karvers, Joyners, and Masons, doe willingly acknowledge that they can worke nothing without reason of Geometrie, in so much that they challenge me as a peculiare science for them. But in that they should do wrong to all other men, seyng everie kynde of men have som benefit by me, not only in building, whiche is but other mennes costes, and the arte of Carpenters, Masons, and the other aforesaid, but in their owne private profession, whereof to avoide tediousness I make this rehersall."[85] Nothing about this was unique to Recorde or his English milieu. In 1689 in Brussels, for example, there appeared a translation by the Jesuit Jacob Kresa, in which he emphasized geometry's utility by praising a princely student: "Your Excellency had not reached fifteen years of age, when in geometry he had already studied the *Elements,* in geography he had explored the gulfs, in astronomy had completed starry courses, in military architecture had penetrated fortifications."[86] The student to whom Kresa refers began with Euclid and then moved to a host of disciplines that were, in turn, united by a common sense of space. Although Recorde and Kresa's texts referred to Euclid from very different cultural contexts, they converged on utility as a justification for the intense study of his greatest work. In the early modern period, Euclid's space truly became ubiquitous.

The Missing "Ism"

Although the literature on Euclid's influence on mathematics is ample, the full breadth of the *Elements*' effects has yet to be explored. I address this lacuna by introducing two concepts, Euclidism and *space making.* Since the term "Euclidism" has not been used outside the history of architecture, I propose a working definition: Euclidism constitutes, first, the projection of homogeneous space onto the *entire* world and, second, the cultivation of mental tools with which to manipulate that space.[87] Anchored historically to the study of Euclid's *Elements,* Euclidism grew in the course of the early modern period to include multiple disciplines, most of which not only used geometry's concepts but also proselytized on the discipline's behalf. Astronomical textbooks, for example, often encouraged their readers to begin with geometry.[88] Indeed, some works even included geometric primers—and much the same was true among texts that related to geography, to optics, and to a host of other disciplines, including now-defunct ones, such as cosmography.[89]

Euclidism's most important contribution is the manner in which it bound together multiple disciplines, anchoring each of them to a common system of thought. Here, I turn to a seventeenth-century example, Père Bernard Lamy's *The Elements of Geometry, or the Measurement of Bodies, Which Contains All that Euclid Taught*. Published in 1685, it was a well-known, widely used geometry textbook.[90] It is, therefore, significant that its introduction offered a list of the disciplines that relied on geometry:

> The utility of this science is evident in its furnishing of the elements of astronomy, gnomonics, navigation, optics, architecture, fortification, mechanics and, generally, of all the sciences that have bodies as their object—and by consequence of physics, too, which has been thoroughly examined, as several philosophers have claimed that [the latter] is nothing more than geometry. Nevertheless, this alone would not persuade me that one should teach it in the public schools, were it not [also] suited to forming the mind, to making it more exact, [more] extended and [more] penetrating.[91]

Lamy's list obviously includes a variety of disciplines. The issue, however, is not the list's comprehensiveness, but what the whole highlights. First, Lamy's list reveals a growing tendency to see geometry as the most fundamental of all disciplines—that is, it was the one from which all others drew their sustenance. Second, it underscores why geometry had such profound anthropological effects. If geometry truly formed human minds, then to read Euclid was, implicitly, to reform humanity.

Space making takes off from this point. Early modernists have long been attuned to the historical significance of printed texts. They have paid less attention, however, to how texts aided readers in *producing* the space in which they understood humanity to be living. I illustrate this point via two images. The first we have already seen; it is the image of Creation in figure 6, in which humanity, the cosmos, and God are put into a fractured space. With this image in mind, I turn to figure 7, which contains a portrait of the geometer Luca Pacioli that was done in 1495 by the Italian artist Jacopo de'Barbari. In it are depicted the mathematical friar himself, a young man who may be the Duke of Urbino, an open book, a closed book, two geometric solids, and a collection of writing implements and instruments. The open book is Euclid's *Elements* and the closed one is

Figure 7. De'Barbari, *Portrait of Luca Pacioli*, 1495. Naples, Museo di Capodimonte © 2017. Courtesy of SCALA Archives, Florence, and the Ministero dei Beni e Attività Culturali e del Turismo.

a mathematical tome by Pacioli. Rather than discuss the entire painting, I concentrate on its presentation of the *Elements*.

The Pacioli portrait illustrates how Euclidism aided *space making*. De'Barbari included a real edition of the *Elements*, namely the 1482 version that I have mentioned. (One scholar has identified the specific page to which the work is opened.)[92] Thus, in addition to its obvious endorsement of geometry as a foundational discipline, the portrait memorializes an important figure's relationship *to a specific edition* of the work. (Significantly, Nicolaus Copernicus owned a copy of this edition, too.) In this context, Pacioli's actions are especially important. With his left index finger de'Barbari's Pacioli directs the viewer's attention to a specific point in the *Elements*, Proposition 8 in Book 13. I do not analyze this proposition except to note that it is from the advanced spherical geometry that appears at the end of the text, well after the first six books that the medieval world knew so well. The text's significance as a *space maker* becomes clearer when we consider that Pacioli's right hand draws the same forms that appear on the opened page. This painting illustrates, in short, how the Renaissance understood Euclidean space's reception: it entered the mind through the eyes and, in turn, exited it via the hands.

Neither Euclidism nor *space making* occupies a prominent position within the contemporary intellectual historical literature. Instead, when tracing classical influences on early modern thought, scholars have emphasized what I will call the "Big Four + 1": Aristotelianism, Epicureanism, Platonism, and Stoicism, on the one hand, with Skepticism identified as a universal challenger, on the other. However, as I have explained, early modern adherents to the "Big Four" were firmly united by their interest in (and respect for) Euclid. And much the same holds true for seventeenth-century "isms," such as Newtonianism and Radicalism, the latter of which has become especially important recently.[93] Without getting into a detailed discussion of either of the latter two "isms," I note that both of their progenitors, Isaac Newton and Benedict Spinoza, were tied directly to the Spatial Reformation's space. Newton was Isaac Barrow's mentee and studied Euclid at his behest, while Spinoza was a lens grinder, which required him to have a command of optics, one of the many disciplines that was being subordinated to geometry. Overall, the conclusion is clear: by the mid-seventeenth century, no major early modern thinker could have been ignorant of Euclidean geometry's basics.

Mere Past Versus Essential Prologue

If we take stock of the history of mathematics, it becomes evident that our own modernity is something to be overcome. Mathematics is a progressive discipline and seeks always to reach beyond its existing bounds. As a result, old ideas and approaches are continually tested and, if found wanting, jettisoned by the newest incarnation. The history of mathematics has absorbed this model and is therefore intrinsically a chronicle of progress, with much of that progress being understood, since the early twentieth century, via the supersession of Euclidean space through the rise of modern mathematical thought.[94] However, if the progressive emphases that govern the history of mathematics provide our primary frame of reference, we risk writing the early modern encounter with geometry's space completely out of the historical record.

As an intellectual historical phenomenon, the Spatial Reformation demands a more differentiated approach. In order to lay a foundation for such an approach, I recount briefly geometry's history from the Renaissance to the nineteenth century. With respect to the Spatial Reformation's earliest phases, it is important to note that Euclid remained a dominant force, as his geometry formed the core of mathematics throughout the quattrocento and into the sixteenth century—and this dominance goes a long way to explaining Euclidean space's cultural ubiquity into the seventeenth century. Both Nicholas of Cusa and Philipp Melanchthon were intensely interested in geometry—and their support lent great prestige to the study of the *Elements*.

In the course of the sixteenth century, however, Euclid's geometry yielded pride of place to newer mathematical currents. During that century's initial half—above all, in Italy—mathematicians began to investigate the power of algebra and trigonometry, with people such as Benjamin Cardono and Niccolò Tartaglia exploring algebraic equations as a means to calculate motion, while still others investigated the possibilities of finding one's place on a sphere via trigonometry.[95] Euclid, of course, had nothing to say on either subject. Moreover, the growth in nongeometric areas of mathematics derives additional significance, if we superimpose onto it the history of the seven liberal arts. This group has comprised, since late antiquity, grammar, rhetoric, and dialectic (the Trivium), as well as arithmetic, geometry, astronomy, and music (the Quadrivium). Although the number of the arts waxed and waned, initially this knowledge system never included mathematics as such, but only geometry and arithmetic.[96] Thus, when the mathematical discipline took its first steps beyond Euclid, it fatally undermined what was left of the Quadrivium's foundations.

With respect to the liberal arts' early modern fate, I am not arguing anything new. Instead, I am simply calling attention to the profoundly anthropological nature of the conceptual reordering that succeeded the original Trivium and Quadrivium. In this instance, Lamy's work is particularly illustrative, as it shows explicitly how, after 1650, geometry came to be considered a source of mental tools that any student could apply to any given body of human knowledge. Geometry, in this sense, preceded both the world and its subsequent study, precisely because it was a human phenomenon. In this sense, space's inherent humanization was an essential precursor to the rise of a "scientific" culture.

With this change in mind, I turn to some of the implications of the move of mathematics beyond geometry. After 1550, two things increasingly defined geometry. The first was the growing recognition of its limits when analyzing the physical world, with the second being the application of ever more sophisticated mathematical tools to that world. In this context, I turn to one towering figure, Johannes Kepler. Kepler was, as I have already noted, a fan of geometry and sought, initially, to explain the planets' motions via geometric analyses. He failed to achieve this goal, because, as we now know (indeed, thanks to Kepler himself), planetary orbits are elliptical rather than circular.[97] This discovery meant that Euclid's spherical geometry could not be applied fully to the cosmos, since the two did not coincide. In this respect, Kepler broke Aristotle's philosophical stranglehold on the heavens, and this achievement enabled other scientists, such as Galileo Galilei, to use mathematics in ways that fundamentally redefined the relationship of Heaven and Earth by subordinating both of them to a humanly projected space.

Within the emerging tradition of "modern" philosophy, meanwhile, the ability to command space put the subject into a dominant position vis-à-vis the experiential world. As I suggested in the introduction, the finest example of this trend is René Descartes, who not only invented analytical geometry but also wrought a new cosmology, a new approach to the philosophical subject, and a new philosophical path to God. (I return to this issue in the conclusion.) Not coincidentally, while Descartes was using mathematics to rethink humanity's relationship to both the cosmos and God, his countryman Blaise Pascal pursued the discipline into a blunt confrontation with the problem of God's presence.[98] I cannot explicate all the details, but I suggest that, if taken collectively, early seventeenth-century thought produced a profound change in homogeneous space's philosophical status, as new departures in mathematics resituated geometry, with respect to humanity, the cosmos, and God.

Mathematics after 1650 was, in turn, characterized by various attempts to analyze the natural world through a suppler way of representing continuity. This system was the calculus.[99] Invented independently by Isaac Newton in England and Gottfried Leibniz in Germany, the calculus had two distinct advantages over geometry.[100] First, it allowed mathematicians to plot a specific point on a curve, which was not possible within traditional geometry. Second, it established the continuity of our number system in a less cumbersome way than Euclid had done.[101] In the course of Books 7–10 of the *Elements*, Euclid borrowed heavily from Aristotle and defined numbers as units that extended over a line.[102] In doing so, he moved the vision of space beyond an older Pythagorean tradition that saw numbers as discrete items (dots) that could be arranged in a geometric pattern, with the lines that produced geometric figures being drawn between the dots.[103] Although Euclid's approach had distinct advantages over Pythagorean approaches to space, it also had limitations. For example, Euclidean magnitudes cannot be multiplied or divided and all numbers are "natural," that is, they are positive integers. This latter characteristic, in particular, limited geometry's utility when it came to understanding natural processes, such as motion. Hence, geometry, although still very useful for imagining space, was no longer to be understood as universal.

The rise of calculus cemented a shift within mathematics that established that the relationship between numbers was real, that our number system was continuous and, thus, applicable to the physical world.[104] The advantages of this shift for natural science are obvious, as one could now mathematize every physical phenomenon. From the perspective of the Spatial Reformation, however, this development also constituted yet another step away from traditional geometry. On the one hand, as I noted with respect to Kepler, Galileo, and company, circles and spheres were becoming ever more artifacts of the human mind, as opposed to the underpinnings of the cosmos. On the other hand, the calculus is more abstract and, therefore, less intuitive than Euclidean geometry, which means that it requires more instruction. Mathematics was, in effect, slowly leaving behind both the Euclidism and the *space making* that had been fundamental to de'Barbari's portrait of Pacioli.

Along these lines, it is significant that eighteenth-century mathematics was primarily an extension of the calculus's original invention. In general, enlightened mathematicians were more concerned with exploring the possibilities that the calculus offered for understanding the physical world than they were in probing the limits of existing geometric methods. Handed a triumph of western thought, they applied their new tools to diverse phenomena, ranging from planetary motion to flowing water, and so on. Some mathematicians did make significant theoretical advances, although the primary tendency was toward application. A good example is the Swiss mathematician Leonhard Euler, who showed how many natural phenomena could be subjected to the calculus, while making important theoretical contributions.[105] Euler illustrates better than any other mathematician of his age how, although Euclid's geometry had not been dethroned, the discipline's seat was nevertheless losing its central position.

The final shift toward a post-Euclidean spatial sense was incubated during the first half of the nineteenth century.[106] During this time, many mathematicians, working independently of each other, turned a critical eye toward the parallel postulate. The *Elements* begins with twenty-three definitions, five postulates, and four common notions. The fifth postulate is the most famous: "That, if a straight line falling on two straight lines make the interior angles on the same side less than two right angles, the two straight lines, if produced indefinitely, meet on that side on which are the angles less than the two right angles."[107] This assertion had long been the subject of doubt, as repeated attempts to prove it had ended in failure. Indeed, already in 1621, Henry Savile became the first mathematician to note that objects drawn in planar space were incompatible with those drawn in spherical space.[108]

A radical change in direction came only when mathematicians managed to develop logically consistent, non-Euclidean geometries. The process took decades. Early in the nineteenth century, some mathematicians argued that Euclid's chief assumptions were unjustified, including the parallel postulate.[109] The reconstruction of geometry along new conceptual lines began with Nikolai Lobachevski and János Bolyai, who developed their ideas independently and published them in 1829 and 1832, respectively. The fundamental rethinking of geometric space's foundations reached its apex, however, with Bernhard Riemann.[110] In 1854, he delivered a talk at the University of Göttingen under the title "On the Hypotheses That Underlie Geometry." This work, published in 1867, laid the foundation for a vision of space that included more than just three dimensions, as Riemann demonstrated that geometry could be logically consistent even if it broke with Euclidean assumptions about space's fundamental three-dimensionality.[111]

Riemann and the others undermined what came to be seen as a naïve belief in geometric space's regularity. The broader cultural significance of this development becomes apparent when we consider that exactly this space had been incorporated into both the Newtonian physical revolution of the late seventeenth century and the Kantian philosophical revolution of the late eighteenth century. Against this backdrop, I underscore two issues. First, one cannot *see* any of the additional dimensions whose existence Riemann managed to justify.[112] Thus, the world in which Pacioli saw things (and was seen by others) had been rent asunder by a vertiginous form of abstraction. Second, Riemann opened the door to a new kind of physics, as Albert Einstein's twentieth-century revolution in physics drew inspiration from Riemann's contributions to geometry.[113] As the Spatial Reformation reached its definitive end in the decades after 1850, another cosmovision began to emerge, with the result that entirely new visions of humanity emerged, too.

Conclusion

Euclid's rise and decline offers an alternative frame for organizing European thought's history between the fourteenth and the nineteenth centuries. First, it allows us to rethink the transition from late medieval to Renaissance thought, which began after 1350 and continued up to 1650. It was not simply that Aristotelian thought waned, as other ancient systems rose, but that a new way of understanding space diffused to the point that Euclid *resituated* all ancient belief systems, including the Christian one. Second, this frame provides a new way of looking at the conceptual changes that dominated the period from 1650 until Euclid's demotion after 1850. As space became an artifact of the human mind, even profounder changes in thought became possible, including, as I detail below, new visions of the earth, of the divine, of politics, and of sentient life in general.

If Euclid's rise and decline between 1350 and 1850 invites us to see the early modern period in ample breadth, it also offers a way to see the period in profound depth. Homogeneous space diffused into every conceptual corner and touched everything within European culture. Thus, every major intellectual and cultural figure applied Euclid's thought to his mental world, regardless of the field of endeavor. In the early modern period, whether one was a mathematician, scientist, philosopher, artist, theologian, political theorist, or anthropologist, one could no more avoid the *Elements* than one could avoid the Bible. Along these lines, the chapters that follow explore different aspects of homogeneous space's effects. All of them are tied to the timeline that I laid down in the introduction, even though each drills down into a specific issue.

In this respect, it is important to keep in mind the broad outline that I sketched in the introduction. First, European thought can be characterized as pre-Euclidean (medieval), Euclidean (early modern), and post-Euclidean (modern). Although I concentrate on the middle period in this book, it is crucial to understand that visions of space have always played a fundamental role in European thought's development. Second, the early modern spatial regime can be best understood through a combination of explications of texts and elucidations of material culture. Put another way, since Euclidism and *space making* cut across the entire period in question, we must pay careful attention to how Europeans imagined and produced the space in which humanity, the cosmos, and God were suspended. We cannot understand the tremors that homogeneous space's rise and subsequent decline caused if we fail to consider how Europeans repeatedly learned new ways to situate things unseen.

Chapter 2

The Renaissance and the Round Ball

On a round ball
A workman that has copies by, can lay
An Europe, Afrique, and an Asia,
And quickly make that, which was nothing, All.
—John Donne, "A Valediction: Of Weeping" (1633), in *The Complete Poetry and Selected Prose of John Donne*, ed. Coffin

The earth was invented in 1477. In that year, a cartographer named Nicolaus Germanus completed a little-known terrestrial globe in Rome.[1] It may not have been the first globe produced; another could have been made around 1444 in Brussels.[2] Germanus's terrestrial globe was, however, the first to be paired with a celestial one—and this innovation marks the start of an epoch, as globe pairing became the dominant practice among globe makers, reaching its height in the middle of the seventeenth century, before ending around the middle of the nineteenth.[3] Today, in contrast, one sphere is sufficient for establishing a "global" space, and for that reason terrestrial globes are generally sold singly. It was not always so. From the earliest stages of production until almost four centuries later, terrestrial globes were usually paired with a celestial counterpart. In short, early modern Europeans put Heaven and Earth into a space that, although it remains accessible to us, no longer structures our own global imagination.

The history of globes is a central theme in the Spatial Reformation's rise and decline, because it illuminates how Europe's cosmos was transformed by the whole's submersion into a new mode of projection. Constructed in the course of the fifteenth and sixteenth centuries and perduring into the nineteenth, this mode boasted as its central attribute the spatial homogeneity whose underpinnings I discussed in the previous chapter. The conceptual space that emerged is illustrated in figure 8, which I have taken from *Cosmography, or Description of the Universal Orb*, a work that was originally published in 1529 by the mathematician Peter Apian, before appearing in multiple reprints and translations.[4] (I have taken

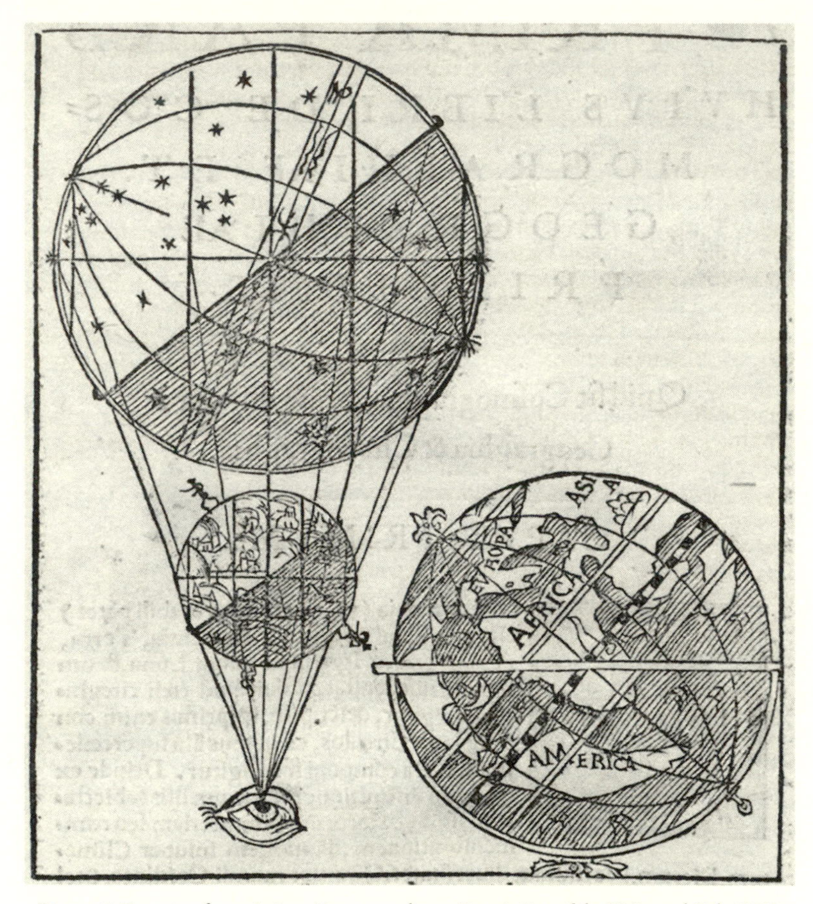

Figure 8. Drawing from Apian, *Cosmography, or Description of the Universal Orb*, 1529. Courtesy of ETH-Bibliothek Zürich.

the image from a 1540 version that was edited by the mathematician Gemma Frisius.)[5] Depicted are a celestial and terrestrial sphere, an omniscient eye, a terrestrial globe, and a homogeneity that permeates and situates everything.

Apian's illustration reveals the vitality of the alliance between the mind and the hands that I identified in de'Barbari's portrait of Luca Pacioli. Consider that Apian's work included inside a single space not only a drawing of a terrestrial globe but also representations of a celestial and a terrestrial sphere. The relative positioning of these things inside this constructed space is significant in three ways. First, the image excludes God—and His absence subtly separated the physical whole that was depicted from the hierarchy of being that I have discussed previously. Second, by subordinating the cosmos to an eye that floats within a generic space, the image implicitly liberated the spheres from their (then accepted) geocentric constraints. Third, by suspending both the cosmos and an object of material culture inside this imagined realm, the image underscored humanity's contribution to this new mode of presentation. Unseen realms could now be encased within texts and reified in material culture for a human viewer who may not have been able to see the *All* but could imagine it.

Historians of cartography have largely overlooked global space, preferring to concentrate on globes' most superficial aspect, their cartographic images.[6] This is true for studies of both celestial and terrestrial globes, but is particularly prominent in the case of the latter. For example, Edward Stevenson's magisterial *Terrestrial and Celestial Globes* details the evolution of drawings on globes of both kinds, as its title would suggest, although it fails to mention globe pairing.[7] Subsequent modern works on terrestrial globes by Oswald Muris, Alois Fauser, Catherine Hofmann, and Dennis Cosgrove cleave to a similar line. A few works, such as the synthetic study *Globes from the Western World* by Elly Dekker and Peter van der Krogt, and the wide-ranging essay "The Image of the Spherical Earth" by David Woodward, mention globe pairing, although without pursuing it.[8] Concerned primarily with establishing a genealogy of the earth's "true" image, the literature has constructed a positivist tale of humanity's progress toward geographic omniscience.[9] The resulting collation of disciplinary milestones assumes, however, that our contemporary globes and their Renaissance predecessors share the same space. This belief cannot, however, be reconciled with globe pairing's decline in the late nineteenth century. Building, therefore, on the previous chapter's discussion of Euclid's rise and decline, I suggest that early modern global space differs in important ways from its modern successor—and this difference has a history.

The contemporary literature's emphasis on globes' surfaces, as opposed to the space that sustained them, has roots in two historical accidents. The first is that the oldest surviving terrestrial globe, the Erdapfel of the Nuremberg merchant Martin Behaim, was built singly and (to this day) stands alone. Completed in 1492, it was immediately dubbed a wonder and, by the nineteenth century, had become so famous that it set much of the literature's trajectory.[10] Germanus's globe pair, in contrast, was probably destroyed in the *Sacco di Roma* of 1527 and, for that reason, had little effect on the literature before the 1980s, when Jósef Babicz published pioneering researches on the matter.[11] As a result, the history of globe pairing and its relationship to homogeneous space's rise has not been the subject of study. Indeed, it is only within the past three decades that scholars have suggested that a full-blown spatial culture may have predated Behaim.[12]

The second (and, in many ways, weightier) accident is the Erdapfel's pre-Columbian date of completion. Behaim knew nothing of the New World, because Christopher Columbus returned to Spain only in March 1493. Influenced, however, by the epochal significance of the Genovese's round trip, scholars have subordinated the history of globes to a timeline that is more appropriate to the

Age of Exploration than to changes in Europe's spatial regime. The latter, as I explained in the introduction, had different conceptual roots and also started earlier.[13] Reports of newly encountered lands were, no doubt, important to globe makers, who incorporated the fresh information into their work. Still, the stream of reports produced neither globes themselves nor the space that sustained them. These things emerged from a different process, in which the mental equipment that allowed human beings to put the earth onto a round ball was recovered and transferred to material culture.

The terrestrial globe's rise and attending pairing with celestial globes was an integral part of the Spatial Reformation's development. As I have already explained, this continent-wide shift had roots in the fourteenth century but reached maturity early in the fifteenth, as homogeneous space was applied liberally to many fields, including architecture, art, astronomy, cosmography, geography, optics, and mathematics. All these fields had important connections to globe making, although the literature has overlooked them. The scholarship has not emphasized, for instance, how globe making and cosmography (of which Apian's text is but one example) shared the same conceptual apparatus.[14] And with respect to the history of art, scholars have read globes' increasing presence in Renaissance paintings as indicating that these items were gaining cultural significance. However correct this observation may be, it misses an important point: globes that were depicted in paintings appeared inside a space that had been created (Apian-like) for a human viewer. In other words, both paintings of globes and individual globes themselves relied on a mode of representation that assumed (and also privileged) a human perspective on everything.

Consistent with what I said about Apian's Eye, paintings of globes *resituated* both the celestial and the terrestrial by subjecting the whole to a humanly projected, homogeneous space. Against this backdrop, I suggest that one can best illustrate this process by analyzing the history of globes in conjunction with the history of art and intellectual history. Put more practically, if one wishes to understand early modern space's rise, one must mix analyses of material objects with explications of both images and intellectual historical texts. Within this inherently multi-disciplinary context, I turn to a Renaissance painting that includes globes: Raphael's fresco, *The School of Athens* (figure 9). Completed in 1511 and located in the Vatican's *Stanza della Segnatura*, which served as a papal study, this work is of incalculable significance for the history of globes.[15] Not only is it probably the earliest depiction of a globe pair, but it also highlights global space's origins in ancient thought.

Raphael's *School of Athens* comprises nothing less than a genealogy of global space. First, it associates each of the globes with a classical authority (see the bottom right of the image). For the celestial realm, this is the astronomer Hipparchus, and for the terrestrial one it is the geographer Claudius Ptolemy (wearing

Figure 9. Raphael, *The School of Athens*, 1511. Vatican, Stanza della Segnatura © 2017. Courtesy of SCALA Archives, Florence.

the crown).[16] Second, by putting these two authorities face to face, the painting highlights global space's anthropological resonances, as both the celestial and terrestrial realms rest in human hands, while the globe bearers look into each other's eyes. Given that God is also conspicuously absent from this image, it seems safe to suggest that ancient spatial thought directly supported the Renaissance era's production of a more human world.

The contributions of homogeneous space's return to the new spatial imagination are, moreover, particularly apparent in Raphael's reference to geometry. Between Hipparchus and Ptolemy is none other than geometry's prophet, Euclid, who has bent over to teach the discipline's subject matter. Both Euclid's location in the image and his engagement in teaching exemplify how geometry's idealized space made possible a human perspective on Heaven and Earth. It was not just that geometry formed the human mind's wings (notice that one student is looking up a Hipparchus's globe) but that the underlying whole allowed the entire cosmos to be presented on this Earth and in fully human terms. In addition, and along more artistic lines, to the right stand the legendary fifth-century painters Protogenes and Appelles, with the latter actually sporting Raphael's visage.[17] That Raphael put himself amid these ancient spatial traditions not only suggests the intellectual complexity of the milieu in which the earth was invented, but also underscores how Euclidean geometry came to pervade all ancient traditions. Once in command of space, fifteenth- and sixteenth-century Europeans fashioned (and continually refashioned) their mental cosmos, with an explosion of cultural productivity being one important result.

Fractured Space

With Raphael's fresco in mind, I turn to terrestrial globes' curious absence from the medieval spatial imagination.[18] The absence is curious, because antiquity boasted a rich globe culture that included the production of both celestial and terrestrial globes.[19] The oldest existing celestial globe, for instance, the Farnese Atlas, dates to the second century AD and is, in turn, a copy of a Hellenistic original that probably dated to the second century BC.[20] Celestial globes moreover, were also painted onto the walls of Roman interiors—and there is evidence that such globes were produced north of the Alps.[21] Concomitantly, terrestrial globes enjoyed great currency in antiquity.[22] In the first century AD the geographer Strabo reported that around 150 BC, a Stoic named Crates of Mallos had built a terrestrial globe ten feet in diameter.[23] Strabo may even have made a globe himself. It is certain, however, that murals of terrestrial globes also proliferated in Roman-era residences.[24] Given what I have said about homogeneous space's role in the earth's invention, antiquity's globe culture suggests the presence of a generalized unity, on the basis of which a spatial whole could be represented.[25] Indeed, a unity was present; it was not, however, the spatial one that dominated the early modern period. Instead, it was primarily a philosophical one that was rooted in Stoicism's chief assumptions about the cosmos. Although I return below to Stoicism and its relationship to homogenous space, I cannot recount this system's entire story. How the Stoic cosmos formed and dissolved—and what this meant for ancient spatial thought—needs to be part of another book.

Antiquity's practice of making terrestrial globes did not long survive Rome's fall, as the empire's successors—Byzantium, Latin Christendom, and Islam—limited themselves to global reconstructions of the celestial realm.[26] Although the evidence is fragmentary, it seems that Western Europeans were making celestial globes by the year 1000, while the same happened as early as the ninth century in the Muslim world and, perhaps, even earlier in Byzantium. Terrestrial globes did make one brief medieval appearance. In the late thirteenth century, the Muslim world produced (as far as we know) a single terrestrial globe, which was shipped to China and promptly forgotten about—by both sides, apparently.[27] If we step back, it is quite startling to behold the global absence of globes. All three post-Roman zones inherited antiquity's intellectual riches. Yet, in all of them, the terrestrial globe became *terra abscondita*.

The earth's reinvention was long in coming, because Rome's successors assiduously cultivated the fractured space whose history I outlined in the introduction. In pursuing the differences between ancient homogeneous space and medieval approaches, I concentrate on Western Europe, because it was in this region that terrestrial globes reappeared. The other post-Roman cultures were more developed than the emerging European one, of course, which makes their failure to cultivate the earth's image all the more striking. Nevertheless, I must place Byzantium and Islam outside this discussion. In order to understand why terrestrial globes reappeared only in the former Roman Empire's backward western part—and only in the fifteenth century—I examine the ideas that the barbarian West borrowed between the seventh and the tenth century from the increasingly moribund Roman Empire.

As late antiquity degenerated into the early medieval period, Western Europeans recovered three doctrines that determined how the terrestrial realm would be encoded for nearly a millennium: (1) a spherical cosmos that was structured by crystalline spheres, (2) geocentricity, and (3) spatial hierarchy.[28] Only a handful of classical texts on space and cosmology remained available in the post-Roman, Latin-speaking world. These included works by Plato (late fourth century BC), Pliny the Elder (first century AD), and Macrobius and Martianus Capella (both fifth century AD).[29] All these thinkers accepted, in some form, the doctrines just mentioned. Their works were read widely and also heavily influenced important early medieval thinkers, including Isidore of Seville (late sixth century), the Venerable Bede (late seventh century), and John Scotus Eriugena (ninth century).[30] The thinkers of the late tenth century, in turn, embraced both this cosmos and its space, as in the examples of Gerbert d'Aurillac, who made the first known medieval celestial sphere, and Notker Labeo, who may also have made one.[31] By the year 1000 Western Europe accepted that humanity resided on an immobile and imperfect *terra*—not a planet in the modern sense, but a blob of substances—while nested, rotating spheres bore the celestial bodies through the perfection above.[32]

The incomplete nature of classical cosmology's return was not, however, the only reason that the medieval world eschewed pasting the terrestrial surface onto spheres. Another was religion. As medieval thinkers digested antiquity's remnants, they grafted Christian assumptions about God and Creation onto the ancient cosmos and, consequently, intensified their commitment to spatial hierarchy.[33] For example, although the Bible relates in the book of Genesis that God created Heaven and Earth—putting the former "above" and the latter "below"—it is vague on His physical location. Classical cosmology proved useful here, because God and His divine entourage could be imposed onto the pagan cosmos's upper realms. (This *location* of the divine is already apparent in Isidore of Seville's works and runs into John Scotus Eriugena in the ninth century.)[34] Thus, when medieval Christians contemplated the cosmos's upper limits, their minds entered not only a different space but also a *holy* realm—and this spiritual dimension proved particularly inhospitable to terrestrial globes.

During the late twelfth century, the medieval emphasis on spatial hierarchy became more intense, after additional classical texts were translated into Latin. Four works were especially important—three by Aristotle (fourth century BC) and the other by Claudius Ptolemy (second century AD).[35] The Aristotelian ones were the *Physics, On the Heavens,* and *Meteorology,* all of which assumed the cosmos to be geocentric and fractured spatially.[36] From among the three, I single out the *Physics,* because this text long directed medieval thought toward non-global ways of encoding terrestrial space. In the *Physics* Aristotle not only explicated the spherical, geocentric cosmos, but also imposed onto it a spatial dichotomy. Dividing this cosmos into sub-lunar and supra-lunar realms, he assigned to each a different physics and, concomitantly, a different space. In effect, Aristotle transferred Plato's distinction between real and ideal to the physical cosmos—and this rendered his cosmological thought incompatible with homogeneous space.

The medieval appropriation of Aristotle's fractured cosmos derived strength, in turn, from the return of the fourth work, Claudius Ptolemy's *Almagest.* Ptolemy wrote this work in Greek in Alexandria during the first century AD.[37] It was then translated into Arabic in the ninth century and into Latin (from Arabic versions) in the twelfth century. Ptolemy was critical of Aristotle's cosmology, although medieval thinkers overlooked this aspect of his thought.[38] In the *Almagest*'s first book, Ptolemy reduced the eighty crystalline spheres in Aristotelian thought to just eight, which bothered no one in the medieval period.[39] He also resisted the radical distinction between sub-lunar and supra-lunar, due to Stoic influences on his thought.[40] This philosophical dimension of Ptolemy's thought remained, however, in Aristotle's looming shadow, with the result that fractured space endured within European thought.

Here, I must make a detour through a theme that will continue to lurk in the background: the continual presence within Western thought of alternative (non-Aristotelian) cosmologies. In Ptolemy's case, the alternative was Stoicism, from which he borrowed heavily. The ancient Stoics saw the cosmos as an organism that was unified by a pervasive substance that they called *pneuma.*[41] This "biological" approach lent to the cosmos a unity that cut across spatial distinctions—and this was, in turn, reflected in Ptolemy's joint application of mathematics to the study of both Heaven and Earth. Mathematics did not itself unify the cosmos within Ptolemy's thought, but its mental tools ran parallel to the cosmos's ostensible pneumatic unity. It is, thus, suggestive that Strabo identified a Stoic—and not an Aristotelian—as an early maker of terrestrial globes. To put the earth's surface onto a round ball required that Heaven and Earth be unified in some, presumably non-Aristotelian, way.[42] In the early modern period, this unity came from projected space.

Medieval interpreters overlooked, however, the tension between Ptolemy and Aristotle on projected space. They did so, in part, because they had no access to Stoic cosmological works, which survived only in fragments and, in part, because Ptolemy also deployed the Aristotelian spatial hierarchy in his discussion of the Heavens.[43] He believed, for instance, that the terrestrial realm was characterized by corruption, while the Heavens were perfect.[44] Thus, Ptolemy could be read (and was read) in the medieval period as confirming Aristotle's physical doctrines, even if his underlying philosophical assumptions remained in tension with them. The Aristotelian reading of Ptolemy's space was reinforced, however, by the *absence* of Ptolemy's main geographic work, the *Geography*. I return to this text below but emphasize here that in it Ptolemy applied to the Earth the same mathematics that he used on the Heavens, bringing in effect longitude down to the terrestrial realm. Prior to the fifteenth century, of course, this practice was almost unimaginable in the West.

While keeping in mind the tensions that pervaded the Aristotelian-Ptolemaic cosmos, one must also recognize the system's utility for organizing celestial phenomena. In no case is this clearer than in the *Almagest*'s most important contribution, its stellar coordinates, which put 1,022 stars on the celestial sphere and organized them into 48 constellations, many of which are still in use.[45] This collection became medieval astronomy's cornerstone, since it allowed heavenly bodies to be put into place. It also meant that the *Almagest* became central to the production of celestial spheres.[46] The oldest existing celestial sphere dates to the first half of the fourteenth century and uses Ptolemy's constellations.[47] (Significantly, it was also the personal property of Cardinal Nicholas of Cusa.) Thus, when framed against Aristotle's philosophy, the *Almagest* could not but accentuate that homogeneous space was reserved to the Heavens alone.

Having sketched what I will call a textual frame for medieval approaches to the terrestrial realm, I turn to an Aristotelian doctrine whose significance for the history of globes has been overlooked, namely the express restriction of geometry to the celestial. In the *Physics*, a core element of medieval education (that remained so into the Renaissance), Aristotle argued that geometry could not be applied to the sub-lunar realm, because it dealt with a perfect, idealized space of the sort that is located "above."[48] Not all ancient thinkers accepted this view, but those to whom early medieval Europeans had access did.[49] Thus, when medieval thinkers applied spherical geometry to the cosmos, they did this in support of the discipline of astronomy—and expressly not of geography. Although they believed the earth to be spherical in shape, they failed to apply to the terrestrial surface the same coordinate system that they applied routinely to the Heavens.[50] For example, the most important medieval work on the spherical cosmos, the thirteenth-century *The Sphere* by John Sacrobosco, contains no specific geographic information, even though it insists that the earth was a sphere.[51] Only one thirteenth-century thinker, the Oxford-trained Franciscan Roger Bacon, applied a system of coordinates to the terrestrial realm. His vision of a terrestrial mathematics withered, however, until fourteenth-century Nominalism enabled homogeneity's descent from the Heavens.[52]

We can gain additional perspective on terrestrial globes' absence by noting that Earth's surface was present within medieval visual culture. When it did appear, however, it was encoded with reference to Aristotelian strictures on space. Here, I call attention to figure 10, which comes from a fourteenth-century *Bible moralisée*. Originating in thirteenth-century France, before spreading around Europe, such religious works combined the biblical text with rich collections of images and goodly doses of pious commentary.[53] Not surprisingly, given the religious backdrop, the earth appears without reference to the amalgamating mode of spatial projections that suffused Apian's work. Instead, this profoundly medieval perspective on the cosmos defines terrestrial space in religiously symbolic and hierarchical terms, with the Christian God cradling our cosmos, as He measures its dimensions with a compass. This starting point for understanding space had a profound effect on medieval traditions in geography.

In spite of God's omniscient perspective (or, perhaps, because of it), the blob on the inside of the image is devoid of geographic features. This detail is significant, because the medieval world had a sophisticated culture of world maps, which are known as *mappaemundi*—and their knowledge could have been deployed here.[54] Nevertheless, the image encoded the terrestrial realm inside a hierarchical space that dismissed "accurate" representations of the terrestrial as insignificant for the divine. God cradled cosmological spheres—and not a terrestrial globe—because the earth's surface features were not *worth* representing. The history of spheres cannot, therefore, be written without reference to the space in which spheres were suspended.

Although there was a religious backdrop to this phenomenon, the earth's absence also had deep roots in astronomy. Astronomers traditionally cultivated an "outside-in" view of the cosmos. This external view was enshrined in a rich array of scientific instruments that included astrolabes, armillary spheres, and celestial spheres.[55] Consistent with the Aristotelian separation of celestial from terrestrial, astronomical instruments generally assumed a supra-celestial perspective from which a terrestrial *sphere* could be imagined, but not a globe. I illustrate the problem through an analysis of the armillary sphere. (In this context, I also call the reader's attention back to figure 4, which contains an image of Nicole

Figure 10. Illustration from *Bible Moralisée,* thirteenth century.
Courtesy of the Warburg Institute Iconographic Database.

Oresme contemplating an armillary sphere.) A tool for teaching astronomy's chief geometric concepts, the armillary sphere had an ancient lineage that reached back to Archimedes in the third century BC and ran up through Ptolemy.[56] Book V of the *Almagest,* for example, includes a discussion of such spheres—and it was via this text that the armillary sphere passed into medieval Arabic astronomical works and, in turn, to medieval Latin ones.

The armillary sphere's geometric structure is an important datum for understanding medieval space. This instrument includes a generic sphere on the inside that represents the terrestrial realm, while assuming another sphere on the outside that represents the Heavens. The inner sphere is a solid body, while the outer one is merely imagined, with its presence suggested by a network of rings (*armillae* in Latin) that is, in turn, affixed to the inner sphere by an axis. An example is illustrated in figure 11, which comes from the frontispiece to *Epitome of Ptolemy's Almagest* (1496), a summary of Ptolemy's great work by Johannes Regiomontanus, who was the fifteenth century's foremost mathematician and also the first person to translate Euclid's *Elements* from Byzantine manuscripts.[57] (Ptolemy is, once again, wearing a crown.) I return to Regiomontanus, below. For now, I concentrate on the armillary sphere.

Regiomontanus's armillary sphere illustrates how, prior to the Renaissance reappropriation of geometric space, homogeneity pervaded the celestial realm alone. The imagined outer sphere is situated by a series of rings, both great and minor, each of which can be divided into equal units. (Great rings divide the celestial sphere in half; minor rings do not.) The most important among the great rings is the celestial equator, because it allows viewers to orient themselves with respect to the celestial sphere's surface—that is, to distinguish its top from its bottom. The space of the outer sphere is, therefore, built on certain assumptions about the geometric stability of "up" and "down" that were deployed to justify an extra-cosmological viewer. (Geometrically speaking, it is easier to justify cutting a sphere in half with an equator than it is to divide the whole with a vertically oriented meridian, since the top and bottom of a sphere do not change, whereas its front and back can.) The result, following this logic, is that the equator justifies celestial latitude, which is understood as the distance from either pole to the sphere's middle and is measured in terms of a ninety-degree arc. Standing on the *inside* of the cosmos, the terrestrial viewer can look up, take a reading, and confidently translate the observations to the outside of his or her material sphere. In this sense, the artificial sphere's surface is subject to a "heavenly" spatial order.

The armillary sphere in the image also calls to mind celestial space's second geometric underpinning, the ecliptic. This ring divides the celestial sphere in half but is 23.5 degrees off, an obliquity that opens the door to celestial longitude, which is celestial globes' *sine qua non,* because it allows the viewer to put things on a sphere with reference to left and right. (The ecliptic itself is the geocentric

Figure 11. Illustration from Regiomontanus, *Epitome of Ptolemy's Almagest,* 1496. Courtesy of the Bayerische Staatsbibliothek München, 1496.08.31.

average of the sun's perceived motion around the earth.) In the image from Regiomontanus's work, the signs of the zodiac are affixed to this line—and this allows people to trace the sun's movement against the constellations. The two celestial intersections between the ecliptic and equator are, in turn, known as the vernal and autumnal equinoxes. These mark the two moments in March and September when terrestrial night and day are equal. The two solstices, which come in December and June, are when night and day are most unequal, and also mark the ecliptic's greatest distance from the equator. From these points the tropical circles extend across the two hemispheres.

Throughout the medieval period, the Heavens were bathed in geometric space, and one key result was the justification of a latitudinal and longitudinal grid that was applicable *above*. Either equinoctial intersection would have constituted a reliable starting point for such a grid, since both are based on regular celestial phenomena. European astronomy begins counting with the vernal equinox and has developed two systems of reckoning, with one based on the celestial equator and the other on the ecliptic. A person can thus drop a line from the observed star down to the projected equator (or the ecliptic) and measure the angle (horizontally) between the point on the line and the vernal equinox's position on that same line, which then yields a longitudinal coordinate.

The ecliptic's role in determining celestial coordinates looms larger if we return to the problem of a sphere's orientation. An equator's relationship to a sphere's surface is more obvious than that of a meridian (which is a great circle that runs through the poles), because "up" and "down" are embedded in three-dimensional space in a way that "left" and "right" are not. Put another way, unlike an equator, a vertically oriented circle has no natural claim to be a starting point on a sphere, because its left/right orientation is arbitrary. If we move inside the armillary sphere, we confront exactly the problem of authorizing a left-right beginning, as our blob of a planet provides no integrated starting point. The heavenly equinoxes cannot perform such a role, if they are transferred to the earth, due to a phenomenon that is called precession, which the astronomer Hipparchus first identified.[58] (The earth wobbles on its axis, which means that the bi-annual ecliptical-equatorial intersections move backward along the terrestrial equator, over time.) Thus, since the earth itself provides no longitudinal warrant, our planet's surface features cannot be put onto a globe without a human—an arbitrary—intervention.

A geographic intervention was, however, precluded by medieval views of the relationship between humanity, the cosmos, and God. I illustrate this point with the image in figure 12, which comes from *Little Work of Geography*, a manual on terrestrial globes that was originally published in 1533 by the mathematician and globe maker Johannes Schöner.[59] (In pursuit of better image quality, I have chosen a later version that appeared inside a posthumous edition of Schöner's

Figure 12. Illustration from Schöner, *Little Work of Geography,* 1533.
Courtesy of ETH-Bibliothek Zürich.

collected works.) The image presents an armillary sphere, in which the earth appears as a sphere that actually has surface features and also boasts an equator, two tropical circles, and two polar ones. Schöner had access to both Euclid's *Elements* and Ptolemy's *Geography,* which is why his terrestrial blob has features. We see this exactly if we compare Schöner's work to the image of Nicole Oresme's contemplating an armillary sphere (see figure 4). The oldest armillary spheres of which we have evidence always represented the earth without geographic features precisely because, before the late fifteenth century, there was no way to locate terrestrial spaces with reference to longitude. Space had to come down from the Heavens.

In this context, it is a significant datum that, after the earth's invention, armillary spheres began to include terrestrial globes *on the inside.*[60] In the course of the sixteenth century, an instrument that had long assumed terrestrial space's exclusion made room expressly for terrestrial globes—and this change, above all, casts light on the fundamental nature of the shift that had occurred.[61] Indeed, the collective reorientation was so profound that the earth's surface also appeared inside armillary spheres that had been illustrated by convinced Aristotelians. For example, in 1570 the Jesuit astronomer/mathematician Christopher Clavius reprinted John Sacrobosco's *On the Sphere,* along with a geocentric cosmological drawing that clearly included a terrestrial globe.[62] Although an Aristotelian, Clavius was also a first-rate mathematician, which is why he used mathematics' spatial uniformity to inject the earth's surface into an instrument that had long excluded it.

Perhaps the most illuminating example of the continental nature of the early modern spatial regime's continued development comes from sixteenth-century Italy. In 1537, there appeared in Venice a book on spheres that had originally been written in the fifteenth century by a friar named Mauro Fiorentino.[63] Published by the printer Bartholomeo Zannetti, this work's full title is *The Common Sphere, Translated with Many Additional Comments from Geometry, Cosmography, Navigation, Stereometry, and [also] Proportion and Quantity of the Elements, Distance, Size and Movement of all the Heavenly Bodies, Things Certainly Rare and Marvelous.* Three aspects of this work are important for contextualizing the Spatial Reformation's progress. First, the text began with a primer on Euclid's *Elements* that explained all of geometry's chief concepts, beginning with points and lines, before discussing "real" things. Second, the succeeding section comprised a translation of John Sacrobosco's thirteenth-century work *On Spheres.* As I noted earlier, this work failed to discuss the earth's surface. Fiorentino's work, in contrast, expressly included the earth's features. Moreover, the publisher also illustrated the underpinnings of Fiorentino's embrace of a fully three-dimensional space by cribbing the illustration that I have dubbed Apian's Eye straight from the original text (see figure 8). A simple case of early modern plagiarism thus illustrates how both Heaven and Earth were now part of a human spatial whole.

With Clavius's inadvertent rebellion and Zanetti's deliberate plagiarism in mind, I now consider another of Raphael's frescos, which is often called (perhaps incorrectly) the "Astronomia" (figure 13). Located above *The School of Athens* in the *Stanza della Segnatura,* this work personifies the celestial science as a woman who looks down onto a milky celestial sphere.[64] Although this visual perspective

Figure 13. Raphael, *Astronomia*, 1510. Vatican, Stanza della Segnatura © 2017. Courtesy of SCALA Archives, Florence.

is consistent with fractured space's notion that perfection lays "above," the rest of the image is not. Much like a "global" armillary sphere, Raphael's celestial globe hosts a terrestrial counterpart *on the inside*. Thus, unlike the medieval God, Raphael's angel looked down onto a globe rather than a nondescript blob—and the inclusion of surface features on an object that should not have borne any indicates the same underlying space that *The School of Athens* illustrated through its assembly of Hipparchus, Euclid, and Ptolemy. Renaissance art's celebration of globality reveals, in short, how all of early modern culture was now submerged within homogeneous space.

Homogeneity's Landing

One way to explore the general nature of the shift that Raphael's work reveals is to consider the textual frame for global space that dominated the fifteenth and sixteenth centuries.[65] Ptolemy's *Almagest* was one element of this frame, as indicated by Hipparchus's presence in *The School of Athens*. At first blush, this suggestion may seem counterintuitive, since it was Ptolemy who wrote the *Almagest*. Nonetheless, it makes sense in two ways. First, Ptolemy already held a terrestrial globe—as is indicated by the crown—which meant that he was unavailable to carry the celestial one. Second, Ptolemy based his *Almagest* on lost works by Hipparchus, especially his star catalogue. For these reasons, it makes sense to read Hipparchus into this image.

Given the *Almagest*'s significance for medieval astronomy, it may, however, appear odd that Raphael's "School of Athens" portrays Ptolemy as a geographer. Although still relatively new, this distinction was not unique to Raphael, as the fifteenth century had already transformed Ptolemy into a geographer—and in direct response to the return of another Ptolemaic work, the *Geography*.[66] As antiquity's embers cooled in the sixth century, Ptolemy's geographic tome withdrew from Rome's Latin-speaking half. Although it is not true that all knowledge of Ptolemy's geographic thought disappeared, manuscript copies of the *Geography* did, as there is no known Latin version from before the fifteenth century.[67] And even if Latin versions had survived, medieval Aristotelianism would have boxed out this work's mathematical approach to terrestrial space. Indeed, exactly this seems to have occurred in the Byzantine and Muslim worlds, since both of them possessed copies of *Geography*, but each failed to incorporate its conceptual apparatus into the regnant geographic thought.[68]

In order to explain the *Geography*'s contribution to Renaissance global space's rise, I consider both the text's specific doctrines and the attending process of recovery. The *Geography* returned in 1406. In that year, the Florentine humanist Jacopo Angelo de Scarperi translated the work into Latin from Byzantine Greek manuscripts. Multiple versions of this translation soon circulated around Italy.[69] One important exemplar, to which I refer again, appeared in Florence around 1470, and includes hand-drawn maps that set the standard for subsequent editions.[70] In the second half of that decade, the process of diffusion accelerated as the work entered print. In 1475, the first edition of Scarperi's translation was published in Vicenza, although without maps.[71] Additional versions with maps were published in Bologna (1477), Rome (1478, 1490), Florence (1482), and Ulm (1482, 1486). The Ulm editions opened the door to the north, as others followed in Strasbourg (1513, 1520, 1522, 1525), Nuremberg (1514), Vienna (1518), Cracow (1519), Basel (1533, 1540), Louvain (1535), Leiden, (1535), Lisbon (1537), Cologne (1540, 1578, 1597), and Paris (1546). And these works came amid other republications in Italy, including a particularly significant one that appeared twice in Rome (1507 and 1508)—and to which I also return.

The Renaissance's broader esteem for the *Geography* is reflected in the eminence of individuals that were involved in the text's publication. The Dutch humanist Desiderius Erasmus wrote the introduction to the Basel edition of 1533.[72] Less well known today but extremely important in his day was the Nuremberg philologist Willibald Pirckheimer, who translated the 1525 edition from Strasbourg and was also Albrecht Dürer's friend.[73] (I discuss Dürer in the next chapter.) Michael Servetus, the unfortunate Spanish humanist who would be burned at the stake in 1553, edited the 1535 version from Leiden, which was based on Pirckheimer's translation.[74] Other important contributors are the cosmographer and cartographer Sebastian Münster, who was involved in the Basel edition of 1540 (and whom I discuss in Chapter 4), and the cartographer and globe maker Gerard Mercator, who contributed to the Cologne edition of 1578.[75] Mercator is, of course, *the* figure in the histories of globes and cartography. His manner of envisioning global space became, in effect, the continent's approach to the world—and, along these lines, it is not coincidental that his teacher was the same Gemma Frisius who had republished Peter Apian's *Cosmography* in Antwerp.

For the course of my narrative, however, the most important person associated with any edition of the *Geography* is the earth's own inventor, Nicolaus Germanus. Not much is known about him, beyond his having worked in Rome as a cartographer.[76] His effect on the *Geography*'s early Renaissance career was enormous, however, as his maps appeared in the 1470 manuscript from Florence, the Rome editions of 1478 and 1490, and the Ulm editions of 1482 and 1486.[77] That the cartographer who first drew Ptolemy's maps also made the first known globe pair could be coincidental. Nevertheless, the *Geography*'s other effects quell any doubts. First, the terrestrial globe that might have been built around 1444 was probably constructed to illustrate the *Geography*.[78] Second, Martin Behaim had a copy of the work to hand, as did Christopher Columbus, who first landed in the New World, and as did Martin Waldseemüller, who named it.[79] Third, the early globe maker Johannes Schöner owned copies of both Ulm editions, which he annotated heavily.[80] Finally, when Raphael painted *The School of Athens*, at his side

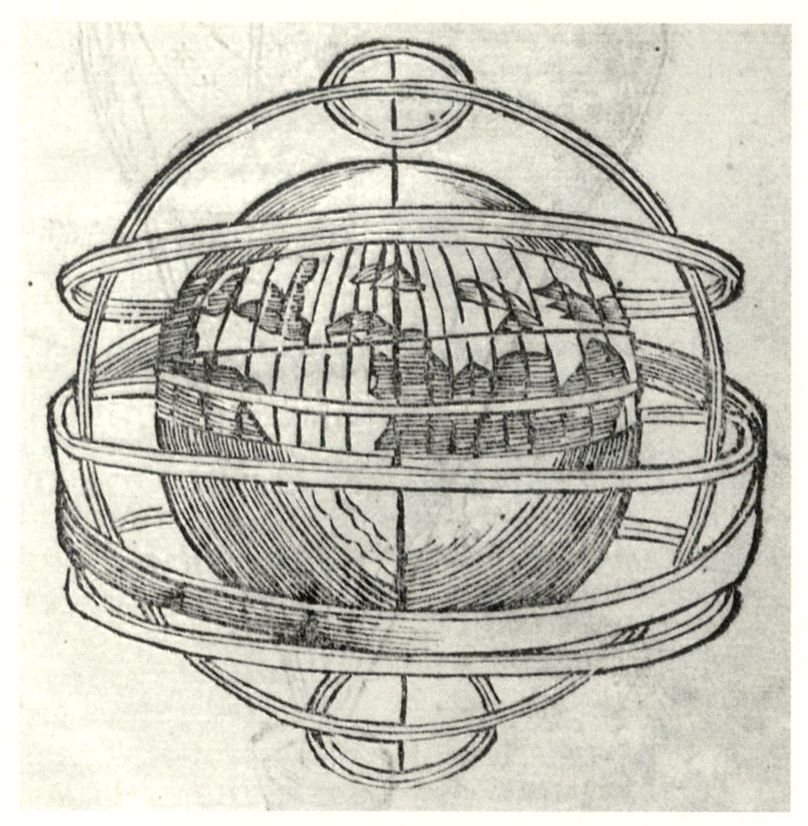

Figure 14. Cover illustration from Apian, *Cosmography, or Description of the Universal Orb,* 1537. Courtesy of ETH-Bibliothek Zürich.

was Johannes Ruysch, a Dutch cartographer who also happened to draw the maps for the two Rome editions of 1507 and 1508, which I mentioned above.[81] Ptolemy's fingerprints are all over the Renaissance globe.

Within this fifteenth-century context, the *Geography* contributed to a new way of encoding terrestrial space, in which celestial and terrestrial were united via mathematics.[82] Here, I call attention to figure 14, which contains the cover illustration from Peter Apian's *Cosmography.* In it we see not only the outline of an armillary sphere, but also how the earth's geography is pasted onto the inner sphere. With respect to what I have already argued about fractured space, two aspects of this image are significant. First, the inner globe has a meridian that authorizes the left-right orientation of Earth's surface features. Second, this meridian is not anchored to a religiously defined space. God does not hold the earth in his hands, nor does the meridian center on Jerusalem, which was often the case with medieval maps. Like Raphael, Apian applied conceptual tools to the earth that the medieval world had reserved to the Heavens—and this tectonic shift was but one reflection of the *Geography*'s return.

Before continuing with the *Geography*, I must integrate this text's arrival with the third (and most important) part of global space's textual frame, Euclid's *Elements*. Although the *Elements* first "returned" in twelfth-century Latin translations from Arabic manuscripts, it did not attain a dominant intellectual position until the fifteenth century.[83] Two obstacles stood in its way. First, Aristotle's views on space limited the *Elements*' potential effects between the twelfth and the fourteenth centuries.[84] Second, when the *Elements* was assigned within medieval curricula, there is little evidence to suggest that people read beyond the first six books—and, as I have noted, these deal exclusively with plane geometry.[85] It was only in the fifteenth and sixteenth centuries that geometry made a full return, including thus the spherical geometry that appears in the later sections.[86] With these things in mind, it is significant that Raphael put Euclid precisely between Hipparchus and Ptolemy, because this arrangement signaled that the *Elements*' space united these ancient thinkers within the Renaissance mind.

Ptolemy's Renaissance legacy cannot be separated from the Spatial Reformation's geometric currents, as I have identified them. In this respect, it is significant that the Renaissance Papacy contributed to the rise of the new space by gathering in Rome people who either understood how to manipulate space or who studied ancient spatial texts. It was, after all, Pope Julius II (1503–1509) who hired Raphael to paint the two frescoes that I have already discussed. Moreover, a series of Julius's predecessors cultivated a mathematical circle that included, at different points, Johannes Regiomontanus and Paolo Toscanelli (it was the latter who taught geometry to the architect Filippo Brunelleschi and also corresponded with Christopher Columbus) and also boasted the likes of Luca Pacioli, who not only was Leonardo da Vinci's friend but who also taught perspective to Albrecht Dürer.[87] Moreover, as I noted in the previous chapter, the *Elements*' entry into print sparked a flurry of editions that probably exceed in number those of every other printed work except for the Bible.[88]

Having completed this sketch, I turn to the *Geography* more narrowly, in order to highlight, first, how homogeneous space suffused this work and, second, to explain why the work fit so well with the Renaissance's reception of Euclid. Ptolemy's great work comprises eight books, among which three (numbers 1, 2, and 7) provide instructions on how to represent terrestrial space in both image and object. Scholars consider these to be the "theoretical" books—the others are lists of place coordinates.[89] For my purposes, the first book is the most important. Comprising twenty-four chapters, it begins with a distinction that was critical to Renaissance geography, the difference between chorography and geography.[90]

According to Ptolemy, the former is based on direct experience, while the latter relates to the earth as a mathematical whole: "For these reasons, [regional cartography] has no need of mathematical method, but here [in world cartography] this element takes absolute precedence. Thus the first thing that one has to investigate is the earth's shape, size, and position with respect to its surroundings [i.e., the Heavens], so that it will be possible to speak of its known part, how large it is and what it is like, and moreover [so that it will be possible to specify] under which parallels of the celestial sphere each of the localities in this [known part] lies."[91] By juxtaposing the vagaries of physical seeing with the regularity of mathematics, Ptolemy justified a new way of *viewing* the earth. Now, its features could be put into a space that could not be experienced, but that allowed real places to be situated on a manufactured sphere. This represents nothing less than the conceptual justification for Apian's Eye.

I illustrate this approach's chief characteristics by comparing Ptolemy's cartographic sense with that of his medieval successors. Figure 15 contains an image of the well-known Hereford world map (*mappamundi*), which was completed in the late twelfth century. It is what scholars of cartography call a T-O map, in which the three then-known continents (known to the Old World)—Europe, Asia, and Africa—were arranged in a T-shape that was, in turn, surrounded by an O of water.[92] Such maps were often centered on Jerusalem, and their "flat" way of encoding geographic space was, in turn, anchored to the limits that Christian Aristotelian thought had imposed on Euclidean geometry. Thus, the continents were not pasted onto a sphere but appeared inside a circle.[93] This difference is profoundly important. Even if one wished to paste a T-O map onto a round ball, there would be no way to "orient" it with respect to a viewer, since a circular map is incommensurable with a spherical projection of latitude and longitude.

Ptolemy's *Geography* suggested that the terrestrial realm could be understood as a mathematical counterpart to the celestial one and, in doing this, made astronomy's spherical grid applicable to the terrestrial. Indeed, Jacopo Angelo de Scarperi was well aware of this change's significance, considering it to have been an "extraordinary discovery."[94] Along these lines, longitude's conceptual underpinnings are also visible in how Ptolemy sandwiched the earth's surface between the Heavens and an imagined cosmological center: "For it has already been mathematically determined that the continuous surface of land and water is (as regards its broad features) spherical and concentric with the celestial sphere [*Almagest* 1.4–5], so that every plane produced through the [common] center makes as its intersections with the aforesaid surfaces [of the terrestrial and celestial spheres] great circles on [the spheres], and angles in [this plane] at the center cut off similar arcs on the [celestial and terrestrial great] circles."[95] Thanks to Ptolemy's embrace of mathematics, the earth's surface could be pasted onto a sphere, even though it ostensibly occupied a different *physical* space. More significantly, this

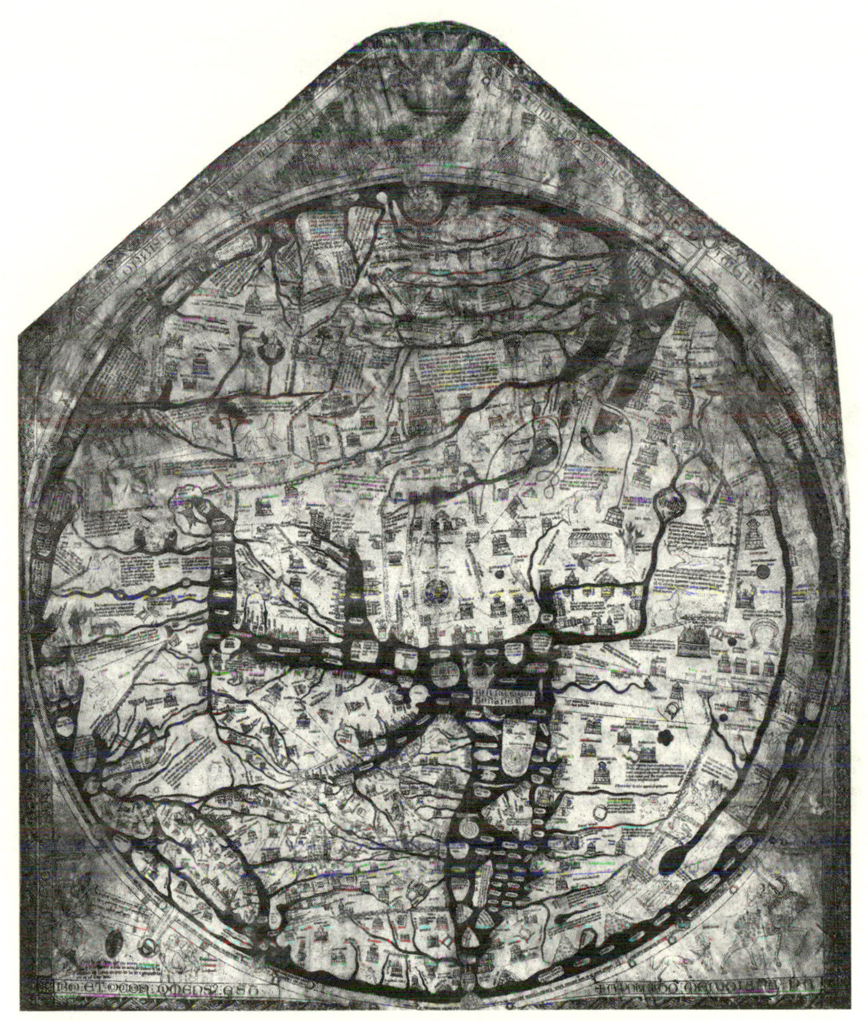

Figure 15. The *Mappamundi* of Hereford Cathedral, thirteenth century. Courtesy of the Dean and Chapter of Hereford Cathedral and the Hereford Mappamundi Trust.

imagined realm's cultivation made it possible for humanity not only to see itself as occupying a position that had been reserved to God, but also to incorporate into its material culture terrestrial features that were, presumably, unimportant to Him. This is the theoretical backdrop to Apian's Eye—and the underlying space that sustained it yielded a fundamental anthropological break.

Ptolemy's role in the rise of terrestrial globes (and globe pairing) becomes most apparent against the backdrop of uniform space's diffusion. The first effect is the simple authorization of a longitudinal starting point on the earth. Ptolemy held that human beings must, as a practical matter, choose a meridian before they can arrange the earth's features in accord with longitude. He wrote: "[By 'locations' I mean] the number of degrees (of such as the great circle is 360) in longitude along the equator between the meridian that marks off [the western limit of the *oikumene*], and the number of degrees in latitude between the parallel drawn through the place and the equator [measured along the meridian.]"[96] Ptolemy's meridian simply coincided with the westernmost region known to the classical world, the Canary Islands, or, as Ptolemy identified them, the Islands of the Blest.[97] This choice was thus an artifact of the Roman Empire's limits.

Ptolemy's second effect takes us more deeply into the human nature of the space that underlay his geography. He did not merely situate the earth's surface on a sphere but made the projection malleable *within the human mind*. Consider that Ptolemy offered not only two different ways to represent our planet's surface on a plane but also another way to put it onto a sphere. Both of the former appear in the final chapter of Book I of the *Geography*.[98] The first projection is conical, and in it latitudinal parallels are represented as concentric arcs, while the longitudinal meridians are straight lines that emerge from the center. (The viewer is assumed to be above the image.) This method works well for spaces north of the equator, with the most significant distortion coming below it. In contrast, the second method represents the earth's curvature better. Here, both the parallels and the meridians are curved, while the viewer is assumed to be at a great distance, situated slightly above the equator. From this perspective, the earth's upper hemisphere appears in proportion, with distortion appearing at the margins.

Ptolemy's projections illuminate the significance of the cosmos's subjection to a human space. When he moved between methods of projection, Ptolemy put his imagined viewer into an extraterrestrial position, but without losing contact with the terrestrial realm. Here, one must recall that within medieval fractured space, to leave the earth signified that one was leaving the terrestrial realm. Inside Ptolemy's mode of thought, however, it was possible for the mind to keep contact with the earth, even when it separated the planet's surface from its underlying sphere. It is therefore important to recall that after deracinating all terrestrial features, Ptolemy simply returned them to a sphere. These mental gymnastics would have been impossible for a doctrinaire Christian Aristotelian.[99]

Having rewoven terrestrial space's early modern tapestry, I return to global space's textual frame. When it was combined with the *Almagest* and the *Elements*, the *Geography* put God's entire cosmos at humanity's service and, in so doing, invited the elaboration of a material culture that included texts, paintings, and celestial and terrestrial globes. I do not delve further into Ptolemy's projections except to note that they inaugurated a continual process of reimagination, as Europeans developed more ways to represent the earth's surface on a plane.[100] Nicolaus Germanus, for instance, improved Ptolemy's methods in his so-called

Donis projection, which was a trapezoidal map, along the borders of which coordinates were inscribed.[101] The method was not especially sophisticated, but it did allow Germanus to put any region into a *global* context, since the grid lines were assumed to extend back through an imagined spatial realm. Later cartographers and globe makers fashioned more effective projections, the finest example of which is Gerard Mercator's terrestrial orange peel, in which the earth's surface is laid flat in a series of lightly attached flaps. The terrestrial globe's surface had become a conceptual veneer that, as we have seen in Raphael's work, was bookended by Ptolemy and suffused by Euclid.

Writing the Earth

The contemporary literature on globes' greatest weakness is its failure to exploit globe manuals. As I noted in the previous chapter, one of the Spatial Reformation's more significant aspects was the way that its texts *made space*. No early modern natural philosophical text was read without reference either to textual discussions of homogeneous space, or to the instruments that early moderns used to represent unseen spaces. It would therefore be distorting to separate globes and global culture from the books into which these things were expressly written. Unfortunately, this is exactly what the contemporary scholarship has done, with its heavy concentration on globes' physical aspects. In this section I offer, in contrast, an attempt to reintegrate three main components of early modern globe culture: books, instruments, and the human user—or what I referred to in the introduction as the practical triad.

Between 1500 and 1850, dozens of globe manuals were published across Europe and North America. An exact number is elusive, since no detailed studies have been done, and many manuals also bleed into other disciplines, such as astronomy, cosmography, and geography. Nevertheless, even a cursory view of this literature enriches our understanding of global space in three ways. First, like their spherical counterparts, globe manuals were often produced in pairs. Second, given that these manuals ran parallel to globes' production, any change in their contents betrays shifts in thought that would be masked by the continuities of globe construction itself. (Beyond the addition of newly encountered geographic details, the design and arrangement of paired globes barely changed over time.) Finally, the manuals reveal that, as we enter the seventeenth century, globe makers expected *users* to bring a different cosmos to paired globes—and the recognition that the broader spatial culture underwent profound conceptual changes remained a prominent theme into the nineteenth century. Thus, even if pairing remained a constant, the mental cosmos that undergirded the use of globes evolved in significant ways.

I illustrate these general themes through an examination of two pairs of manuals, one from the sixteenth century and the other from the eighteenth. The former pair was the work of Johannes Schöner and among the earliest such manuals produced.[102] They were also probably the first to be published as a pair, as they appeared in 1533 in Nuremberg under the titles *Celestial Globes, or Spheres of the Fixed Stars* for the first volume and *Little Work of Geography, Taken from Various Books and Maps* for the second.[103] Beyond their pairing, these works replicated exactly global space's textual frame, as they borrowed liberally from the *Almagest,* the *Geography,* and the *Elements.* The first volume, for instance, used concepts that came straight from the *Almagest,* including geocentricity, the spherical cosmos, and the projection of mathematical regularity onto the Heavens. Consider Schöner's definition of a celestial sphere: "The sphere we understand as a globe; it is a round, solid body. The ancients, however, who explained better the rotation of the heaven, thus set forth that it was directed and turned around as a wheel on its axis, but a certain distinctive feature of this axis is that it is above and below us, or it lies hidden in each direction. Since the position of the earth is settled so strongly, in this case the highest are called poles."[104] Schöner applied concepts to the Heavens that both classical and medieval thinkers had embraced, including sphere, axis, and pole. He then explained how celestial latitude and longitude allowed one to put stars onto a sphere.[105]

If Schöner had not written a second volume, little would have separated him from his medieval predecessors.[106] In this context, Schöner's *Little Work of Geography* confirms the effect that classical thought had on early modern spatial sense. This volume applied to the earth's surface geometric concepts that had once been reserved to the Heavens. Not surprisingly, it recapitulates much of the logic that Ptolemy used in the *Geography,* as it is divided into two parts, with the first part explaining the doctrines of spheres and the second comprising a collection of coordinates that are organized by continent—Europe, Africa, Asia, and the New World. I concentrate on part 1, since it contains the conceptual apparatus for Schöner's terrestrial globes.

Although Schöner's broader sense of space was relatively new, his cosmos remained traditional. In the first seven chapters of part 1, he summarized the geocentric cosmos.[107] Among other topics, he discussed the earth's shape in chapter 1, its geocentric position in chapter 2, and the definition of an axis in chapter 3. Other discussions of ancient geographic concepts, such as climatic zones, followed.[108] In chapter 8 Schöner exposed most clearly, however, the depths of the Renaissance rupture with medieval space, through his justification of terrestrial longitude. He began with Ptolemy's starting point, the Canary Islands (calling them the *insulae fortunatae*) before explaining how longitude was determined.[109] In chapter 9, he added latitude, which he explained is calculated by measuring the angle between the Pole Star and the northern horizon, all the while using terms and methods that reflected his knowledge of geometry.[110] Following Ptolemy, Schöner justified representing the earth's surface on a sphere by putting the whole inside a mathematical continuity. Not coincidentally, in the next chapter he explained how to put Europe's cities onto a sphere, which was something that medieval thinkers failed to do for roughly a millennium.

Schöner's volumes set a baseline against which to judge further developments, precisely because they were not meant to break new ground but recounted established ideas about global space. As one would expect, Renaissance space's textual frame (the *Almagest*, the *Geography*, and the *Elements*) figured prominently in this work. Subsequent globe manuals broke radically, however, with this arrangement. The main force behind this change was heliocentrism, whose rise began with Nicolaus Copernicus around the middle of the sixteenth century and concluded with Isaac Newton at the end of the seventeenth. This first step in the rise of what we call modern physics was, as I suggested in the introduction, a product of homogeneous space. I do not, however, pursue heliocentrism's rise in detail.[111] Instead, I concentrate on chronicling its relationship to global thought's textual frame.

The "global" effects of changes in European views of the cosmos became obvious in the late seventeenth century, when globe manuals began to avoid using Ptolemy as an authority and turned largely to Euclid.[112] As globe manuals entered the eighteenth century, the shift away from Ptolemy became complete, with both the *Almagest* and the *Geography* disappearing from discussions of globes. I illustrate this change via a pair of manuals that appeared in Nuremberg, about 250 years after Schöner's great work. In the years 1791 and 1792, the mathematician Johann Wolfgang Müller published two volumes under the title *Instruction in the Knowledge and Use of Man-Made Celestial and Terrestrial Globes*.[113] The first dealt with the celestial globe, while the second covered the terrestrial one. These tomes were closely tied to globe production, since they were commissioned and published by the globe maker Johann Georg Klinger.[114] I highlight these eighteenth-century volumes because they are as ordinary as those of Schöner. Müller was a competent mathematician but hardly an innovative one, and much the same was true of his globe manuals, especially since they were based on a text that had been written almost sixty years earlier in the sleepy Neckar town of Heilbronn.[115] Müller's two volumes thus incorporated nothing more than the most mundane consensus.

With this backdrop in mind, I examine each of Müller's volumes. In the first volume, Müller began by separating what humans see of the Heavens from how the Heavens are actually constructed: "But here we are only considering how the universe appears to our eyes and not its actual configuration, which must first be explored through profounder study. Beyond that, only this remark needs to be made: that because the universe appears to our senses in a manner distinct from its true configuration, with the stars appearing to be uniformly distant, at an inconsiderable distance, and fixed to a hollow sphere, we should not conclude rashly, perhaps, that it is, in fact, so."[116] In this way, Müller applied to global space the changes in astronomy and physics that had washed across the continent during the seventeenth century. Not coincidentally, neither volume mentioned Ptolemy or the *Almagest*, but each cited only Isaac Newton and an array of eighteenth-century astronomers. The new physics' rise had made Ptolemy's *Almagest* irrelevant to celestial globes, even as it also fixed Euclid's homogenous space within the European mind.

Müller expected globe users to accept a cosmology that differed sharply from what globes actually show. For example, in the first chapter of the second volume Müller explained that although terrestrial globes depict the earth as a sphere, our planet is not a sphere but a spheroid, flattened at the poles.[117] He added that Isaac Newton had predicted this on the basis of his law of gravitation and that other scientists had confirmed it.[118] The earth could thus no longer be understood as a sphere. A brief reflection on the *Geography* underscores this point. Ptolemy's belief in the earth's perfect sphericity justified his application of mathematics to the entire cosmos. Müller, in contrast, began his second volume by undermining exactly this justification for the earth and instead highlighted that perfect spheres existed only in the human mind. It is therefore no accident that Müller failed to cite Ptolemy, since, beyond the limitations in the Alexandrine's geographic knowledge, his philosophical assumptions had become unsustainable. Thus, the terrestrial element of global space's textual frame dissolved, too.

Müller's sense of global space rested, in the end, on the only remaining component of the original frame, Euclid's *Elements*. Extending a trend that had become apparent in the late seventeenth century, Müller united his volumes via the space that suffused Euclidean geometry. The first volume, for instance, began with a tutorial on basic concepts, including points, lines, and angles—with special attention paid to right, oblique, and acute angles. Next, Müller discussed circles, radii, circumferences, diameters, and midpoints before pursuing the projection of spheres. The reliance on spherical geometry is also apparent in the second volume, where the second chapter carries the title "On the Mathematical Partitioning of the Earth's Surface."[119] Here, Müller applied a fully human mathematical grid to the earth and explained how the equators and meridians that human beings imagine also allow us to situate terrestrial places on a manufactured sphere.

Müller's work exemplifies how, by the end of the eighteenth century, global space had been reduced to one remaining text, Euclid's *Elements*. Of course, this reduction only occurred after seventeenth-century mathematicians, such as Johannes Kepler, had characterized Euclid's thought as an artifact of the mind, and perdured only until nineteenth-century mathematicians pulled the rug out from under Euclid's most important assumptions about space. Thus, although globe makers continued to produce paired globes—the discipline's most glorious examples stem from the early nineteenth century—the resulting objects were situated within a mental context that was becoming increasingly hostile to the core assumptions of globe pairing itself. When Euclid's geometry finally dissolved, there was little justification for going to the trouble of pairing Heaven and Earth.

Conclusion

In this concluding section I explain why the history of globes must reach beyond the narrow search for "who got what right" and recognize how geographic knowledge has always been situated by a larger spatial regime. Above, I examined how changes in astronomy and geography altered the space that users brought to paired globes, even if globe pairing endured as a manufacturing practice. Here, I go deeper and sketch out how developments in geometry coincided with the most fundamental change in the material culture of globes since the earth's invention, the dissolution of globe pairing. In effect, shifts in the broader culture of space changed the meaning of a venerable historical practice so thoroughly that the practice itself dissolved.

Early modern global space lost its underlying conceptual supports, when post-Euclidean geometries arose. As I explained in the previous chapter, early in the nineteenth century, mathematicians criticized geometry's foundations, holding that some of Euclid's assumptions were unjustified, especially those that undergirded his parallel postulate.[120] Then, in the second half of the nineteenth century, Bernhard Riemann established that a logically consistent, non-Euclidean geometry was possible. Once this geometric turn had been made, a new context for space began to diffuse throughout European culture—and this was bad news for paired globes.

Euclid's demotion by Riemann accelerated a trend that the great mathematician Augustus de Morgan had already lamented before mid-century. De Morgan held that all the new knowledge about the cosmos was ruining what had been a very useful globe culture. In 1845, he published *The Globes, Celestial and Terrestrial* as a manual for globe pairs that were then being sold by London's exquisitely named "Society for the Diffusion of Useful Knowledge."[121] He highlighted, first, how modern science had undermined celestial globes: "The time was when the only elementary work on astronomy was the description of the celestial sphere. The quotation in my title page is a passage of commentary inserted near the commencement of one of the earliest printed editions of the celebrated work of Sacrobosco, the astronomical Euclid of the middle ages. As the Copernican theory gained ground, the globe, which represented the *real belief* of the falling sect, was no longer judged essential to the instruction in the science."[122] He added that this globe culture should be considered useful precisely because it *only* dealt with appearances:

> The time was when every educated person was tolerably well acquainted with the appearance of the heavens, and came to the discussion upon Ptolemaic and Copernican explanation with a clear idea of what it was about. It is not so now: few persons look at the heavens, while most learn a certain astronomy from books. The books teach the true or real motions of the heavenly bodies, those in particular of the solar system, so well, that if their reader could take up his residence in the sun, he would find himself quite at home, and would have daily illustrations of the Copernican theory passing before his eyes. But, confined to this earth, he does not see written in the heavens the astronomy that he learns: no wonder, then, that he does not care to look at the heavens, or feel much interest in the globe.[123]

De Morgan was swimming upstream. Nevertheless, he persisted through his critique, even adding that globe makers' open embrace of heliocentrism had undermined the culture of globes. And he put the blame for this squarely on the seventeenth century, even citing a "heliocentrist" globe manual that first appeared in 1627 as an example of the problem.[124] Human beings had, in short, learned so many new things by looking out into the cosmos that a venerable tradition of celestial globes was withering away.

De Morgan's lament about the decline of terrestrial globes was less extensive but equally revealing. When turning to the terrestrial he noted: "In the terrestrial globe the position is reversed. Here we have an imitation of the reality, in opposition to our senses, and to the maps, which rather represent appearances."[125] De Morgan underscored here the slow disappearance of the conceptual unity inside which globe culture had persisted. Thus, as he saw it, terrestrial globes were useful for the study of geography only if their use was separated from other aspects of human knowledge. He accentuated the point by writing, "With these astronomical systems we have nothing to do: as far as power of using and understanding the globes is concerned, it matters nothing whether their owner imagines the sun to move round the earth or the earth round the sun."[126] To maintain a radical separation between what one *saw* and what was *known* was now an essential if quixotic mission.

De Morgan wrote his globe manual before Riemann and company shredded what was left of global space's traditional underpinnings. If we look back to Germanus and his globe pair from this modern perspective, the history of globes takes on a new character. Rather than being the story of inexorable progress in geography, it becomes a tale of decaying cosmography. This discipline, which arose in the sixteenth century and disappeared in the course of the seventeenth, placed both Heaven and Earth into a projected space that offered profoundly new perspectives on the cosmos. In this context, I return to Apian's Eye (see figure 8). A second look reveals that Apian did exactly what de Morgan suggested was essential to the use of globes: he transferred his *experience* of the relationship between Heaven and Earth to globe pairs by putting them into a space that made it possible to "see" both of them. However, as I noted in the introduction, the same space that Apian applied to globes not only allowed Nicolaus Copernicus to rethink the relationship between the Earth and the sun, but also impelled Isaac Newton down a path that ultimately vindicated Europe's first heliocentrist. Thus, the human embrace of a space that made terrestrial globes possible also augured paired globes' ultimate decline.

Although one can still buy globe pairs today, the lone terrestrial globe is the dominant market presence. As de Morgan suggested, globes' fate was tied to changes in how the unseen was known. Thus, it is apparent that globe pairing's final dissolution is a consequence of profound changes in an underlying spatial sense. This point, by itself, challenges the geographic positivism that suffuses the contemporary literature. It is true that globe makers "got the earth right." Nevertheless, the slow process of terrestrial discovery was but one part of a complex history of space's effects on European thought and culture. The rise of a new space not only made the earth's invention possible, but its continued development also changed how human beings understood their own relationship to Heaven and Earth—and this change had consequences not only for our planet, but also for the entire Western triad.

Chapter 3

Divine Melancholy

> [Philosophy] excludes the doctrine of angels and of all those things that value neither bodies nor the physical state of bodies; because place is not joined in them, nor distributed, as in nothing is greater or lesser; that is, place is not reckoned.
>
> —Thomas Hobbes, *Elementorum philosophiae sectio prima de corpore* (1655)

Albrecht Dürer's *Melencolia I* is the most analyzed engraving in the history of art (figure 16).[1] Almost since its completion in 1514, scholars have attempted to divine the meaning behind its arresting depiction of lassitude and disgruntlement.[2] The first commentary was printed in 1541 and, over the centuries, the body of analysis has become as voluminous as it is inconclusive.[3] Considering only those works published since 1900, it remains unclear whether one should read the *Melencolia* by itself, or as one in a pair, in a trio (as one of the so-called *Meisterstiche*), or as one of four works, which is the approach that I favor.[4] More significantly, there is no agreement on the backdrop to the *Melencolia*'s seeming paganism, as it is uncertain whether Dürer borrowed more from Aristotelianism, Galenic medicine, Hermeticism, Neoplatonism, Platonism, or Stoicism.[5] Regardless, the scholarship remains steadfast in insisting that the *Melencolia*'s interpreters must forever concentrate on illuminating the ancient lineage of the philosophical ideas and the artistic motifs that Dürer appropriated.

For all its underlying erudition, the literature's historical imagination operates within a narrow analytical band, as it effectively oscillates between the *Melencolia* and the ancient world.[6] In this respect, the most important scholarly work remains *Saturn and Melancholy: Studies in the History of Natural Philosophy, Religion, and Art* by Raymond Klibansky, Erwin Panofsky, and Fritz Saxl.[7] First published in 1964, this indisputably brilliant analysis of melancholy's history from antiquity up through the early modern world reads the *Melencolia* as a culmination of classical prototypes, or *Urtypen*. As Saxl characterized his

Figure 16. Dürer, *Melencolia I*, 1514. Courtesy of the Metropolitan Museum of Art, New York.

perspective in another venue, "The problem is antiquity's legacy (*Nachleben*)."[8] Following this lead, most of the standard interpretations of the *Melencolia* emphasize Dürer's reliance on ancient ideas and motifs, as opposed to concentrating on his connections to the medieval world, or even to exploring his own early modern context.[9] Thus, as the extant literature would have it, Dürer's greatest work signifies nothing less than that the Christian God, ultimately, came to terms with the pagan Saturn.

Recent Dürer scholarship has shifted some key historical reference points by emphasizing that one must read Dürer's oeuvre through his own historical context.[10] The scholarship on the *Melencolia,* however, has yet to embrace using Dürer's life and culture as an analytical backdrop. Still, such an approach holds great promise for opening new vistas on Dürer's greatest engraving, because it reduces the reliance on antiquity and pulls the analysis of Dürer's work into his own mental world. In appropriating these trends in the Dürer literature, I foreground two themes that figured prominently in Dürer's personal development. First, Dürer remained a convinced Christian and continually infused his creations with religious leitmotifs and symbols.[11] In the case of the *Meisterstiche* this undercurrent suggests that the problem is not simply to trace antiquity's legacy, but to identify a transformation in Christianity's dialogue with the divine. Second, the mature Dürer was a committed Euclidean, as he not only studied the *Elements* in the years prior to executing the *Melencolia,* but also celebrated the geometer in his later writings.[12] Thus, it may be that the most important thing that Dürer borrowed was neither antiquity's philosophy nor its artistic motifs, but its manner of projecting space onto the world.

Dürer's application of Euclidian geometry to his later work recalls our attention to homogeneous space's incompatibility with medieval approaches to the divine. Consider that each of the three *Meisterstiche*—the *Melencolia, Jerome in His Study* (1513), and *Knight, Death, and the Devil* (1514)—subjects the divine (both good and bad parts) to spatial homogeneity (see figures 17 and 18, for the latter two of the *Meisterstiche*). Thus, although spiritual beings are depicted in each one, in no case does the divine seem to have a legitimate *place.* In the third of the three engravings, for instance, a knight is accompanied by an array of demons, none of which has an obvious physical effect. In the second, meanwhile, Saint Jerome, the Bible's first translator, sits in his oddly Germanic study and contemplates the divine, albeit (and much like Dürer's knight) without actually *seeing* it. Indeed, the divine seems to be exclusively *within,* as the light that emanates from the saint's head provides no actual illumination; that function is assigned to the rays that stream in through the window.

The literature has pursued ancient philosophies and motifs so zealously that it has overlooked how the *Melencolia* imposed Euclidean space onto the Christian cosmos.[13] Thus, it is significant that the *Melencolia*'s terrestrial world is permeated by a uniform space that was inherently hostile to spatial hierarchy. As a preliminary example of what a fully spatial analysis can do, I again turn to Dürer's brooding angel (figure 16). Within medieval Christianity angels long enjoyed a secure place *in Heaven* that was anchored to their position within the Christian Aristotelian hierarchy of being.[14] As high-ranking creatures, angels

Figure 17. Dürer, *Jerome in His Study,* 1513. Courtesy of the Metropolitan Museum of Art, New York.

Figure 18. Dürer, *Knight, Death and the Devil,* 1514. Courtesy of the Metropolitan Museum of Art, New York.

resided in the perfection *above,* although they occasionally descended (miraculously) to the imperfection *below.* Nevertheless, Dürer's illustration expunges this hierarchy, as it presents the angel as being present on Earth and inside a space that, before the fifteenth century, applied only to the Heavens. The implicit extension of homogeneity to the cosmological whole is, I argue, why the angel gazes so morosely into the distance. Suffused by a humanly imagined space, this divine being is now *out of place.*

Dürer's use of space in his *Melencolia* marks him as a scion of the Spatial Reformation. Thus, if we analyze this engraving against the backdrop of Euclid's post-1350 rise, we can see yet another way in which spatial homogeneity torqued the Western triad. The ever-growing ability to imagine space and to represent it within books, artwork, and objects of material culture required a thoroughgoing reappraisal of the relationship between humanity, the cosmos, and God. Given that the *Melencolia* included divine figures, this work reveals with particular clarity the effects that homogeneous space had on early modern conceptions of the divine. By executing an image of a divine creature inside a Euclidean space, Dürer participated—if not altogether knowingly—in a broader rethinking of humanity's relationship to God and the cosmos.

Divine Melancholy

I have argued throughout this book that the Spatial Reformation drove a realignment of Europe's triadic reference points. Some of the changes altered how Europeans saw their cosmos, with the earth's "global" invention being one result. Dürer's vision of melancholy was but one aspect of this much broader shift in thought. I sketch out the context for understanding Dürer's contributions to this change by reaching back to ancient Greece. The oldest discussion of melancholy appears in the *Problemata,* which was probably written by one of Aristotle's students.[15] This lineage makes it certain that the *Problemata*'s melancholy was suspended inside the fractured space whose history I have already discussed. It is, therefore, significant that the contemporary scholarship—with the work of Klibansky and company heading the list—has overlooked the effects of homogeneous space's rise on a melancholy that was rooted precisely in fractured space.[16]

I illustrate space's potential for rereading both melancholy and the *Melencolia* by evaluating one of the contemporary scholarship's unexamined assumptions, namely that Dürer's melancholy pertains to humanity. The consensus on this point is curious, given that Dürer depicted no human sufferers, but only an emaciated dog, a banner wielding bat, a gloomy cherub, and a dyspeptic angel—none of which seems to have a history of representing humanity in art.[17] Even more significant is that by excluding humans Dürer broke with the tradition from which he had ostensibly drawn inspiration, since from the *Problemata* onward, melancholy was usually ascribed to human sufferers.[18] As a result, by not confronting humanity's absence from the image, scholars have foreclosed the possibility that the *Melencolia* actually illustrated then-recent changes in humanity's vision of the divine.

Dürer's application of homogenous space to the eternal malady disrupted a tradition that dated back almost two millennia. By executing his *Melencolia* entirely within this space, he contested (implicitly, albeit) the spatial hierarchy on which Aristotle had built his cosmology. The tension that arose from homogeneous space's application, in turn, undercut not only the *Problemata* but also the angels that medieval theologians incorporated into their mental cosmos. From this perspective, Dürer's exclusion of humanity from the *Melencolia* was deliberate—and this intentionality was related closely to the *Melencolia*'s injection of homogenous space into a terrestrial setting.

The history of space thus opens to scrutiny how the appropriation of *nothing* in the *Melencolia* forced a break with both the medieval and the ancient the past. Beginning with the *Problemata* and extending through medieval and early Renaissance discussions of melancholy, the disease was *situated* by the Aristotelian spatial dichotomy. Thus, it is important to underscore how Christian views of the divine—including the nefarious parts—were integrated into this space. Medieval theorists inscribed Heaven and Hell directly onto the cosmos, burying the latter below Earth's surface and perching the former above Saturn's sphere, with both humanity's physical position and its final end being ensconced inside a heterogeneous whole. This fissured backdrop exerted a powerful influence. For example, Henry of Ghent, a prominent late thirteenth-century scholastic, wrote of melancholics:[19] "[T]hey make the best mathematicians, but the worst metaphysicians, because they are unable to extend their minds beyond location and magnitude, on which mathematics is founded."[20] According to the contemporary scholarship, Henry's approach stemmed from a suggestion in the *Problemata* that creative people were susceptible to melancholy, because a surfeit of ideas overheated their brains.[21] (A glass of wine was the preferred remedy; it probably still is.) Henry's appropriation of this idea reflected the Christian Aristotelian cosmos's chief assumptions, given how he not only rejected homogeneous space's general applicability but also saw its pursuit as an impediment to the search for God.

If we keep Ghent in mind, it becomes significant that the same mathematics that inspired medieval melancholy also permeated European thought after 1350. Thus, to juxtapose the medieval period with the early modern one (and along these lines) suggests that melancholy changed its *place* as European culture moved from fractured space to homogeneous space. The tectonic depth of the resulting shift is evident in the *Melencolia*'s sufferers, as those obviously stricken by the disease—the angel and the putto—are divine creatures rather than human ones. Thus, when we read the *Melencolia* through the intellectual context in which the image's space existed, we can see how geometry's diffusion inverted melancholy's most basic reference points, with the once-human disease becoming an affliction of the divine.

The *Melencolia* is a particularly revealing example of a broader reassessment of the divine that was especially important to the culture of the fifteenth and sixteenth centuries. As Charles Trinkaus has noted, Renaissance thinkers remained profoundly interested in God and the divine, even if they were critical of medieval approaches to both.[22] It is thus important that the stricken beings in the *Melencolia* are angelic, because angels occupied a prominent position within medieval anthropology and theology. Nestled between God and humanity in the great hierarchy of being, angels had served, since at least the ninth century, as both anthropological and theological boundary markers.[23] While keeping humanity at both a physical and an ontological a distance from God, they also guaranteed a *philosophical* connection to Him, thanks to reason's supposed ubiquity.[24] Thomas Aquinas, for instance, defined the limits of human reason with respect to what angels could know and also limited angels' knowledge by outlining what God could know.[25] The resulting great chain of knowing was, in turn, woven into Aristotle's physical cosmos.[26]

Keeping this angelology in mind, I turn to a detail in Dürer's image that relates to my earlier discussions of God's physical position in space, the compass in the angel's hand. This instrument lays bare the problems that homogeneous space represented for the medieval divine. In Dürer's day, a compass was an essential tool for measurement, whether of length or angles.[27] Yet, unlike most human beings, who had grown comfortable with measuring both things and spaces, Dürer's angel appears overcome at the prospect of taking a reading. I demonstrate the significance of this act of non-measurement by returning to the fourteenth-century *Bible moralisée*, which I discussed in the previous chapter (see figure 10). The image shows God holding the cosmos in one hand and a compass in the other. Note, here, that the whole assumes no underlying spatial unity, as the stylized sphere rests in the Lord's hands, while the earth lacks all geographic features. The *space* of this image is thus religiously symbolic rather than geometrically uniform, with one important result being the terrestrial surface's effacement.

If we return to the *Melencolia* with the medieval God's compass in mind, the disjuncture between God's activity and the angel's passivity moves to the fore. The medieval God could measure the cosmos, because (in addition to being God) he looked *down*. However, Dürer's angel encountered the terrestrial realm from within a space that *should not have been where it was*—below the lunar sphere—and the outcome was the failure to measure. This angelic incapacity was not, however, the only evidence of a shift in the divine's context, as the putto hardly appears to be energized. Like the angel, the putto is passive, although in this case the little one's somnolence directs our attention to terrestrial artisanal production. Rather than simply fail to measure the world, the putto eschews all creative activity, as evidenced by its holding idly tools that artisans readily employed—such as the burin, which was essential for engraving, and a plate (probably made of copper) on which engravings were made. (Indeed, these are exactly the tools that Dürer used to make his *Meisterstiche*.) Veritably paralyzed by space, this diminutive divinity could not bring itself to create—and in Dürer's day, this was almost sinful.

Dürer's image contains additional details that underscore the medieval divine's retreat before the new space. Here, I turn to the two animals, the bat and the dog. The former completes the outline of the traditional divine hierarchy, insofar as bats had long represented the netherworld within art.[28] Thus, if taken together, the bat and the two angelic figures evoke the *located* Hell and Heaven of the Christian Aristotelian cosmos. It is not clear whether the bat suffers from melancholy, but its nunciatory role underscores how homogeneous space threatened noncorporeal entities, given that even Hell was falling victim to space's depredations. The dog, in turn, connects the *Melencolia*'s representation of the hierarchy of being to those people who cultivated it, scholastic thinkers such as Henry of Ghent. Within European artistic traditions, dogs usually represented loyalty but especially the loyalty of scholars to textual authority.[29] For many fifteenth- and sixteenth-century humanists, the reflexive loyalty to texts connoted scholasticism, the school that gave a place to everything, including angels and demons.[30] That Dürer's melancholic canine is starving says all that needs to be said—especially if the poor animal is compared to St. Ambrose's well-fed companion.

Before exploring space's relationship to the *Melencolia*'s physical cosmos, I reflect on the broader triadic implications of homogeneity's diffusion. In this context, I turn to a key figure in the history of space, Galileo Galilei. As Alexandre Koyré has explained, the great Italian scientist united celestial and terrestrial physics inside a common space—and this made him an important figure in heliocentrism's victory.[31] However, rather than retell the same story via Galileo's physical writings, I highlight a neglected text that relates the problem of space directly to the divine, namely the record of lectures that Galileo delivered in 1588 under the title, "Regarding the Form, Position and Extent of Dante Alighieri's Hell."[32] In these talks Galileo examined the location and area of Hell as Dante had imagined this place in his *Inferno*. While evaluating Dante's views on Hell's dimensions, Galileo did not exactly state that the damned could not be both beneath our feet and also remain within homogeneous space, but the conclusion was clear. Toward the end of the sixteenth century Galileo had the temerity to suggest openly what had been implicit in Dürer's work: geometric space humanized the *entire* cosmos, evicting God's entourage from the Heavens above, while also clearing His fallen angels from below.

Dürer's Mental Cosmos

Dürer's reassessment of the divine was not the only evidence of a spatially induced tension within the Western triad. Having reread the *Melencolia*'s relationship to early modern anthropological and theological currents, I turn to the image's relationship to early modern cosmology. The *Melencolia* cannot be understood without reference to homogenous space's implications for the medieval cosmos. I do not mean by suggesting this to limit myself to the most crucial change in sixteenth-century views of the cosmos, the rise of heliocentrism. Instead, I wish to emphasize that homogeneity introduced a more general instability into a culture that had long combined geocentrism with fractured space. Consider that Nicolaus Copernicus's initial ideas on heliocentricity, which appeared definitively in 1543 in *On the Revolutions of the Heavenly Spheres*, were already circulating in textual form by 1514, just as the *Melencolia* appeared.[33] And we know that Copernicus was fully aware of Euclid's *Elements*, since he owned a copy of the print edition that appeared in Venice in 1482. I would suggest, therefore, that we understand Dürer's art and Copernicus's science as comprising different aspects of the same reformation in space.

With this spatially induced instability in mind, it is significant that the physical cosmos that Dürer assumed within his *Melencolia* was orthodoxly Aristotelian. The traditional cosmology's outlines are most apparent at the back, where a comet and rainbow are situated. With respect to the former, some scholars have suggested that Dürer included a comet that he himself had seen.[34] Although biographically convenient, the explanation is unnecessary, since Aristotle's thought assumed the comet's presence. In his *Meteorology*, for example, the Stagirite defined comets as sub-lunar phenomena that appeared at the edges of the earth's atmosphere, just below the sphere that carried the moon.[35] Hence, although comets were often understood as portents from Heaven, within Aristotelian science they also marked the terrestrial realm's limits. If we read the comet with respect to Dürer's use of geometric space, it becomes apparent that the *Melencolia* puts spatial homogeneity inside an expressly terrestrial realm.

The space in Dürer's *Melencolia* offers additional perspective on the burgeoning conflict between spatial homogeneity and Aristotelian cosmology.[36] From the mid-thirteenth century onward, there was no avoiding the Stagirite's thought, as his corpus was central to university education and, consequently, to medieval philosophy and theology's main currents. And with the rise of print in the fifteenth century, many of Aristotle's ideas about the cosmos diffused even more broadly. An example, which may have influenced Dürer himself, was *On the Meteors of Aristotle* (1512), an epitome of Aristotle's *Meteorology* that was published by the luminary Jacques Lefèvre d'Étaples, to whom I return.[37] Figure 19 is taken from this text, and a comparison with the *Melencolia* reveals important parallels: both images depict a comet, a rainbow, a city on a shore, a body of water, and floating vessels. The major difference between them is that Lefèvre d'Étaples's image includes celestial phenomena, whereas Dürer's does not. It is difficult to say whether Dürer had access to this text, although it is not improbable that he did. (In any event, he could not read Latin, so any relationship to the

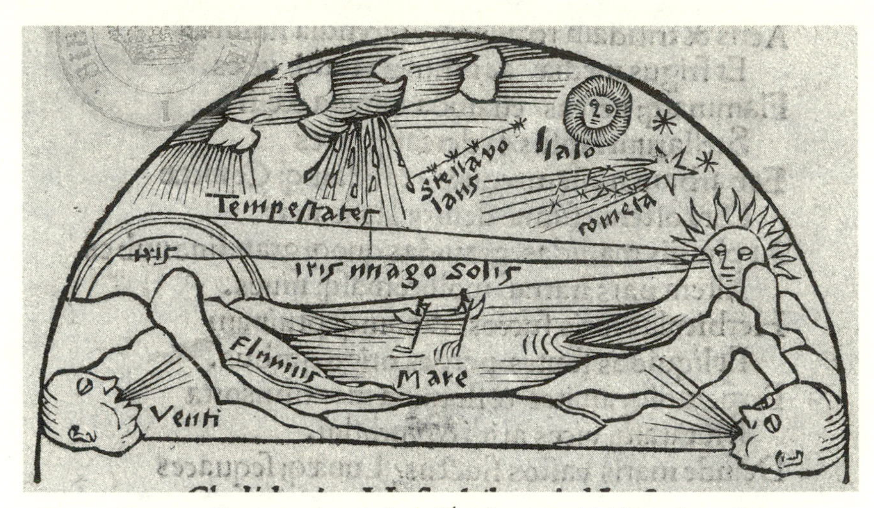

Figure 19. Illustration from Lefèvre d'Étaples, *Aristotle's Meteorology*, 1512. Courtesy of the Bibliothèque nationale de France.

work would have come through interpreters, of which there was no shortage in Nuremberg.) It is certain, however, that Dürer understood Aristotle's cosmos—and this is what makes the celestial realm's absence from the *Melencolia* significant. Following Aristotelian precepts, the realm that Dürer depicted is wholly terrestrial, while the geometry that undergirded the image was, by traditional definition, otherworldly.

If we move into Dürer's image itself, then the rainbow further illustrates the tension that emerged from the injection of Euclid into Aristotle's terrestrial world.[38] The physics of the day dictated that the sun must be at a right angle to the arc itself.[39] (This is illustrated in figure 19, with the straight lines that run from the sun to the rainbow.) Within the *Melencolia*, however, given the rainbow's perpendicular staging, the sun would have to be understood as lurking somewhere on the other side of the image's surface, behind the viewer's perspective. We do not know where Dürer acquired this knowledge, although it is safe to say that it was ubiquitous. This point is exemplified in *Pearl of Philosophy*, a popular encyclopedic text by Gregor Reisch that first appeared in Freiburg in 1503 and circulated widely.[40] In this text one finds a discussion of exactly the physical doctrines that I just mentioned, as well as an Aristotelian explication of the terrestrial world. (Looking ahead to the next chapter, it is also significant that Reisch was one of the cosmographer Sebastian Münster's teachers.)[41] Therefore, with all this in mind, the sun's absence from the *Melencolia* underscores the image's terrestrial nature and also reveals why an older divine was becoming unstable.

Dürer's application of homogenous space to the terrestrial realm was transgressive, even if the most explosive intellectual consequence, heliocentrism, had

Figure 20. Cover illustration from ibn Athari, *On Knowledge of the Motion of the Orbs*, 1505. Courtesy of the Houghton Library, Harvard University.

yet to manifest itself. In this context I turn to figure 20, which contains the frontispiece to *On Knowledge of the Orbs' Motion*, a Latin edition of a late eighth-century work by the Persian astronomer Masha'allah ibn Athari that appeared in 1505 in Nuremberg.[42] Dürer was certainly aware of this book, since its publisher, Johannes Stabius, hired him to engrave the frontispiece.[43] However, given that Dürer could not read the Latin text, it is likely that the image illustrates Dürer's views of space more than those of Masha'allah, which were wholly Aristotelian.[44] More significantly, the work's date of publication establishes that Dürer was already applying homogeneous space to depictions of the cosmos a decade before he completed the *Melencolia*.

The frontispiece to *On Knowledge of the Orbs' Motion* reveals how deeply homogeneity penetrated the early modern mental cosmos. The image presents Masha'allah in the guise of a wise man who is seated on a throne that, in turn, incorporates half of the celestial sphere in the hemisphere above while evoking the other spheres through the symbols on the throne's back. These represent six of the (then known) seven heavenly bodies, beginning with Saturn at the top and ending with Mercury at the bottom. Without a legend, it is difficult to be certain of what the circle at the top right represents. Nevertheless, since Masha'allah was an astronomer and would necessarily have been located on the earth in order to look into the night sky, I suggest that this circle probably represents the moon. Taking into account what I have already argued, Dürer's image broke with medieval attitudes in two ways. First, he placed a man—and a Muslim, at that—onto a throne that belonged to God. Second, he presented the terrestrial realm in the context of a three-dimensional space that should have pertained only to the Heavens.

Dürer's transgressions appear even greater when we add the compass and the sphere, both of which are in Masha'allah's possession. With reference to the compass, if we compare this image to the *Bible moralisée,* we now see a human being taking a measurement of the sort that had long been reserved to God. Moreover, this human did so in a context that was shaped by recent innovations in material culture. Thus, with respect to the sphere itself, it is important to note how generic it is, because the round ball itself could have denoted either the celestial or the terrestrial realm. This ambiguity is particularly significant, given the time period. As I explained in Chapter 2, it was during the late fifteenth century that latitudinal/longitudinal grids were becoming equally applicable to Heaven and Earth. Although this point is speculative, I submit that if we understand the sphere as generic, then its presence in the image reflects exactly the new mode of space inside which terrestrial globes could be paired with celestial ones.

Along these specifically "global" lines, I consider two additional pieces of evidence. First, ten years after *Of Knowledge of the Orbs' Motion* was published, Dürer joined Stabius once again to engrave a terrestrial map, whose image was executed in accord with latitude and longitude.[45] Put simply, in order to complete the Stabius map, Dürer had to assume and to apply a fully three-dimensional space to the terrestrial realm. Second, Dürer had likely already been exposed to this way of imagining the terrestrial realm, given that Martin Behaim's Erdapfel was completed in 1492 in Nuremberg, and it is virtually certain that Dürer had seen this globe. Behaim's work became famous immediately, and it is inconceivable that Dürer, as a local notable, would have failed to admire it. Moreover, Dürer also had personal connections to this work, insofar as the house in which he grew up sat next door to the Behaim family home. Although Behaim and Dürer never met—the former was a half-generation older and emigrated to Portugal before

Dürer's family moved in next door—there can be little doubt that Dürer would have known of Behaim's achievement and admired it.

The Magic of Number

With Dürer's connections to the new space in mind, I return to the *Melencolia*. The engraving harbors two different methods of foisting homogeneity onto the cosmos: measurement within space and the application of numbers.[46] Although I have emphasized geometric space throughout this chapter, I make a detour through number in order to reconsider a venerable thesis from the Renaissance scholar Frances Yates, which holds that the *Melencolia* was rooted in ancient Hermeticism.[47] In this context I call attention to the top right of the image (see figure 16), where Dürer put a magic square. This item arranges integers in a grid so that each column, row, and diagonal produces the same sum (in this case, 34). Knowledge of magic squares entered Europe as far back as the thirteenth century, via astronomical and astrological works that had been imported from the Muslim world. Such squares were popular in the fifteenth and sixteenth centuries, as both mathematical exercises and tools for dabbling in the occult. Two noted early modern figures, the geometer Luca Pacioli and the occultist Heinrich Cornelius Agrippa—both of whom I discuss below—included such squares in their written works.

Before evaluating Yates's approach to the *Melencolia*, I offer additional background on the Hermetic tradition. Hermeticism was an ancient system of ideas that saw magical correspondences between the cosmos, human beings, and the divine—and that believed these correspondences to be subject to manipulation by humans.[48] Its cornerstone is a collection of texts known as the *Hermetic Corpus*, which was ostensibly penned by the mythical Hermes Trismegistus ("Thrice-great Hermes"), roughly at the time of the biblical patriarch Moses.[49] (The writings are, in fact, from the third century AD.) In the fifteenth century, prominent thinkers such as Marsilio Ficino saw the *Hermetic Corpus* as a primordial work that contained *prisca theologia*, or first theology. This knowledge's proximity to the earliest phases of Judeo-Christian history signified (at that time, at least) that it was extremely powerful. Concomitantly, fifteenth- and sixteenth-century thinkers hoped that by artfully using Hermetic knowledge, they could influence events on Earth.

In Yates's view, a direct line runs from the Hermetic corpus up to Ficino and Agrippa and then into the *Melencolia*.[50] There are reasons to be skeptical.[51] First, Dürer's square probably did not come from Agrippa's great work, *On Hidden Philosophy*, since the number pattern in the *Melencolia* differs from the ostensible original.[52] One may imagine that Dürer saw the square in Agrippa's work and fiddled with the pattern in order get the year 1514 to appear in the bottom row.

The problem, however, is that Dürer was no mathematician, so it is unclear that he would have pursued this avenue alone. Along these lines, help would have been available to him in Nuremberg's rich cultural community, of course. As we will see, however, if Dürer sought any help in mathematics, he probably did so while he was traveling through Italy from 1505 to 1507—and this possibility suggests both a substantially different intellectual genealogy and a potentially different view of number.

There is a better explanation for the *Melencolia*'s magic square. Consider that its number pattern matches exactly that of a magic square that appeared in *On the Powers of Quantity*, a text that Luca Pacioli completed around 1508.[53] This peculiar document covers many topics, including arithmetic and geometry, magic tricks, and magic squares, of which it includes seven examples, with each paired with one heavenly body. The square that Dürer used was allied in the text with Jupiter—and this is an important datum in itself, since this planet was not traditionally associated with melancholy. (Indeed, Agrippa's square is associated with Jupiter, too.) Thus, contrary to what the literature has assumed since Klibansky and company published their *Saturn and Melancholy*, Saturn may not lurk in Dürer's work at all. Instead, it may be Jove who looks down—and this opens quite a different avenue back to the ancient world.

The evidence in support of the idea that Dürer borrowed primarily from Pacioli goes beyond the similarity of their respective magic squares. Dürer and Pacioli were probably acquainted, as it seems that in 1506 the former traveled to Bologna, in order to study perspective with the latter.[54] (I emphasize here that the evidence on this score is suggestive rather than definitive, since we know only that Dürer had arranged to go to Bologna—and not whether he actually arrived.) Given how close *On the Powers of Quantity*'s completion date was to the time of Dürer's possible encounter with Pacioli, it seems plausible, however, that the two men would have discussed the manuscript's chief ideas, although one can also imagine Dürer having read parts of the text itself. In any event, Dürer would have had no other access to Pacioli's text, since the manuscript remained unpublished until only recently.[55]

With Dürer and Pacioli's possible meeting in mind, it will be worth considering the significance of the title to Pacioli's manuscript. The Latin title, *De viribus quantitatis*, is occasionally translated as *On the Power of Numbers*.[56] This is problematic in two ways. First, *viribus* is a plural form of the Latin word *vis* (power, force) and should probably also be translated in the plural, especially given the hodge-podge nature of the work. Second, *quantitatis* is the genitive singular of *quantitas*, which means "greatness, extent, or quantity," as opposed to "number." It would, thus, be more appropriate to use any of these meanings in a translation, since all of them connote measurement in ways that are consistent with Pacioli's approach to mathematics. Finally, if Pacioli had meant to say "number," he had a proper Latin word for it, *numerus*. Indeed, when Pacioli referred to numbers in the manuscript, he used *numeri*, which is the plural form in both Latin and Italian. For these reasons, I have suggested the translation *On the Powers of Quantity*, since it not only fits better with the Latin but also incorporates the era's growing emphasis on measurement.

Next, I deepen our understanding of measurement's philosophical significance with a return to Euclid's *Elements*. I have already noted one major difference between medieval and early modern approaches to geometry: medieval readers seem not to have gone beyond the *Elements* first six books. In that context, as I have explained, when the *Elements*' final books (11–13), which explicate spherical geometry, were studied in detail, their approach to space opened new ways to imagine unseen things. (Johannes Kepler's thought is a primary example of this trend's effects.)[57]

With the vastness of projected space in mind, I would like now to cast light on the significance of the *Elements*' middle books (7–10). Within the history of mathematics these were extremely important, because in them Euclid conceived of number as a segment of a line, rather than as a series of discrete items, which was how the ancient Pythagoreans (sixth century BC) saw things.[58] Thus, with these books Euclid addressed a potential philosophical incompatibility that resulted from the mixture of a system of idealized space that was rooted in *nothing* with a system of number that was based on discrete things.[59] What Euclid seemed to see, in short, was that number had to be associated with spatial extension, in order for the measurement of physical things to be justified. As the compasses that I have highlighted in different Renaissance-era artworks would indicate, this lesson was not lost on the fifteenth and sixteenth centuries.

In the context of the Western triad, the consequences of the *Elements*' justification of number were as jarring as those wrought by spherical geometry's return. With his inclusion of Books 7–10, Euclid infused the *Elements*' spatial homogeneity with number, and this exacerbated the conceptual problems that space presented to the divine. God had obviously *created* number, which meant that Books 7–10 of the *Elements* did not threaten his existence directly in any way. Nevertheless, the question naturally arose: What of both His and humanity's relationship to a cosmos that was based in number? In this context, the express connection of number to measurement threatened the medieval divine's noncorporeal entities, including most especially, angels and cherubs, since they were expressly understood as intermediary beings. It would, therefore, be difficult to overstate Euclidean geometry's philosophical significance in the early modern period. First, the generic nature of Euclidean space justified the human application of mathematical forms of analysis to both celestial and terrestrial realms. Second, Euclidean thought justified human efforts to gather knowledge both *down here* and *up there*, insofar as the measurement of all things within Creation was now warranted. Put another way, within an early modern context, Euclid's inclusion of number in the *Elements* vindicated the wresting from God's hands of the compass that He wielded in the *Bible moralisée*.

Along these numerical lines, I offer one final point that directs us back to Dürer and the *Melencolia*. For simplicity's sake, I have used a shortened version of the engraving's original title; it is, in fact, *Melencolia I*. Here, I propose that Dürer's inclusion of the "I" represents exactly the association of number with the measurement of space that the full reception of Euclid's *Elements* had made possible. Thus, it is essential for my analysis of Dürer's work that ancient geometric thought backstopped artisanal productivity by affording it an independent, human dignity. With a copy of the *Elements* in hand, à la Pacioli, any prominent fifteenth- or sixteenth-century figure could celebrate the power of human creativity by imagining, producing, or using an object of material culture.

Now I return to the distinction I drew earlier between the early modern emphasis on number and the foundations of Hermeticist magic. First, it is important to underscore that Dürer could not have seen Agrippa's *On Hidden Philosophy*, as he died in 1528, whereas the text rolled off the presses in 1531. Frances Yates has argued, however, that manuscript versions of Agrippa's great text were circulating by 1510 and, building on the work of art historians Erwin Panofsky and Karl Giehlow, concluded that Dürer must have seen one of these copies before he executed the *Melencolia*.[60] Other than some parallels in Agrippa and Dürer's respective cosmologies—which are to be expected, given Aristotle's dominance—Yates's argument is supported only by Giehlow's supposition that the *Melencolia*'s magic square is inscribed on a metal plate.[61] The question of the metal's type is important, because Agrippa held that a magic square should be engraved on the metal that corresponded to the planet with which it was associated. The *Melencolia* does not, however, indicate what the underlying metal was, if it depicts any metal at all. In the end, therefore, the only evidence from the *Melencolia* that we have for the magic square's provenance is its number pattern—and, given that this pattern can plausibly be ascribed to Pacioli, Dürer's magic square would seem to undermine Yates's thesis.

Here, I reinforce my argument in favor of Pacioli from the other side, with respect to Agrippa's *On Hidden Philosophy*. Another reason to be skeptical of the Yates thesis is that Agrippa used concepts in his work that were conspicuously absent from the *Melencolia*. He included, for instance, instructions on the use of magic squares in magical ceremonies, of which there is almost no trace in the *Melencolia*. The closest one can come to this is Dürer's inclusion of the flaming crucible in the center left of the image, which some scholars believe refers to alchemy.[62] There is, however, an alternative explanation for this item to which I return below. Finally, Agrippa also translated the Arabic numerals within the magic squares into Hebrew characters and presented the new grids next to the old ones.[63] The *Melencolia*, in contrast, included no Hebrew characters—and this absence is critical, because it means that the image lacks a direct reference to the Mosaic era, which was the essential backdrop to the embrace of Hermeticist doctrines. In the end, the *Melencolia* seems to associate itself more with Euclidian geometry's constructions of continuity than it does with the Hermeticism of Heinrich Cornelius Agrippa.

Dürer's Euclid

Here, I return to the parallel that I set up in Chapter 1 between Euclid's *Elements* and the Bible. Early modern Europe witnessed a massive upsurge of interest in and access to published versions of the *Elements*, and the *Melencolia* was directly related to this upsurge's early years.[64] The first Latin translation of the *Elements* appeared in 1482 in Venice. This edition was, as I have noted, the one that Copernicus owned and that also appeared in de'Barbari's portrait of Luca Pacioli. Another important early edition appeared in 1505 in Bologna, and in the context of this chapter it is very significant, because we know that Dürer bought a copy of it in 1507.[65] This edition was in Latin—the earliest German version appeared only in 1555—so that Dürer had to ask his friend Willibald Pirckheimer to translate it for him.[66] Although we do not know when the translation was completed, the evidence suggests that by 1513, at the latest, Dürer was studying Euclid's great work intensely.[67]

The effects of Dürer's geometric studies are reflected not only in his artistic production but also in his written works. Dürer penned two books, both of which applied geometry to artisanry in a way that allowed the reader to *make space*. For example, the first line of Dürer's initial text, *Instruction in Measurement* (1525), reads: "The wholly astute Euclid constructed geometry's foundations, and who understands these well has no need of the following written work, but it is [written] exclusively for the young and for those who have nobody to teach them faithfully."[68] Dürer intended this work to be both a tutorial in geometry and a manual for creative work. Thus, it is particularly significant that he began with points and lines, agreeing with Euclid that they constitute nothing. On points, for example, he wrote, "But a point is a thing that has neither size, length, breadth, or thickness, and is nonetheless the beginning and end of all the bodily things *that we may fashion, or that we may imagine in our thoughts.*"[69] The highlighted text further draws attention to the anthropological effects of Euclid's rise: geometry provided a path by which human thoughts could become reality, that is, by which they could not only *make space* but also put things into it.

Dürer's geometric studies return us to space's broader philosophical backdrop. Unlike our contemporary culture, fifteenth- and sixteenth-century culture was fully aware of homogeneity's philosophical implications. Luca Pacioli, for instance, associated geometry directly with philosophy. His *Principles of Arithmetic and Geometry: Proportion and Proportionality,* which appeared in 1494, celebrated Euclid not simply as a geometer or a mathematician, but as a philosopher.[70] And Pacioli's own translation of the *Elements* appeared in 1509 under the title *Euclid of Megara, Sharpest Among Philosophers and of Mathematicians.*[71] The identifier "of Megara" was the product of a common confusion. Euclid of Megara was not the author of the *Elements,* but a student of Socrates. This confusion highlights, nonetheless, the chief reference point for Renaissance-era interpreters (including Dürer): Renaissance Platonism's rise and its corresponding challenge to medieval Aristotelianism. Thus, we see from another perspective how Euclidism cut across everything that had ostensibly divided early modern European culture. To study Euclid was to open an alternative perspective on the entire Western tradition.

During the Spatial Reformation's initial phases, Euclid's *Elements* mediated stealthily between pagan and Christian thought. If we are to understand how this worked, we must keep in mind that its space threatened primarily those ideas that few people seriously doubted—and to which they clung for decades—namely, the belief in the hierarchical (and geocentric) physical arrangement of Heaven and Earth and its accompanying chain of divine beings. Raphael got it exactly right in his *School of Athens* when he not only depicted Euclid as standing between Ptolemy and Hipparchus, but also put the group off to one side, with Plato and Aristotle depicted in the middle. This flanking arrangement is crucial, because it illustrates how Euclid's effects came from an oblique angle. Euclid's rise did not erase the thought of Plato and Aristotle, but instead chipped away at their respective views on space's philosophical status. As I have explained, Plato concentrated on geometry's forms and embedded them within the world of ideas, while Aristotle limited geometry's applicability to the Heavens above. Euclid, in sum, threw both these traditions into disarray by putting space into an altogether different conceptual place.

I illuminate the changes that Euclid heralded via the cover image to an important mathematical work, Niccolò Tartaglia's *The New Science* (figure 21). First published in 1537, this text explicated the mathematics of ballistics, for which it is rightly famous, and did so by mixing algebra with geometry.[72] (This process had been going on since the Middle Ages and would reach its height in the first half of the seventeenth century with René Descartes's invention of analytic geometry.) Tartaglia was particularly well versed in geometry, as he emerged from the mathematical culture that Pacioli had helped to cultivate.[73] Indeed, like Pacioli, he proselytized on behalf of Euclid, as he published in 1543 the first Italian edition of the *Elements,* under the title *Euclid of Megara, Incomparable Philosopher and Guide to the Discipline of Mathematics.*[74] If the pervasive nature of Pacioli's influence is apparent in the reference to Euclid as a philosopher, the mathematical friar's

Figure 21. Cover illustration from Tartaglia, *The New Science*, 1537.
Courtesy of ETH-Bibliothek Zürich.

emphasis on Euclid's fundamental utility is unavoidable in *The New Science*'s frontispiece.

The frontispiece to Tartaglia's work underscores how Euclid resituated *everything*, including ancient philosophical debates. It depicts two realms, the respective entrances to which are guarded by classical figures. Before the outer entrance stands Euclid, who allows the select few to pass into the realm of the liberal arts, where Tartaglia himself is ensconced and flanked by Arithmetic and Geometry. At the inner entrance, in turn, stand the two greats Plato and Aristotle, beyond whom sits Philosophy herself. I highlight two details. First, Plato stands a bit higher than Aristotle, which suggests that although Tartaglia held the Stagirite in high regard, he ranked him slightly below Plato. Second, if we view the image as a whole, we see not only that number and space come before everything within the image itself, but also that three-dimensional space suffuses the entirety of the depicted realm. In this sense, the unity that Tartaglia cultivated pervaded and *situated* all philosophical systems—and this is exactly what we see in Dürer's work, with equally stirring effects.

The New Science's frontispiece illustrates that those scholars who have pursued the Platonist backdrop to the *Melencolia* are correct, but in a way that needs refinement.[75] Euclid's rise profoundly affected how Renaissance thinkers read both Plato and Aristotle. The existential clash between Aristotelian and Platonic philosophies in the fifteenth century is well established in the literature, even if the battles between them were never as absolute as once believed.[76] In this context, I highlight two points that reflect back on Dürer's *Melencolia*. First, Renaissance geometry undermined Aristotle's spatial dichotomy by applying homogeneity fully to his cosmos. Second, the very idea that nothing (i.e., points and lines) could unite sensible and intelligible worlds democratized Plato's austere dualism. Now anyone who could get a copy of Euclid's work had the keys to a bigger cultural world, including artisans who lived in Nuremberg and had copies of Dürer's printed works at hand. Thus, as I have suggested previously, it may be better for historians to consider Euclidism as having stealthily enveloped all other ancient systems of thought, leaving them, in the end, with no other place to go.[77]

With the early modern reception of Euclidism in mind, I return to Dürer's world via Jacopo de'Barbari and his portrait of Pacioli (see figure 7).[78] Although Dürer never saw this painting, he did come into contact with its maker, who worked in Nuremberg from 1500 to 1503. At that time, Dürer observed how de'Barbari applied geometry to his professional work, even if the established master refused to teach the rising artist anything.[79] Indeed, de'Barbari's demurral may have spurred Dürer's subsequent pursuit of geometry in Italy, both in terms of his meeting with Pacioli and his subsequent purchase of the *Elements*. (In the course of his life, Dürer took two trips to Italy, the first between 1494 and 1495 and the second between 1504 and 1507.) I suggest, therefore, that Dürer's headlong pursuit of three-dimensional space marked a watershed in his artistic career—and one result was his new vision of melancholy.

Against the backdrop of Dürer's personal development, de'Barbari's rendition of Pacioli highlights two important aspects of geometry's effects. First, the canvas encapsulates how homogeneous space justified a *human* world. De'Barbari not only used perspective, which was essential to painting, but also combined his work's space with various geometric figures, whose presence in the painting is, at once, real and ethereal. Geometry, in this sense, not only provided the work's conceptual frame, but also permeated the image's interior. Second, the image illustrates geometry's path to cultural ubiquity. In the painting, Pacioli holds open a copy of Euclid's *Elements* (the first edition of 1482) while he draws onto a slate the figures that are described in the text.[80] This transfer—abstract space passing from a book into the mind and out through the hands—epitomizes Euclid's early modern career, as generations of students learned geometry by consulting texts and scratching their lessons onto slates. Put more broadly: the *Elements*' print diffusion allowed Euclidean space to ensconce itself between theory and practice in a way that begat a new mental world.

Geometry's God

Having explored space's effects on humanity's relationship to both the divine and the cosmos, I circle back to an issue from the introduction: the problem of God and his physical location. It is not coincidental that the Christian God is as conspicuously absent from the *Melencolia* as is humanity, given that both absences are direct products of homogeneous space's rise. As I have explained, space threatened traditional visions of the divine, because its diffusion came in the wake of bitter medieval debates between what can loosely be called Thomism and Nominalism. Nominalism's emphasis on God's will—as opposed to Thomism's emphasis on His reason—shook the alliance between anthropology, cosmology, and theology that had dominated medieval thought and, in doing so, allowed Euclidean space to dominate European thought.

In order to connect Dürer's *Melencolia* more fully to geometry's theological effects, I look back to a figure whom I have already discussed, Nicholas of Cusa. One would be hard-pressed to find anyone more in tune with the speculative currents that Nominalism had unleashed. Born in 1401 in the German city of Cusa, which lies along the Moselle, the young Nicholas studied in Heidelberg, Padua, and Cologne, which were then centers of Nominalism, Latin Averroism, and Thomism, respectively.[81] After completing his studies, he worked professionally on both sides of the Alps, representing the interests of a German noble at the Council of Basel in 1432, before embarking on a clerical career that terminated with a bishop's miter and a cardinal's cap.[82] Industrious and intellectually curious, Cusa was also a noted plunderer of classical manuscripts in monasteries, as well as the author of important theological, philosophical, and pastoral works. When he died in 1464, few would have disputed his intellectual significance.

Cusa's influence is well established, although amid much disagreement.[83] It is generally accepted, for example, that Renaissance Nurembergers, such as Dürer's friend Willibald Pirckheimer, read Cusa's works with interest, even if the specific effects remain unclear. I set aside the disputes over the nature of Cusa's influence and highlight, instead, its breadth. Beyond his personal contacts and written correspondence, which were extensive, Cusa's influence is exemplified in the repeated publication of his works.[84] The first edition of his *Opera omnia* appeared in 1488 in Strasbourg, which at the time was a thriving center of the German Renaissance.[85] This edition was copied and republished (with some changes) in 1502 in Piacenza. In 1514, another edition appeared in Paris, and this one is particularly reflective of the background from which the *Melencolia* emerged, because its editor was none other than Jacques Lefèvre d'Étaples.[86] Finally, in 1565 another version appeared in Basel, the busy publishing hub that united Europe's northern and southern worlds. This edition is, of course, not relevant to Dürer's work, but it establishes that interest in Cusa's thought extended into the second half of the sixteenth century.

Although scholars agree that Cusa's thought was important, they have yet to agree on what it signified. The philosopher and historian Ernst Cassirer considered him to be the first "modern thinker," with the historian of science Alexandre Koyré following this line, while other scholars have emphasized the medieval and mystical aspects of his thought.[87] These differences aside, it is safe to say that the scholarship overall sees Cusa as a transitional figure whose thought hovered somewhere between the medieval and early modern, although it remains unclear whether he looked forward or backward.[88] This Janus-like reputation is, in turn, anchored to the year of Cusa's birth. Because he was born in 1401, Cusa could scarcely have avoided contact with the issues that had dominated the late fourteenth century, even as his education and experience exposed him to new intellectual currents, including the continental rise of Euclidean geometry.[89]

With this point in mind, one backdrop to the scholarly ambivalence becomes especially important, and that is homogeneous space's effects on Cusa's thought. The scholarship has yet to concentrate on this issue—and this is significant, since it was exactly Cusa's use of space that made his thought early modern. Fortunately, Cusa's professional career allows us to cast light on his process. The church exposed Cusa to both a mature Aristotelianism and a rising Platonism, with the result that the two currents were mixed liberally within his own thought. More important, however, the church also nurtured Cusa's profound interest in mathematics. While living in Rome, he joined Pope Nicholas V's circle, which comprised a glittering collection of the mathematically inclined, including Leon Battista Alberti, Basilios Bessarion, Johannes Regiomontanus, Paolo Toscanelli, and George of Trebizond.[90] Cusa's immersion in mathematics offers a different perspective on his relationship to both the medieval and the early modern worlds. By injecting geometry into theology, Cusa applied ostensibly new methods to what seem to be old questions, with the result that his speculations appear odd to us.[91] This is the case precisely because Cusa was among the first to wrestle with a problem that wracked early modern thought (but that no longer bothers us), namely, space's relationship to both the divine and the cosmos.

To present an adequate overview of Cusa's oeuvre is impossible in this space. Instead, I concentrate on two themes that figured prominently in his thought—his doctrine of "the coincidence of opposites" and his embrace of geometry. The former contains an important theological insight. According to Cusa, God is so transcendent that he incorporates everything within himself, including things that contradict reason. This approach was not wholly new, as it dated to at least the eleventh century, when Anselm of Canterbury made God's existence into a rational imperative in his "ontological proof." (The proof went as follows: God is the greatest of all beings that the mind can conceive; to exist in reality is greater than to exist in the mind; therefore, God exists.) However, the force of fourteenth-century Nominalist critiques undermined this "rationalism," and one result was that Cusa turned away from the pursuit of certainty through logic and toward what he called "learned ignorance."

In 1440, Cusa completed a work entitled *On Learned Ignorance*, in which he held that humans could only examine their relationship to God after coming to terms with their own limits.[92] This approach effectively derailed Christianity's millennium-long dialogue with the divine, as the Platonist, Neoplatonist, and Aristotelian strains of Christian thought had all foisted a rational hierarchy onto humanity's relationship to the divine and the cosmos. Cusa detected, however, a problem with this rationalist tendency in ontology: "Created being derives from the being of the First in a way that is not understandable."[93] Thus, human reason simply could not bridge the gap between the divine and the mundane in anything but a *human* way, which meant that Christianity's hoary emphases on both a pervasive rationality and the hierarchy of being were rent asunder by a brusque turn inward.

Cusa did not lift his mind up to God by ascending a rational stepladder, but he invented oblique ways of backing into His transcendence. Put another way, he sought paths to God that imposed no limits on Him, while also justifying a more assertive sense of human agency within a context that was becoming ever more human. Geometry seemed a promising discipline in both respects, because the purity and unity of its space illustrated God's majesty, even if its doctrines yielded imperfect knowledge of God. Indeed, all too conveniently, perhaps, geometric theology ended in absurdity. In his *On Learned Ignorance*, for example, Cusa argued that although we can think of God as an enormous circle, we cannot limit Him to the relationship between center and circumference, because God exists beyond all these concepts.[94] In this sense, God was leaving geometry behind, while (to an equally profound degree) the reverse was happening, too.

Cusa's geometric approach to God was not new, as geometric theology dated as far back as the fourth century AD and extended into the thirteenth.[95] One important example from the fourth century runs, "God is an infinite sphere, whose center is everywhere, whose circumference is nowhere."[96] Cusa did not use these exact words in his own work, but a similar spirit pervades his thought, as he groped for God geometrically, while propounding theological doctrines that, when understood in geometric terms, are nonsense.[97] Indeed, we can detect an echo of this approach to the divine in chapter 14 of *On Learned Ignorance,* where Cusa argued that "an infinite line is a triangle," adding in the next chapter that "the maximum triangle is a circle and a sphere."[98] For a geometer this was absurd; for a theologian it was a beginning. This problematic application of space to the divine marked, nevertheless, a theological watershed, as it soon became clear that geometry's doctrines were as incompatible with the divine as they were consistent with the human mind. Thus, having placed God just out of geometry's reach, Cusa used geometry itself to emphasize continually, in his later writings, the almost-divine aspects of human creativity.

Given my attempt above to distinguish the *Melencolia* from early modern Hermeticism, it is important to note that the fourth-century phrase that I quoted in the previous paragraph comes from a Hermetic text. (To this day, the author of the text is listed as Hermes.) Preserved under the title *Twenty-four Books of Philosophy,* this work draws near to God through the application of geometric doctrines to His majesty.[99] It does this, however, not to establish the validity of Euclid's geometry, which is what early modern thinkers did. Instead, it applies geometry symbolically—that is, it illustrates humanity's journey toward the divine via geometric concepts, as opposed to celebrating humanity via its ability to produce space. Given this ancient text's very particular characteristics, anyone who studied the *Elements,* which Cusa surely did, would come to understand that geometry's own logic was attuned to the human mind's needs, while remaining wholly inadequate to the needs of the divine. Geometry, in this sense, became essential to the rise of learned ignorance, insofar as it highlighted both human reason's glory and its inherent limits.[100]

Cusa's theology was both related to and also becoming increasingly independent of an ancient tradition of associating homogeneous space with the divine. His own riff on this issue is crucial to my broader argument, because it wrought a break that was essential to heliocentrism's subsequent rise, namely, the transfer of infinite homogeneous space from God to the physical cosmos.[101] In *On Learned Ignorance* he wrote: "Thereupon you will see—through the intellect, to which only learned ignorance is of help—that the world and its motion and shape cannot be apprehended. For [the world] will appear as a wheel in a wheel and a sphere in a sphere—having its center and circumference nowhere, as was stated."[102] He continued: "Hence, the world-machine will have its center everywhere and its circumference nowhere, so to speak; for God, who is everywhere and nowhere, is its circumference and center."[103] Space now provided more than a symbolic path to God, as it also brought homogeneity to His cosmos in a way that justified increasingly detailed human analyses of its physical functioning. Moreover, the concomitant shift in infinite space's *location* had additional implications, as it suggested that human beings could know the cosmos in a way that was not only non-Aristotelian but also non-Platonic. In this respect, early modern analyses of the cosmos still maintained a tenuous contact with God, even as they pushed Him toward Euclidean space's margins. (The angels' fate, of course, was sealed.)

In addition to nudging God away from the cosmos, Cusa justified a new respect for humanity's intellective and creative power. In *On Conjecture*, which he completed three years after *On Learned Ignorance*, he humanized both human knowledge and human reason by limiting it to experience: "Therefore, no other end exists for the active creativity of humanity than humanity. For it does not proceed beyond itself when it creates, but when it unfolds its virtue, it reaches toward itself alone. Nor does it bring about anything new, but discovers that everything that it creates through its unfolding already existed within itself, for we said that all things existed in a human way."[104] These rather elliptical assertions illustrate how space's progress reconfigured the relationship between humanity, the cosmos, and God. And with this in mind, I refer back to another phrase from *On Conjecture*, which I have quoted previously: "Thus, man is God, but not absolutely, because he is man; humanly, therefore, he is God. Furthermore, man is a world, but not comprising All, because he is man. Therefore, man is a microcosm, or certainly a human world." Cusa added: "Therefore, man is able to be a human god and, as god, is able humanly to be a human angel, a human beast, a human lion or bear, or whatever else. For everything exists in its own fashion inside the power of humanity."[105] Taken together, these quotes throw into relief a pending change: humanity's mastery of space was making it into a little God.

Having explained how Cusa used space to reposition humanity with respect to both the cosmos and God, I turn to his work *The Layman on the Mind*. Completed around 1450, this collection of dialogues between a layman and two academics on knowledge's foundations harbors the same subversive currents that I have identified within the *Melencolia*. First, it embeds reason in the terrestrial world through geometry's core doctrines. Cusa wrote: "Layman: The mind makes the point into the end of a line and the line the end of a surface and the surface [the end] of a body; it also makes number, whence multitude and magnitude come from the mind, and it measures all things."[106] Second, geometry justified human productive activity via number. Cusa added: "Layman: Similarly, I say the exemplar of our mind is the concept of number. For without number nothing can be made; neither similarity, nor an idea of a thing, nor difference, nor measurement [would be possible] were number not to exist."[107] Building on number's connection to measurement, Cusa then reconstructed the relationship between God and Man by defining measurement's philosophical limits. He noted, for example, that all earthly things are subject to measurement, while adding that the spiritual realm, which was beyond humanity's reach, could not be measured. The divine had clearly lost its *place* within humanity's cosmos.

Cusa reveals how geometry's rise resituated traditional conceptual boundaries and, in so doing, impelled European thought along its new path. Spatial homogeneity and number became so pervasive that both the human and the divine were redefined in ways that demanded an ever-greater conceptual distance between them. For example, presaging the words from Thomas Hobbes that appear in this chapter's epigraph, Cusa separated God's world from ours when he noted in *The Layman on the Mind* that God can count spiritual beings, but humans cannot.[108] Space had, in effect, cut humanity off from the divine—and it was now left to humanity to find meaning both within its own sphere and through its own abilities. Homogeneity had defined, in short, the mental world inside which Dürer's *Melencolia* became thinkable.

Conclusion

Dürer's *Melencolia* is endlessly fascinating, precisely because it is filled with tensions that appeared after 1350 and whose partial resolution came only around 1650, when geometry became an artifact of the human mind. The engraving is, in this respect, only one echo of a continent-wide change that transformed almost every aspect of early modern thought. Much of the tension in the image was due to Dürer's artistic genius, which was sufficiently capacious not only to associate contraries, but also to give them free rein. Consider, along these lines, why anyone would give a burin to a putto. Dürer did this (and other odd things) because he was groping for ways to illustrate the effects of an intellectual revolution that, although underway, remained inchoate in many respects.

Dürer was, therefore, one among many sensitive souls who recognized that homogeneity caused problems for traditional approaches to the divine. To this point, the scholarship has read the *Melencolia* in different terms, arguing that Dürer's greatest work must be understood as representing a shift in melancholy's definition away from the physical and toward the psychological.[109] As I have tried to show, however, if any beings were brooding in this image, they were divine, rather than human—and this makes the whole question of modern human psychology's rise quite beside the point.

Moreover, as my discussion of Nicholas of Cusa has established, after the fifteenth century a profound relationship emerged between the cultivation of space and the justification of human creativity. Thus, it is hardly coincidental that the *Melencolia*'s divine entities, the angel and the putto, are not simply lost within homogeneous space, but also sit amid tools that were utterly useless to them.

Some of the tools relate to the measurement of quantity, such as the hourglass and the bell, which defined the early modern workday, while the scale refers to a form of measurement that was fundamental to economic life.[110] In addition, the other tools are related directly to production, such as the millstone—on which the putto has languidly placed its posterior—the clamp in the angel's lap, the tongs, the saw, the planing instrument, the straightedge, the hammer, and the nails, all of which are arrayed around both a sphere of unknown material and a decorative piece of wood. It is especially through these latter items that Dürer highlighted artisanal productivity. Decorative items were a normal part of architecture and interior design, while the sphere was the sort of thing on which the earth's image might have been pasted.

Thus did Dürer underscore how artisans produced things, and with this emphasis on production in mind, I turn to the last of the image's tools, the ladder and the crucible. The contemporary literature has invested much energy in finding reasons to ascribe symbolic or otherworldly meaning to these items. Frances Yates, for example, held that the ladder behind the putto represented Jacob's ladder, which is not altogether plausible as the incident is not very important within the Old Testament, being mentioned all of once.[111] Yates and others have also suggested that the flaming crucible represented alchemy, which was closely tied to Hermeticism. I do not doubt that the *Melencolia* had religious overtones, nor do I discount entirely Hermeticism's influence on Dürer's work. In my view, however, it makes more sense to emphasize the human and this-worldly nature of both the ladder and crucible.

In this context, we must note, first, that Dürer was an artisan and would, therefore, have encountered many ladders and crucibles in his professional life. He had spent years studying in foreign parts, where he acquired a variety of skills that included drawing, painting, engraving, and woodcarving—all of which are represented in his oeuvre. Given his varied experience within early modern Europe's artisanal traditions, it would have been difficult for him to avoid seeing and using either of the tools in question. With respect to ladders, they would certainly have been present within his workshop in Nuremberg, let alone the other workshops in the other cities with which he was familiar. With respect to the crucible, it is important to note that Dürer's father not only was a goldsmith but also directed his son's early training in that trade. In Dürer's experience, therefore, crucibles were tools that trained artisans could use to produce something of value—and this was exactly what angels and cherubs had become incapable of doing. There was still magic within Dürer's mental world, but I suggest that it did not flow primarily from the Hermetic tradition. Instead, it emerged from the discovery of the amalgamating power that lurked within the human mind's growing command of homogeneous space.

Figure 22. Dürer, *Adam and Eve,* 1504. Courtesy of the Metropolitan Museum of Art, New York.

I would thus not disagree with the notion that the *Melencolia* transferred melancholy into a new intellectual realm. I would underline, however, that this realm was not a psychological phenomenon, but a spatial one—that is, it was an effect of homogeneous space's diffusion through both the Western triad and early modern culture more broadly. In this context, I turn, finally, to another of Dürer's engravings, *Adam and Eve*, which was completed in 1504 (figure 22). The contemporary scholarship has long excluded this engraving from the *Meisterstiche*. More than two decades ago, however, the scholar Peter-Klaus Schuster argued that *Adam and Eve* should be added to the group of three.[112] I agree with Schuster's suggestion, although for reasons that differ from his. In Schuster's view a stylistic unity cuts across all four images, and he held, concomitantly, that the resulting aesthetic parallels mandated the *Meisterstiche*'s expansion. I hold, however, that what unites *Adam and Eve* with the other works is the space that pervades all of them.

It has escaped notice to this point, but *Adam and Eve* was completed the year before the publication of the Latin edition of Masha'allah's *Of Knowledge of the Orbs' Motion*. Based on this sequence, I argue that, at the very least, the spatial sense whose evolution I have traced both into and beyond the *Melencolia* is likely also present in *Adam and Eve*. With this idea in mind, I call attention to how this image of humanity's first breeding pair handles physical space. On the one hand, the image was executed using the same mental tools that Dürer applied to the *Melencolia*, insofar as the objects are depicted within a uniform space and with little distortion. On the other hand, the space in which Adam and Eve are placed is not real in any terrestrial, that is, geographic sense. The mixture of tropical flora and fauna with European ones and the hodgepodge of geographic zones (note the mountain goat on the top right) all deny to Eden an identifiable *terrestrial* location. Thus, although sixteenth-century culture routinely projected detailed images of unseen terrestrial spaces, Dürer's work denied to Adam and Eve a terrestrial *position*. Mankind may have passed its salad days somewhere on Earth, but this history could not be integrated with the new spatial thought. Much like Isaac Newton's God, Eden was moving into a "not nowhere."

Dürer's Adam and Eve were as out of place as were his angel and putto. The growing gap between humanity and the divine that this spatial turn implied thus directs our attention to the truly profound nature of the underlying change. Angels brood and puttos snooze, while Adam and Eve have no location. If we take these things together, we see that the increasingly sophisticated understanding of Euclid that characterized intellectual life in Dürer's day continually contorted traditional ideas about humanity, the cosmos, and God. In this respect, where the *Melencolia* directs our attention to changes in the general approach to the divine "above," *Adam and Eve* calls attention to a fundamental shift in the mundane—the view of the Earth "below." Older categories were collapsing and a new *All* was emerging.

Chapter 4

Eden's End

> The Great Map of Mankind is unrolld at once; and there is no state or Gradation of barbarism and no mode of refinement which we have not at the same instant under our view.
>
> —Edmund Burke, "To Dr. William Robertson" (1777), in *The Correspondence of Edmund Burke, III*, ed. Guttridge

The cosmographer and geographer Sebastian Münster was a prominent figure within sixteenth-century Europe's increasingly Trans-Alpine culture of space. A German Rhinelander who became a professor at the Swiss University of Basel, he, like many others, borrowed from Italy's advanced mathematical culture and applied its achievements to multiple areas of thought. Although he contributed to the development of many disciplines, including lexicography and philology, it was through his effect on geography and cartography that he has gained particular renown. In this context, his greatest work was the *Cosmography.*[1] Published in 1544 in Basel, it is notable for, among other things, being the first work to present both written descriptions and cartographic images of Germany's geographic diversity. Republished for almost a century, the *Cosmography* secured for the German-speaking region the geographic *place* within the European mind that it occupies to this day.[2]

Histories of geographic thought often use Münster's *Cosmography* to frame the contemporary geographic discipline's rise (and vice versa), with Münster embodying, in effect, an early step in humanity's journey toward global omniscience.[3] Although not incorrect from a disciplinary perspective, this view is misleading from an intellectual historical one. The emergence of what modern geography considers to be its conceptual apparatus is a product of changes in spatial thought that occurred in the first half of the seventeenth century—and this particular development was punctuated by the appearance in 1650 of Bernhard Varenius's *General Geography.*[4] Published in Amsterdam, Varenius's text is generally deemed to be the first example of modern geography.[5] Nevertheless, the fixation on chronicling the geographic discipline's *progress* toward modern ways of knowing terrestrial space has obscured a shift that separated Münster from Varenius: the former was a product of Euclid's early days, when geometry seemed to coincide with the cosmos; the latter, meanwhile, matured as European thinkers restricted geometry to the human mind. Thus, although both men were children of the Spatial Reformation, each was formed by a different phase in homogeneous space's broader unfolding.

A profound shift in early modern spatial sense drove both cosmography's decline and geography's rise. In this chapter I trace the history of these two phenomena in separate sections that analyze Münster and Varenius's respective contributions to the European project of imagining the whole earth, while also considering the implications that each man's work had for the triadic backdrop that I have already discussed. In the first section I explicate Münster's *Cosmography* and show how its mode of representation not only *resituated* the earth, but also nudged aside traditional visions of the divine. Like Dürer's *Melencolia*, Münster's *Cosmography* applied homogeneous space to terrestrial knowledge in a way that denied to the divine its accustomed place. In the second section I use Varenius's *General Geography* to illustrate how innovations in spatial representation induced modern geography to exclude from its portfolio not only God but also the Heavens. In contrast to the progressive reading of geography's history, my juxtaposition of these celebrated figures reveals that cosmography and geography's histories comprise more than the discovery of the earth's true appearance, but were part of a more profound change in the earth's spatial context.[6]

With this alternative narrative in mind, I return to two concepts that I have previously mentioned, decaying cosmography and spatial secularization. As to the former, I have already explained in Chapter 2's discussion of globes how my emphasis on decay contrasts with contemporary celebrations of disciplinary progress. Where other scholars have chronicled how Europeans got the earth "right," I explore how changes in the spatial imagination also required the dissolution of older (once innovative) ideas about terrestrial space's *position* within European thought. By extension, in this chapter I stress that cosmography's descent into geography, although it was a progressive phenomenon from our perspective, was characterized by, among other things, a loss of conceptual richness—and it was this richness that long underwrote the divine's terrestrial presence within European culture. Thus, whereas cosmographers continually balanced competing anthropological, cosmological, and theological commitments within the new mode of spatial thought, geographers made little effort to include "extraneous" themes within their emerging discipline, but they went all in, as it were, on space. Put another way, cosmographers were ingenious thinkers whose minds worked on multiple levels simultaneously; geographers, in contrast, were sophisticated calculators whose full embrace of mathematics demanded homogeneity's victory over *All*.

Against this backdrop, the history of spatial secularization becomes critical to understanding European thought's development after 1650. If, following Henry of Ghent, the best mathematicians made the worst metaphysicians, then by tracing space's progression through European thought we can divine exactly the mathematical reorientations between 1350 and 1850 that governed the shift from cosmography to geography. Whereas medieval philosophy hovered above mathematics, both early modern and modern philosophy are tied to mathematics' continual disciplinary renovations. When framed against this backdrop, the early modern period becomes a time when homogeneity's incompatibility with traditional assumptions about humanity, the cosmos, and God moved to center stage.

From Mathematics to Cosmography

Cosmography's withering into geography highlights an inflection point in the early modern encounter with homogenous space. If we look backward from the seventeenth century, Münster's thought highlights two crucial themes. First, it exemplifies the continued vigor of spatial thought, as it journeyed over the Alps and into not only the German-speaking regions but also the Netherlands, France, and England. Second, it directs our attention to the peculiar fertility of the soil on which this new space landed. In these respects, it is significant that Münster was a trained Hebraist and classicist, in addition to being a competent astronomer, geographer, and mathematician. The former group of disciplines had become the North's forte, while the latter had been renovated most impressively in the South. Münster mixed these traditions and, in turn, produced an extraordinarily broad body of work that included, among other things, a Latin translation of a medieval Hebrew grammar, a Latin-Greek-Hebrew dictionary, and (most important) a version of Claudius Ptolemy's *Geography*—whose "return" I discussed in Chapter 2.[7] The breadth of Münster's thought, therefore, cannot be understood separately from the richness of the culture from which he emerged.[8]

Münster's *Cosmography* was not simply a collection of new and more comprehensive maps but constituted a particular kind of intellectual whole.[9] It is important to underscore this point, because homogeneous space's expansion through the seventeenth century repeatedly altered how the whole was defined. In this context, I turn to a relatively overlooked detail in Münster's *Cosmography* that will accompany us both through this chapter and into the next one: the text not only included a careful description of humanity's pre-civilizational State of Nature, but also boasted an illustration of this human idyll (figure 23). Similar to what we saw in Luther's Bible, Münster juxtaposed his discussion of humanity's murkiest past

Figure 23. The "State of Nature," in Münster's *Cosmography*, 1544. Courtesy of ETH-Bibliothek Zürich.

with a specific illustration, with the resulting combination of text and image marking the moment when a fully human history became possible.

When analyzed in conjunction with the accompanying text, Münster's State of Nature reveals how the projection of space drove the rise of a secular world.[10] I pursue this point, first, by analyzing the how Münster's image presented the State of Nature (figure 23 again). Judging by the woodcut, this state was a placid affair, with men and women living under trees, eating their meals at wooden picnic tables, and suspending their cooking implements in the branches above. (Evidently, as in Southern California, it never rained in the State of Nature.) In this context, a comparison of the image to Dürer's (four) *Meisterstiche* underscores how space continually reshaped the Western triad's context. First, this State of Nature is definitely not Eden, as there is no attempt to relate this generic couple to Adam and Eve's biblical world. Second, this fully human couple is located within a regular terrestrial space, insofar as the geographic features are presented consistently. (There is, for example, no mixture of tropical flora with mountain goats, as we saw in Dürer's work.) Finally, foreshadowing what I argue in the next chapter on the rise of secular political thought, the people seem well adjusted and rational, living happily within their own world.

Münster's *Cosmography*, therefore, should be understood as one contribution to a continental project of imagining the world in ever more human ways. First, his cosmographical study extended the "global" trend of putting the *entire* cosmos into homogeneous space, as he assumed that both Heaven and Earth were equally subject to mathematics and (not coincidentally) equally knowable. Second, this resituating extended into theological issues, as Münster's space ushered the most mythical aspects of the biblical past to human history's margins. As Münster saw it, although History began with Creation, history dated only to the receding of the biblical Flood waters—and he punctuated this transition by situating postdiluvian humanity within a natural state. Münster was not the first sixteenth-century figure to rethink humanity's past, nor was he the last to grapple with Natural Man. His *Cosmography* reveals with particular clarity, however, how homogeneity's rise drove a broader recalibration of the European sense for both space and time.

Along these lines, Münster's career also calls our attention to the growing self-confidence of the Central European cultural sphere.[11] By Münster's day, Central Europe had grown wealthy enough to boast numerous, prosperous urban centers, such as Basel, Nuremberg, Prague, and Strasbourg, as well as a burgeoning collection of universities that cultivated the knowledge that had been pouring in from older universities in Italy, France, and England. In the course of the fourteenth and fifteenth centuries many new institutions appeared, including in Prague (1347), Krakow (1364), Vienna (1365), Heidelberg (1382), Cologne (1388), Erfurt (1389), Leipzig (1415), Rostock (1419), Louvain (1425), Greifswald (1456), Freiburg im Breisgau (1457), Basel (1460), Ingolstadt (1472), and Tübingen (1477). This growth does not imply a massive expansion in the number of intellectuals, relative to the population. The web of learning spread thinly and was not anchored deeply, as universities remained small.[12] Nevertheless, the network of newer institutions integrated itself into the existing medieval educational system and thus accelerated the pace of intellectual change.

In the course of the fifteenth century an increasingly effervescent "northern" intellectual world appeared—and this world produced Sebastian Münster. Born in 1488 in Ingelheim, a town near Mainz, he began his studies at a Franciscan school in Heidelberg before joining the Franciscan Order in 1505.[13] He pursued his interest in ancient languages at the universities in Louvain and Freiburg im Breisgau before heading to Rouffach in Alsace, where, under the guidance of the noted humanist Konrad Pelikan, he studied astronomy, geography, mathematics, and ancient languages. In 1511, he followed Pelikan to the University of Basel, where he pursued all these fields. Three years later, he moved to Tübingen, residing at the Franciscan school in that town while pursuing advanced mathematical studies with Johannes Stöffler, the University of Tübingen's renowned mathematician and astronomer.[14] In 1518, he returned to Basel before settling down at the University of Heidelberg in 1521, where he occupied the chair in Hebraic studies. In 1529, in that other reformation's wake, Münster left his order and returned to Basel, where he remained a professor until his death in 1552.

The rise of a Central European intellectual infrastructure casts light on two aspects of Münster's thought on which the literature has yet to concentrate. First, he lived mostly in German-speaking cities that were located on a river, especially Heidelberg (the Neckar), Freiburg im Breisgau (the Dreisam), and Basel (the Rhine). (The Flemish-speaking Louvain constituted the extent of his foreign experience; and Louvain was, like the other places mentioned, a small city that sat on a river, the Dijle.) Münster's experiences in a world of urban islands made him an archetypical *Bürger*, as he saw life from the perspective of a city dweller whose activities required a network of cities that was connected via rivers.[15] This urban-riparian perspective is evident in figure 24, which presents Münster's view of the Rhine basin. The map is obviously not drawn to scale, as the cities and rivers are larger than they ought to be. This prominence is, nevertheless, significant, because it highlights what Münster held to be civilization's true foundation.[16]

The second important aspect of Münster's thought is the northern humanist character of his education. His classroom studies coupled historical and linguistic research on the ancient world (both biblical and classical) with burgeoning traditions in spatial thought, many of which had bubbled up from Italy or had migrated from England and France.[17] This mix of traditions yielded new ways of situating humanity within not only space but also time. In this sense, Münster's many works, including his edition of Ptolemy's *Geography* and his works on ancient Hebrew, underscore his cultivation of a highly developed sense for how place and history interacted. Moreover, this sense worked entirely within Münster's mathematical world, because his domination of mathematics' projective techniques allowed him to put both ancient and contemporary peoples onto a globe that, unlike Eden, existed within historical time. By putting humanity onto a round whole, Münster found a new way to separate its experiences from both the cosmos and the divine.

In this respect, it is important to note that Münster's State of Nature reflected the prejudices of his untraveled cosmopolitanism. From his urban perspective, the precivilized world appeared thoroughly human, but utterly disorganized—or, put another way, the State of Nature hosted rational people but no cities. I put the matter this way for two reasons. First, human rationality's status had been (and would remain) a central issue in the broader discussions of space. (I return to this problem in the next chapter.) Second, Münster emphasized what people in the State of Nature did *not* have. He began: "They had no documented coins in circulation. There was neither manufacturing nor commerce. Instead they bartered and repaid one good deed with another. No one owned anything in particular or [had] property. Instead, just as the air and the sky above were held in common, so too were the soil and the waters free for every man."[18] This State of Nature was thus simply the inverse of city life. Consider that Münster began with the absence of currencies before turning to the absence of markets. These specific

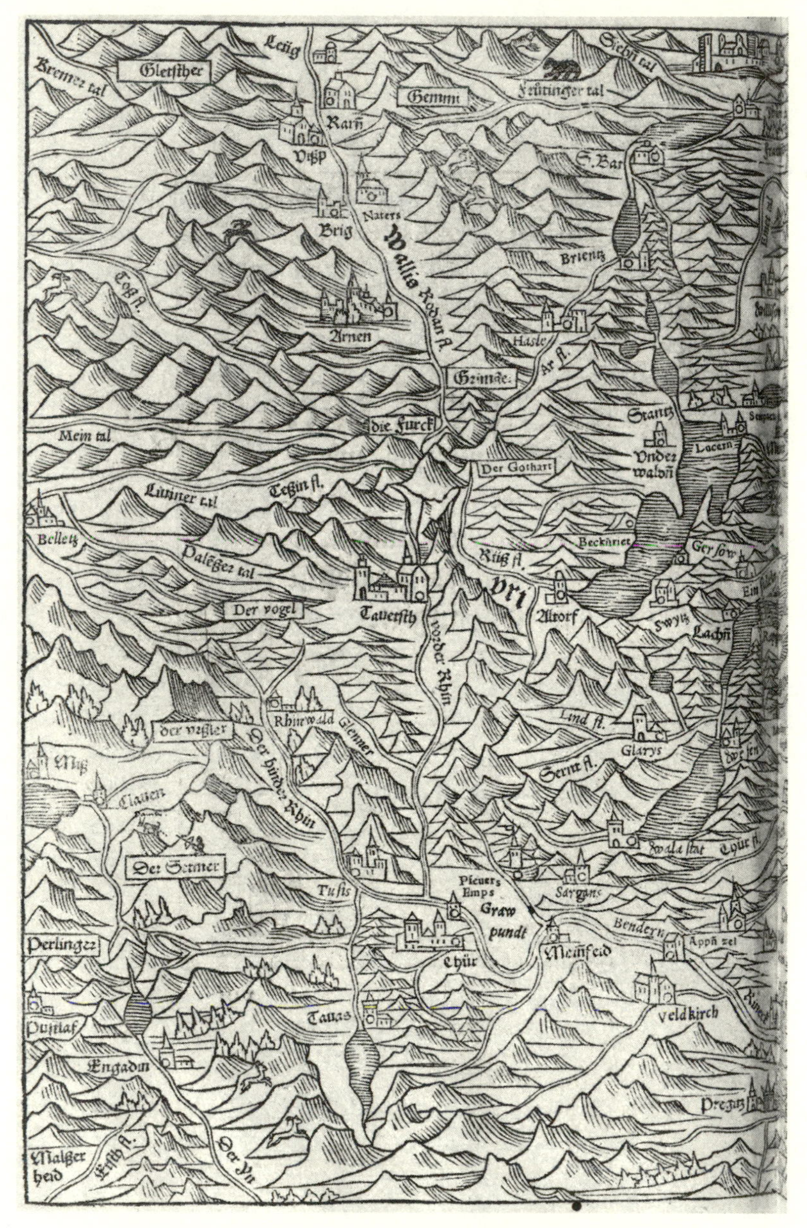

Figure 24. Map of Germany from Münster's *Cosmography,* 1544.
Courtesy of ETH-Bibliothek Zürich.

observations must be understood against two additional backdrops. First, some of Europe's most important currencies were issued by cities, including the Florentine florin and the Venetian ducat. Second, the Continent's most prominent markets were urban. The Frankfurt book fair, for example, was central to intellectual life in the sixteenth century—and given that Münster grew up not far from Frankfurt, he would have at least known of the fair.[19] Moreover, civilization's urban origin is underscored by Münster's juxtaposition of what humans in the State of Nature lacked with what they possessed: "They had, moreover, neither curtain wall surrounding them nor any moat, but they rambled freely among the free animals. And where night seized them, there they lay down to sleep. They did not fear murderers or thieves, but knew nothing of these things."[20] Natural Man did not live behind city walls, buying and selling things, but picked nature's fruits directly.

Of course, if this natural life was so idyllic, why did humanity leave it? This question injected Christianity into the new spatial context in a way that emphasized humanity's contributions to history, as Münster relied on his religion's anthropology, as the key explanatory variable. As Münster saw it, over time, the human population became too large for the food supply, with the result that some people began to steal, that is, to sin. Cities were founded, in turn, as a means to preserve life and to protect property. He held: "At the time when the soil and agriculture and work did not produce sufficient food and other multiple shortages arose among, the wild animals and foreign people began to turn to robbery. Then the people joined together and organized mutual aid, with which in unison they offered resistance."[21] He added: "[T]hey began to build their cottages next to each other and increasingly lived together as burghers and countrymen and to turn toward each other more humanely. Thereafter, when they confronted additional emergencies, they protected themselves with walls and moats, made rules and established authorities, so that they were able to live with one another in peace." It is important to recall that in the sixteenth century even small cities boasted defensive walls, given that chaos was but one invading army away. Indeed, Münster continually associated urban spaces with battlements, as every illustration of a city in his *Cosmography* includes a defensive wall. Thus, humanity's transition from biblical time into the State of Nature and up through civilization's emergence reached fruition in the proliferation of fortifications.

Münster pursued his emphasis on cities as oases of civilization by examining the rise of the urban economy. After providing a refuge from the increasingly dangerous State of Nature, the network of fortified islands was, in turn, nourished by commerce: "And finally, they subsisted not from agriculture and animal husbandry alone but also from invented trades and other work. They bound trees together and sailed across seas and began to engage in commerce, also using horse-drawn wagons."[22] The resulting economic development not only sustained cities' growth but also brought positive cultural and moral externalities. Münster added: "They transformed ore into coins, dressed themselves more delicately and suavely, spoke in a friendlier manner, built more ornately, repulsed murder and cannibalism, prevented robbery and open promiscuity."[23] This quote takes us deeper into the urbanity of Münster's thought. In addition to ushering humanity from the State of Nature, cities provided the disciplining that kept everyone from leaving civilization's fragile state, on account of how their walls enabled the policing of behavior.[24] Put another way, cities offered no space for crime to enter and left no place for it to go.

Before I continue with space, I consider the vision of historical time that Münster justified via his State of Nature. Humanity's precivilized state was expressly postdiluvian, that is, it came only after the earth's surface had been radically changed by God's wrath. This approach was significant, because it suggested that between humanity's Edenic leisure and the end of Noah's zoological cruise, no biblical figure lived in a fully historical period. Consider that prior to the Flood, human beings not only lived on a surface that had yet to be washed away, but also remained in relatively closer proximity to God. (Judging by the texts, God spoke more frequently to early figures, such as Adam and Eve, than he did to the urban people of the postdiluvian world.) Thus, the State of Nature's rise marked the end of not only biblical space but also biblical time, as humanity was now on its own recognizance within the realm that it would, one day, prove capable of imagining. In this vein, it is significant that human beings built cities as a response to other humans' sinfulness, as finding the distance between cities was one of cosmography's essential tasks.

Münster's conceptual reframing of History's biblical timeline had three consequences. First, it made the rise of contemporary civilizations a human process that functioned at an ever-greater distance from God. (After all, the State of Nature yielded the criminality that drove human beings to live in communities, in the first place, which meant that God had to be farther away.)[25] Second, with reference to the post-Columbian context that I discussed in Chapter 2, Münster's reframing of historical time allowed the peoples of the New World to be integrated into a fully "global" narrative. On the one hand, the newly encountered populations could plausibly be identified as descendants of one of Noah's sons, who then went their own way on the earth's surface. On the other hand, this postdiluvian genealogy meant that none of the New World's indigenous peoples could be characterized as living in Edenic innocence. Instead, having first located them within space, Europeans placed the peoples of the New World into a specific stage of *historical* development, through which the former had (conveniently) already passed. This change in perspective justified, in the end, exactly the ranking of peoples that is obvious in this chapter's epigraph from Edmund Burke. Everything and everyone would be subjected to a European mind that commanded both celestial and terrestrial space.

Having completed these preliminary discussions, I turn to the *Cosmography* itself. At over 750 pages, it is a large work, full of information about unseen places and peoples. The contemporary scholarship concentrates on the maps that Münster inserted just after the text's preface and evaluates these items against modern notions of accuracy. To ponder what Münster got right, however, is not nearly as interesting as explicating the mental structures that allowed a *Bürger* on the Rhine to put the whole world into place. The *Cosmography*'s maps are, therefore, not simply pre-positivist drawings of the earth's surface, but are examples of how space undergirded the elaboration of a new mental world.

Along these lines, it is significant that Münster's maps follow the same historical logic that sustained his State of Nature. Here, I turn to *Cosmography*'s first "map," which depicts Adam and Eve in Eden (figure 25). This image has been overlooked by historians, largely because they see it as representing Münster's piety. It constitutes, however, much more than a mere nod to religion, and is also a profoundly spatial statement. Consider that the image's space *situated* Eden within biblical space and time, and thus also separated it from the geographic space that sustained the *Cosmography*'s maps. Along these lines, it is significant that the image is presented without reference to the projective techniques that undergirded early modern maps, including, especially, longitude. For that reason, the geographic zone depicted is fundamentally irregular when compared to the regularized space that dominated Münster's cosmographical thought. What we see, therefore, is how the imposition of a specific kind of space shifted Eden into a conceptual nowhere.

With this point in mind, it is worthwhile to explore the space that Münster imposed. Inside the image itself this ostensibly geographic zone appears within a roughly drawn circle that is permeated by the cosmos's four elements. (Earth, fire, and water are depicted directly, while the flying birds suggest the air.) This expressly religious realm thus has no *global* location but is terrestrial only in a philosophically schematic sense. (I return to this theme in Chapter 6, where I discuss sixteenth-century anthropology.) Consistent with what I argued in Chapter 2, Eden's roughly circular space cannot be transferred to a sphere and, moreover, cannot be subjected to the kind of mathematical grid that was cosmography's *sine qua non*. By extension, consider also that before Adam and Eve stands an angel (perhaps Jophiel, who expelled them from paradise—although the image is too jumbled to be certain) who, as an angel, was rapidly losing his *place* within space. Given that angels are also conspicuously absent from the *Cosmography*'s terrestrial maps, Münster's Eden underlines the growing separation of humanity from the divine, in the wake of homogeneity's arrival.

We can understand the depths of the conceptual reordering that was underway by looking to the work of Münster's younger contemporary, the French explorer and royal cosmographer André Thevet. In 1575 in Paris, he published

Figure 25. "Creation," in Münster's *Cosmography,* 1544.
Courtesy of ETH-Bibliothek Zürich.

Universal Cosmography, in which he situated his empirically based global knowledge within a spatially ordered cosmos.[26] In most ways, this work was a standard cosmography, as it put knowledge of the earth's various places into a conceptual space that also incorporated the Heavens. In the preface, however, Thevet included a revealing comment about cosmographical knowledge's broader status: "Therefore, this cosmographical discipline serves to reveal the vanity in which we languish, then to lessen our pride. It directs our spirit to that which is great and no longer permits the binding to that which is nothing. And for this reason, I think that there is no science, after theology, that has greater virtue in introducing us to divine grandeur and power, and of holding it in admiration, than this one."[27] For Thevet, the construction of a whole that answered to the human mind allowed the viewer to admire not only the cosmos as a thing but also God's wisdom in having designed it as He saw fit. Humanity's vanity was thus put in check by its ability to imagine by itself and for itself the physical whole that had been created for its benefit. This approach to the natural world was not new, insofar as Creation had long been seen as another sacred "book." My interest here is in Thevet's express ranking of cosmography just beneath theology, when it came to humanity's search for God. Here, a traditional hierarchy clearly perdures. Nevertheless, the anthropological changes that space's rise had set in motion would soon produce a more broadly wrenching reordering, insofar as space continually elevated the human mind in a way that undermined exactly the longstanding theological valuation of space.

Returning, now to Münster and his production of what I am calling a biblical spacelessness, I consider the *Cosmography's* collection of terrestrial maps. The first two that follow on the image of Eden would have been recognizable to any educated person, as each was based on one of Ptolemy's two methods for representing a curved surface on a plane. Unlike the *Cosmography*'s image of Eden, however, these maps make use of latitude and longitude. (As I noted in Chapter 2, fifteenth- and sixteenth-century geographers distinguished themselves by applying latitude and longitude to the terrestrial realm.) Münster's Ptolemaic projections, when they are understood against the earlier image of Adam and Eve, underscore that although Eden had no location, the many places of post-Noachian Earth did—and this distinction was essential to the rise of secular time. Although Münster was hardly alone in separating the biblical past from the human one, his work illustrates space's particular role in driving an even broader secularization.

Once he put the earth's entire surface into his maps, Münster turned to the specific spaces of his home continent, beginning with regional depictions of Western Europe, before detailing other parts of the globe. Not coincidentally, many of Münster's maps indicated latitude and longitude at the margins, especially the ones that illustrated European regions. In this way, unseen spaces became known by their being submerged within the new projected space. Beyond partaking of the new mathematics, however, Münster's maps of Europe also share a peculiar characteristic: each is "upside down," with north and south inverted. With this maneuver Münster underscored his transalpine origins. Rather than replicate Ptolemy's perspective—in which Northern Europe existed only on the margins of a civilized realm that centered on Egypt's Alexandria—Münster imposed his own perspective, drawing every region to the south, as it would have appeared from the eminently civilized Basel. Moreover, in doing this, he also highlighted homogeneous space's malleability, as the viewer's specific orientation with respect to the globe did not really matter, given that the physical whole was bathed within a homogeneous space.

Although the better part of the *Cosmography*'s maps relates to Germany and Western Europe, the text also presents a "global" view of geography. This part of the work's cartographic collection is remarkably comprehensive, as it includes (in the order presented) views of Sweden, Norway, Denmark, Italy, Russia, Poland, Transylvania, Greece, the Levant, the Middle East, India, China, Japan, Africa, and the New World. Münster, of course, had never visited any of these regions. Yet he dared to put all of them into an imagined space. Whence did Münster's "global" knowledge come? Münster was quite clear that it he got it from books—and this revelation underscores his participation in *space making*. Münster obviously used books and objects of material culture to put unseen things into the uniform space that suffused his maps and their attending textual descriptions.

Scholars' narrow concentration on the *Cosmography*'s maps has led them to see Münster proleptically, that is, as an incipient geographic thinker. Although he was exactly this in many ways, a cursory glance at his work's title casts light on the non-geographic themes that coursed through the background. Here is the full title: *Cosmography: A Description of All Countries by Sebastian Münster, which includes Peoples, Regimes, Cities and Notable Regions: Customs, Practices, Organizations, Religions, Sects and Occupations, throughout the World, but especially for the German Nation. Also the Unique Things To Be Found in Each Country and Bestowed thereon. Everything Illustrated with Figures and Nice Maps and Designed for the Eye.* By starting with the term "cosmography," Münster embedded every concept in the work's title within the sixteenth century's mathematical unity. Not coincidentally, he situated the earth's diversity of places and peoples on a sphere, as every postage stamp of soil could now be given a meaningful location. Finally (and most revealingly) the title notes that the figures and maps were arranged *for the eye*—an eye that existed within the same space that, as I explained in Chapter 2, Peter Apian had projected so firmly onto the cosmos just over two decades prior. By the mid-sixteenth century, at least for the European mind, every part of the terrestrial realm was subjected to homogeneous space inside which an imagined eye reigned supreme.

From Geometry to Geography

In this section I turn from the cosmography of the sixteenth century to the geography of the seventeenth century, with special emphasis on Bernhard Varenius.[28] Scholars have long recognized Varenius as a geographic pioneer, with his chief contributions being *Description of the Kingdom of Japan*, which appeared in 1649, and *General Geography*, which was published the following year.[29] The first work did what its title suggests: it described Japan in detail while putting the island inside a global understanding of space. The second work, in turn, reduced geography almost entirely to geometry and related mathematical fields. Thus, rather than posit any relationship between Heaven and Earth of the sort that suffused Münster's work, the *General Geography* concentrated on locating things on an imagined sphere while mostly avoiding discussions of where that sphere may actually be. (And this went doubly for God.) In short, Varenius isolated geography by fixing it within a system of uniform projected space that was becoming wholly independent of all older traditions of thought.

Varenius's intellectual formation throws into relief how the Spatial Reformation's center of gravity shifted to different points on the European continent. Although Varenius was a German, like Münster, his thought was heavily influenced by an increasingly sophisticated Dutch and Flemish culture, including especially the Lowland's reception of mathematics. Varenius was born in 1622 in the town of Hitzacker, which sits on the river Elbe about 100 kilometers inland from Hamburg; he received his primary education in a town called Uelzen before heading to the University of Helmstedt. When Varenius attended this institution, it was an important avenue into Germany for Lowland approaches to the natural world. In the course of the late sixteenth century, the Dutch and the Flemings had become leaders in globe making, cartography, geography, and mathematics.[30] Some of this glory was built on the achievements of German cosmography.[31] Nevertheless, in the second half of the sixteenth century, the Lowlanders rapidly set new standards in the representation of global space, some of which can still be admired in the globes and maps of Gerard Mercator, who, like Münster, studied in Louvain and finished his professional career in the German-speaking lands.

Netherlandish culture's increasingly global knowledge base suffused every part of Varenius's world. At Helmstedt he studied with the noted scholar of everything Hermann Conring, who, like many of his colleagues at the university, had been educated at the relatively new and strikingly effervescent University of Leiden.[32] (Leiden had been founded only in 1575 and, significantly for European thought's development, it hosted René Descartes at the same time that Conring was there.) Conring, in turn, influenced multiple academic disciplines in Germany, including theology, history, law, and politics by insisting that both they and all other disciplines rely on a truly general method, which he associated with mathematics. Moreover, at Helmstedt Varenius also encountered the geographic work of a German-Dutch thinker, Philipp Clüver, a native of Danzig who had studied geography and history in Leiden before finally settling there. Clüver's life and work are particularly illustrative of the spatial trends that produced Varenius. In 1624, he published *Introduction to Universal Geography*, in which he distinguished geography from both cosmography and astronomy:

> Geography, however, describes solely the earth's setting. In geography, on the contrary, we do not accept with the appellation Earth the separation

> of the four elements, as among the natural philosophers. But we understand a complex earth, one with waters infused, as the center of the entire universe, and which has the shape of an orb, or globe, and is called Earth. A globe, as one might expect, is a solid body, rounded on all sides, and enclosed by one surface. In the middle, it has a center or a point, from which all lines leading to the surface are equal. Therefore, from every element a whole is put together.[33]

Clüver reveals a subtle shift in the earth's spatial underpinnings, as he put geography on a foundation that was conceptually independent of the Heavens, even though there was no *physical* reason to do so. He remained a geocentrist, which meant that the cosmos's physical structures still coincided with spherical geometry. Nevertheless, that he separated Heaven from Earth so sharply indicates how the new emphasis on mathematics opened the door to further changes in geographic thought's *situation*.

It is thus particularly significant that mathematics dominated Varenius's education even more fully than it had in either Münster or Clüver's case. After spending four years in Helmstedt, Varenius moved to Hamburg, where between 1640 and 1642 he attended the famed "Johanneum" Gymnasium and studied under the polymath Joachim Jungius.[34] The latter's career casts additional light on the continental nature of mathematics' upsurge. Jungius had studied at the University of Padua (founded in 1222), where he pursued mathematics and medicine, taking his degree in the latter before occupying chairs in mathematics at the new University of Giessen (founded in 1607) and, later, at the Renaissance-era University of Rostock. He was a vociferous proponent of intellectual reform, stating in 1629, "Following the example of arithmetic and geometry, all sciences should be reorganized, in accord with the increasing complexity of their subject matter."[35] Mathematics was thus becoming the foundational model for all knowledge. And here, he merely voiced an emerging consensus. Two decades later, René Descartes wrote: "Those long chains composed of very simple and easy reasonings which geometers customarily use . . . had given me occasion to suppose that all the things which fall within the scope of human knowledge are interconnected in the same way."[36] The system of idealized space that had made the unseen imaginable also reordered early modern Europe's entire way of thinking, along with the attending emphasis on logic.

Varenius remained in this leavening environment on the Elbe for four years before heading to the University of Königsberg (founded in 1544), where he again pursued mathematical and astronomical studies.[37] In 1645, after spending two years in Königsberg, he moved to Leiden, where he studied medicine and mathematics, taking a degree in the former in 1649. (His time in Leiden also coincided with that of the mathematician and natural philosopher Christiaan Huygens, whom I discuss

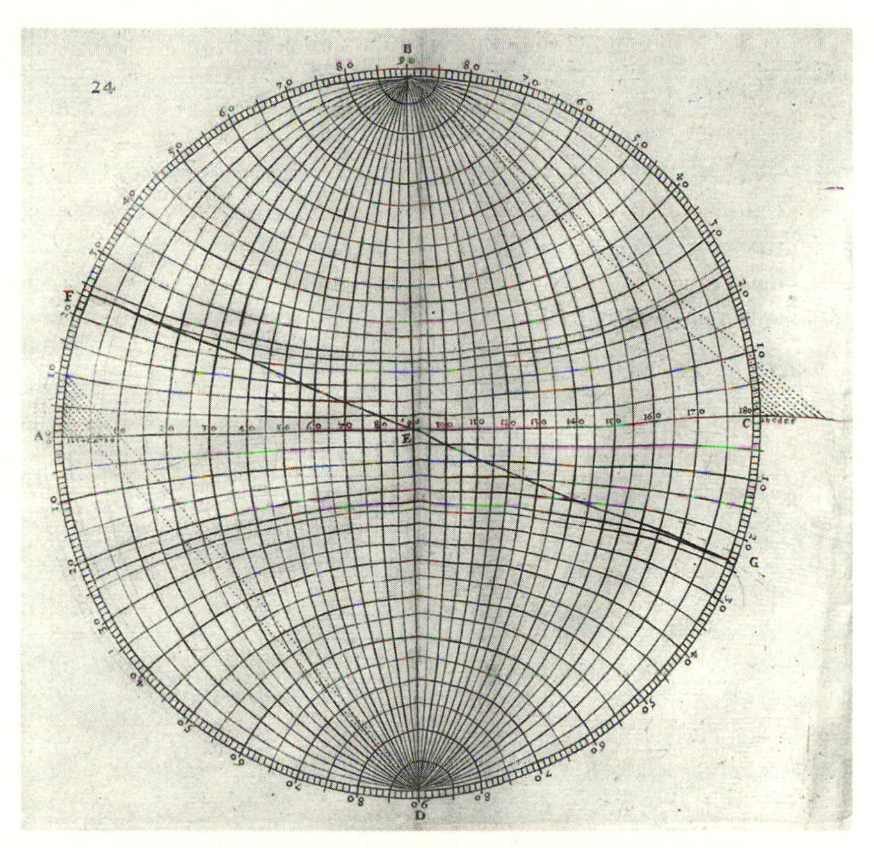

Figure 26. Hemispheric projection in Varenius, *General Geography*, 1650. Courtesy of ETH-Bibliothek Zürich.

in Chapter 6.) Varenius ultimately settled in Amsterdam, where he published his geographic works and died at only twenty-eight years of age.

The *General Geography* pinpoints geometry's role in accelerating cosmography's decay into something new. Along these lines, let us consider an aspect of this work that has received insufficient commentary: although it is considered to be a work of geography, it does not include any maps. This would seem to be a striking omission, especially when juxtaposed with either Münster's *Cosmography* or Clüver's *Introduction*, as both of them incorporated such illustrations. Significantly, however, Varenius's text did include a number of geometrical drawings that explicated the *space* of all maps (figure 26). As is apparent from the image, the specific knowledge of things was now less important than the projection of a conceptual realm inside which places and peoples could be *located*.

Looking ahead, I underscore that the emphasis on conceptual rigor guided European thought on spatial matters, up through the seventeenth century's end.

In 1690, for example, the French court creature and dramatist Samuel Chappuzeau published in the German court city of Celle *Idea of the World, or Easy Introduction [to] and Method [for] Cosmography and History.*[38] The work's contents are clear from the title. Most significant for anchoring our understanding of space's agenda-setting influence, however, is how Chappuzeau subordinated his work's contents to abstract space. On the first page, there appears what will be a familiar definition: "The sphere is a solid body, that contains one surface, in the middle of which there is a point, called the center, from which all the lines drawn to the circumference are equal to one another. Flat spherical figures, such as those representing geographic maps and world maps are only an imitation of a solid sphere."[39] Following what was already standard practice, Chappuzeau began his discussion of the cosmos with idealized spheres before putting both human populations and the history of their attending civilizations into the global space that this geometric space made possible.

With this additional context in mind, I return to Varenius's geographic thought and concentrate on the *General Geography* itself. In my analysis of this work, I examine only the preface and the first two chapters, because it was in these parts that Varenius constructed his system's foundations. (Everything else serves to elaborate the work's early insights.) Varenius made clear from the outset that he intended to break with tradition, holding that although many of his contemporary geographers associated Heaven and Earth, he would not do so: "We are accustomed to refer to everything in the universe as a world [*mundum*], and for a long time now, according to the common opinion of all men, it was divided into Heaven and Earth. This arrangement, seems consistent with the rules of the Logicians [*Aristotelians*] and nature itself, partly because the essential difference between Earth and Heaven is that one is beneath, partly because a small Earth compares to the entire heaven as the center of an orb."[40] Instead, he argued for a space that was rooted not in our experience, but in mathematics—and also added that mathematicians can readily understand the position of things on a sphere because they have access to the requisite mental tools.[41] Geography's adherence to mathematics, it would seem, had finally dissolved any remaining link between a divine Heaven and a mundane Earth.

I have emphasized the *General Geography*'s preface in order to highlight, as I did with respect to Münster's Eden, how the things that some scholars see as textually marginal can nevertheless be used to identify important changes. (In the history of geography, for example, the scholarship has been interested mostly in the maps and the descriptions that geographers included in their works, rather than how religious themes mixed with space.) It is, therefore, significant that the Varenius's preface included a definition of geography, because this definition offers a sense for how he *situated* his knowledge of the earth's surface: "Geography is a two-fold [science]: one [part is] general, the other is special. The former considers the earth in general, explicating the variety of its parts and their general relations. The latter, however, the special obviously, observes the general rules, surveys the locations, divisions, limits and other suitable known things of individual regions."[42] Varenius distinguished between geographic knowledge that was gathered via experience and mathematically organized knowledge of things unseen. This distinction dated back to Ptolemy, but it was particularly significant for Varenius, because mathematics had advanced considerably since ancient Alexandria's heyday. Consider, along these lines, a statement that he later made on geometry's relationship to nature: "In fact, in all of nature there is no body that has exact and geometric roundness."[43] Thus, if we look from Varenius back to Münster, we can see how mathematics changed both the earth and the viewer, because unseen realms now answered to a mind that commanded mathematics. The cosmographer André Thevet would probably have been shocked by such an idea.

The extent and power of Varenius's mathematical commitments are apparent in the *General Geography*'s first two chapters, "On the Fundamentals of Geography" and "On the Fundamentals of Geometry."[44] In the former, Varenius explained geography's fundamentals and offered another definition: "Mathematical Geography is called a mixed science, which teaches the parts of the earth and its attending relations through quantity, namely in its form, location, size, motion, the appearance of the Heavens and other associated qualities."[45] The geographic discipline was built on the consistent application of mathematics to all aspects of human knowledge, with a newly delimited vision of the earth being the chief result. Put another way, once the earth was sealed inside a larger mathematical envelope, it could be studied separately from the Heavens. Thus, as did Clüver, Varenius denied any link between Heaven and Earth beyond their both being subject to mathematics. This latter emphasis became specific in the second chapter: "And these tables are called the Mathematical Canon and they have infinite uses in all mathematical and physical sciences. And on that account I was determined to teach a little of geography to the eager few. The first particular use, however, is in the measure of triangles, *both spherical and planar*. But because the measure of spheres poses some difficulties, it seems necessary for those desiring this discipline to immerse themselves profoundly."[46] Here, we reach the core of Varenius's thought. Given that geography was about measurement within an abstract space, prior training in mathematics had become a prerequisite for any serious study in the new discipline. In this context, it was crucial that Varenius also mentioned both planar and spherical triangles, because, as I have explained previously, Euclid did not discuss the latter kind of triangle in the *Elements*. Thus, with this stipulation Varenius made clear not only that mathematics was evolving, but also that terrestrial knowledge's *situation* was changing along with it.

Although Varenius separated Heaven from Earth, it would be incorrect to separate his geography entirely from the history of cosmology, given that mathematics was revolutionizing both fields. Instead, it would be better to say that changes in mathematics altered the relationship between the two fields of endeavor by putting the human mind in a dominant position over the respective objects of study. This point becomes clearer through a comparison of Varenius with Münster and Clüver. Although all three men framed the earth mathematically, Varenius's vision of the cosmos had been reshaped by heliocentrism, while Münster and Clüver's views remained geocentrist. In the seventeenth century, it became increasingly difficult for a geographer to avoid the debate about heliocentrism. In this context, it is particularly significant that Varenius's own beliefs remain unclear, as he took no definitive stand on the cosmos's structure, but explicated each system: "And of the two, the Ptolemaists [Ptolomaici] affirm the motion of the stars themselves to be in the orbs. For the Copernicans, however, such motion comes not from the circling of the earth fixed in place, but asseverate that it [the earth] in its own place maintains its rotation around its own axis from the west to the east (as seems to be set in the stars.)"[47] From there, he explained the reasons that each side gave in support of its hypothesis and also included presentations of the criticisms that each directed at the other.[48] Varenius's official presentation was, in this sense, both fair and noncommittal.

Even though Varenius declared no opinion on the debate, there is ample reason to believe that he was a heliocentrist. Consider that he failed to cite a single contemporary geocentrist as an authority but referred repeatedly to this group as the "Ptolomaici." In itself the collective nature of these references mean very little. When we consider, however, that he also explicitly named multiple heliocentrists, his reticence to name geocentrists becomes important. For instance, Varenius identified prominent members of the new school, including Johannes Kepler, Galileo Galilei, and Philip van Lansbergen. Even more significant, he described the first two as mathematicians, with Kepler identified specifically as "Imperial Mathematician" and Galilei as "Mathematician of the Grand Duke of Tuscany," while Lansbergen appeared as "the Belgian." This moniker was probably sufficient, however, as Lansbergen was both a respected mathematician and vocal supporter of heliocentrism.[49] That Varenius cited only famous heliocentrists therefore suggests where his sympathies lay.

Here, we reach an important issue, namely that Varenius's position on geocentrism (or heliocentrism) was irrelevant to his geography. Indeed, his cosmological agnosticism was important to his project, as it signified that mathematics alone situated human knowledge of terrestrial things. (Within a mathematical geography, it made no difference whether the earth revolved around the sun or vice versa.) Homogeneous space had thus resituated the world by subjecting it to a mode of spatial projection that not only answered primarily to the human mind, but also could be cut up by that mind for its own independent purposes. Thus, by presenting both sides evenhandedly, Varenius made explicit a change that had been implicit in Clüver's work: within mathematical geography only the arrangement of the earth's surface features mattered. Indeed, Varenius spent only fourteen pages of the *General Geography*'s almost eight hundred in explicating the differences between Ptolemy and Copernicus before simply dropping the matter.[50] The earth's *actual* location was unimportant, because knowledge of this whole was ensconced within a capacious human space.

By remaining agnostic on heliocentrism, Varenius achieved three things. First, he cast doubt on geocentrism without making his reservations an issue, which meant that both geocentrists and heliocentrists could read his book. Second, he reinforced the centrality of mathematics to knowledge of the terrestrial realm. Astronomers could go whatever way they wished; geographers would put terrestrial spaces into mathematically defined places, regardless. Finally, Varenius reinforced the growing distance between geometry and the cosmos that was obvious in the work of Kepler, Galileo, and others. In this sense, Varenius's geography marks another waypoint in the human mind's journey away from both the cosmos and God.

As I argued in this book's introduction, prior to 1850 any change in one of the Western triad's concepts invariably affected the other two. With this backdrop in mind, I consider a theological implication of geography's separation from cosmography by examining Varenius's attitude toward the biblical Flood. Unlike Münster, who strove to associate biblical and human history, Varenius simply dispensed with the issue, holding that the Flood could not have happened as the earth's mountains were too high for the seas to cover them fully.[51] He added: "But the truth of the matter is one has taken refuge in miracles without sufficient cause in this matter. As we have tried to show in chapter XVIII, inundation by water or of all land is prohibited by the height and consistency of the soil."[52] The last sentence is particularly significant, because it underscores how problematic miracles (of whatever type) had become for a culture that was committed to homogeneous geometric space. The divine, in all its aspects, was being marginalized from the physical realm by homogeneity's remorseless expansion.

We can glimpse this spatial commitment's broader implications in a subsequent chapter, "On the Change in Place of Water and Earth, or On the Changing of the Surface from Water to Earth and the Opposite."[53] Varenius's chief insight here was that although the specific relationship between land and water was variable in its details, the broader physical relationship between them *has always been constant*.[54] This is homogeneous space's true legacy, as it allowed the reconceptualization of physical relationships in a way that undergirded an expansive vision of historical time. Varenius noted, for example, that water and land were roughly equal in quantity, which meant that although they can exchange places, neither substance can overwhelm the other—and he also added that one can see in the dry places of the earth exactly how water was once located there.[55] Therefore, Varenius reduced all geography—whether it related to contemporary or biblical times—to a spatial uniformity, in which physical changes occurred over such a great period that humanity may not even have been around to witness any of them. (This point becomes clear in my discussion of the Comte de Buffon's relationship to the nebular hypothesis in Chapter 6.)

With this fully "spatial" view of the earth's surface in mind, I compare Varenius to Münster more narrowly. Thanks to changes in mathematics, the relationship between the earth's soil and water held a different meaning for Varenius than it had for Münster. First, in Varenius's view rivers could not serve as a conceptual anchor for either geography or anthropology because each one merely reflected an inveterate "global" interrelationship between soil and water. Second, Varenius's spatial view of natural knowledge excluded the miracles to which Münster had granted an important (albeit circumscribed) place within his cosmography. Finally, using space's vast reach as his philosophical backstop, Varenius simply erased Münster's biblical time, as all events were now reduced to a common timeline that organized the terrestrial globe's entire past.[56] There was no longer any place within European geographical thought for the Bible's most mythical of tales—and this meant that human space undergirded and justified all historical time. Eden was over.

Varenius's embedding of terrestrial changes inside homogeneous space echoed through the seventeenth century's end and across the continent. In 1686, the astronomer Edmond Halley wrote a letter to the geometer John Wallis, a figure who appears in the next chapter. In this letter, Halley included a view of the earth's past that could have come straight from Varenius:

> [A]t present I can inform you thoroughly of the Hypothesis of Mr. Hook[e], concerning the changes which seem to have happened in the earth's surface, from the shells in bedds, found unpetrified in the Alps and other hills far from and above the sea, and again sea sand and shells found at great depths under ground. Upon this subject Mr. Hook[e] has read severall lectures before the Societie; and it seems that he hath shown how the superficies of the earth may have been frequently covered with water, and again dry, so as to answer to all the appearances.[57]

Halley's mention of Robert Hooke opens another door onto how space produced a larger past. Hooke clearly knew his geography; but he also knew geometry very well, as from 1664 until his death in 1704 he was Professor of Geometry at Gresham College in London. I return to Varenius's broader influence below. Before doing so, however, I underline that Halley returns us to two themes: first, geometry's space continually shaped humanity's relationship to its Earth; second, space pushed the biblical past beyond European thought's increasingly terrestrial margins.

We can also observe this spatialization of historical time within French thought, most especially in the work of the geographer Pierre Duval. In 1672 in Paris, he published *Treatise of Geography, which Provides Knowledge of and Use of Globes and Maps,* and included in the work's preface the following reference to history:

> Those who know geography have great advantages in the reading of history and they recognize that maps render to them great clarity in all matters. They confess that it is not less fitting for men, than for beasts, to know the location of one's dwelling, and to receive pleasure from travelling without peril in distant regions. Chronology and genealogy are truly necessary for grasping history well, but only owing to geography, because one normally asks where, when, and by whom things were done. Besides, with God having created the earth [first], rather than time and man, according to common reckoning, it seems that one must first know the places and only after this can one know the succession of time and [of] illustrious men.[58]

By the end of the seventeenth century, the increasingly sophisticated tradition of geography allowed the early modern mind to put every aspect of humanity's world *into space*. Europeans came thus to "know" places and peoples that few of them had ever seen—and they did so via the material culture of globes and maps.

I now pursue Varenius's influence with an eye toward the issues that drive the next chapter. As Halley and Duval attest, Varenius's approach seems to have diffused widely. The *General Geography* was originally published in Amsterdam in Latin, which made it transferable to every part of Europe, and multiple new editions appeared. For instance, the first edition was reprinted in Amsterdam in 1664 and 1671, while revised ones (also in Latin) appeared in Cambridge in 1672, 1681, and 1712.[59] Moreover, translations were published in Italian (1713), English (1733), Dutch (1750), and French (1755).[60] It would be difficult to find a geographer who had a greater influence on Europe's understanding of terrestrial space than did Varenius.[61]

In order to illustrate, from another angle, space's significance for the tradition that began with Varenius, I concentrate on England, for two reasons. First, the English geographical community embraced Varenius's work more completely than did any other one. Second, England was seventeenth-century mathematics' epicenter. Thus, the *General Geography*'s publication history in England affords a view of how homogeneous space permeated whole cultures. It is particularly significant, along these lines, that the first person outside of the Netherlands to oversee a reprint of the work was a certain professor of mathematics at the University of Cambridge named Isaac Newton. Newton, whose command of mathematically projected space was second to none, prized Varenius's work highly, even going so far as to oversee two editions. That the inventor of modern physics influenced the reception of Varenius's geography is no mere coincidence but reflects the completeness of homogenous space's victory over *All.*

Newton's role in supporting Varenius's diffusion highlights two additional issues. First, it calls attention to Newton's own connection to the Spatial Reformation. Isaac Barrow, who was Newton's mentor at Cambridge, not only published England's most important edition of the *Elements*, but also seems to have suggested to the precocious undergraduate that he familiarize himself with this work.[62] Obviously, Newton's achievements in mathematics far outstripped those of Euclid, not to mention those of Barrow. Still, Newton's education, like that of all prominent early modern thinkers, began in (and continued to be permeated by) encounters with Euclid's work. Second, the trajectory of Newton's thought was well suited to Varenius's emphasis on mathematics. Here, the chronology is significant. The epochal *Mathematical Principles of Natural Philosophy* appeared in 1687, or, after Newton had already edited and published Varenius's work twice.[63] This progression suggests that within English academic culture, at least, the earth had been *situated* within homogenous space, years before Newton destroyed all previous cosmologies.

Conclusion

Any mention of Isaac Newton's physics directs our attention to the eighteenth century, given Newtonianism's significance for the Century of Light. Thus, with the new physics' rise as a backdrop, we are in position to appreciate the cocktail of spatial knowledge and global arrogance that is revealed in Edmund Burke's words in this chapter's epigraph. Burke's comments about the "Map of Mankind" appeared in a note that he penned to the historian William Robertson, in which he thanked him for having sent a gift copy of his latest book, *The History of America* (1777).[64] Robertson's reading of the New World's history is novel and sympathetic—even the Spanish invaders get credit for not being complete barbarians. Nevertheless, this work also trafficked in the same Eurocentric prejudices that Burke cultivated. One excellent example appears in Book IV, which contained information on the "manners and policy of [America's] most uncivilized inhabitants."[65] For Robertson, not everyone over there was uncivilized; the Aztecs and Incas were quite developed, even if they had not reached European levels of glory. One cannot escape the reality, however, that another European academic who was ensconced in an omniscient institutional perch (this one located in Edinburgh) had imposed distinctions on "others" whom he could not actually see, but whose *place* he could imagine.

Taken together, Burke and Robertson illustrate how Europe's confidence in its ability to locate things justified the instinct to rank peoples and cultures that were, by comparison, spatially blind. This ranking occurred, not coincidentally, inside a Euclidean matrix. Consider, in this respect, Burke's full description of the "Map of Mankind": "[T]he very different civility of Europe and China; the barbarism of Persia and Abyssinia; the erratic manners of Tartary and Arabia; the Savage of North America, and of New Zealand."[66] The ranking that suffuses these words was rooted in the era's construction of a "global" knowledge that allowed Europe to extend its prejudices (both positive and negative) around the globe. In the seventeenth century, it was generally accepted that the Chinese cultivated an advanced scientific culture (which was true) and that this culture embraced Euclid (which was false).[67]

As I explained in the introduction, the mistaken belief in Chinese Euclidism sustained a symmetry that many early moderns perceived to exist between Europe and China. Both civilizations were held to be equal, because each had the wherewithal to put the other *into space*. Put another way, Europeans could look east and fully expect the Chinese to be looking west. Along similar lines, consider Thomas Hobbes's ranking of world civilizations in *On Body* (1655), an important text in the history of political thought to which I return in the next chapter. In this work, Hobbes noted how useful was the ability to measure bodies and to calculate motion, especially given that architecture and navigation were impossible without them. He added: "Almost all the peoples of Europe take delight in them, and most [of the peoples] in Asia, and some in Africa, but among Americans and those people who are close to the poles, [there is] none at all."[68] For the early modern mind, the ability to put knowledge into space was a hallmark of civilization—and invidious rankings of entire peoples mushroomed from there. Those who looked down from "above" were, in short, authorized to dismiss as barbarous any culture that lacked a spatial foundation.

As we look forward from both Münster and Varenius into the ostensibly enlightened eighteenth century, we see that knowledge of homogeneous space reordered European thought's every aspect, informing its theology, undergirding its cosmology, and transforming its anthropology, up to the point that both foreign cultures and races could be ranked. Put another way, space's continual advance distanced humanity from both the cosmos and God—and the result was the encasement of humanity's populations inside a space that answered to the European mind. One final example from France underlines the profound nature of the change. In 1685, Allain Manesson-Mallet published a work of geography that included these introductory thoughts: "The human spirit being extremely narrow-minded, has not been able to imagine in one view this immense space. In order to facilitate this knowledge, it has composed a machine, or little world [Monde], which is proportional to it, and which in every manner is alike to the great [world], representing both the celestial and elementary parts of this universe. This instrument is called the artificial sphere. We will represent it on the next page."[69] With his reference to an image of a material sphere, Manesson-Mallet illustrates that Euclidean space's passage from books into material culture produced a human vision of the once-divine *All.* Europeans assiduously cultivated this other *All,* if with decreasing reference to the humility that Christianity had long counseled.

Chapter 5

Modest Ravings

> The art and secret of the Sciences consists only in deducing the first truths that God has placed in our heart, that is to say, the principles that permeate natural science; and that is what the geometers do admirably, as we will see, in discovering simultaneously the principles and foundations of geometry.
>
> —Bernard Lamy, *Les Élémens de géométrie: Ou, de la mesure du corps. Qui comprennent tout ce qu'Euclide, en a enseigné: Les plus belles propositions d'Archimède & l'analise* (1685)

In 1651, just over a century after Sebastian Münster's *Cosmography* went to press, and only one year after Varenius's *General Geography* appeared, Thomas Hobbes published his *Leviathan*. This fantastic screed is commonly recognized as the first work of secular politics, because it severed humanity's natural state from the fable of Edenic innocence and, in so doing, justified the modern state's formation.[1] As many scholars have noted, the secularity that was inherent to *Leviathan*'s politics went hand in hand with a dour anthropology. For Hobbes, human beings in the natural state were so ferociously violent that they could not even organize themselves into gangs of thieves. Instead, early humanity comprised a pool of malevolent corpuscles, whose reflexive brutality rendered human life, as Hobbes put it, "solitary, poore, nasty, brutish and short."[2] Hobbes concluded that the only answer to Natural Man's congenital pugnacity was a social contract that granted to the state a monopoly on violence. Both this bleak assessment of human nature and its attending authoritarian politics were, as I argue in this chapter, implicated in homogeneous space's rise.

As a formal matter, the traditional interpretation of the *Leviathan* is not wholly incorrect. Hobbes did elaborate a secularized vision of the state. Nevertheless, consistent with what I have argued throughout this book, this reading is incomplete, because it overlooks the spatial backdrop to Hobbes's views on the nature of the political. Contrary to what one may expect, given the current reading,

Hobbes's secular political thought was constructed on a peculiar reception of geometry. On the one hand, as I noted in Chapter 3, Hobbes accepted the creeping marginalization of the divine, as is evidenced in his expulsion of angels from the physical world. On the other hand, he also rejected Euclidean geometry's main doctrines, including especially the notion that points are indivisible, because he believed all of them to have positive anthropological implications—and a positive anthropology justified at least some human freedom. In response, Hobbes invented a geometry that was wholly compatible with the negative anthropology that his political preferences demanded. Put another way, the same Euclidean spatial sense that had long threatened God and his divine entourage also posed risks to the state's authority. Hobbes was fine with the former; the latter, however, was another matter.

In order to highlight how space shaped both Hobbes's anthropology and his politics, I frame Hobbesian thought against the previous chapter's analysis of Sebastian Münster and Bernhard Varenius. I begin with Münster, who, as I have explained, was a scion of the Spatial Reformation. He studied mathematics intensely throughout his education, which meant that he knew Euclid's *Elements* well.[3] Moreover, as his vision of humanity's State of Nature revealed, knowledge of empty space succored a basically positive anthropology, with Münster anchoring a natural human sociability and rationality to the production of three-dimensional space. Hobbes, in contrast, although he embraced the distancing of the divine that homogeneous space brought to European culture, also abhorred the positive anthropology that was associated with it. Put another way, Hobbes rejected both God and Euclid, precisely because of his doubts about human freedom.

If we read Hobbes via his engagement with Euclid, then it is not coincidental that the *Leviathan* appeared about the time that Varenius's *General Geography* was published. As a product of the Spatial Reformation, the latter work had important theological and anthropological implications. First, Varenius used his command of spatial reckoning to extend humanity's distancing from God, as attested by his assertion that the biblical Flood could not have happened. Second, he also separated the Earth from the Heavens through the emphasis on space's generic nature, which allowed him to make geography into a fully terrestrial discipline. There is no evidence that Varenius influenced Hobbes directly. However, if we consider Hobbes with respect to the progression in spatial thought between Münster and Varenius, we can analyze the Englishman's thought with respect to the trends that drove sixteenth-century cosmography's transformation into seventeenth-century geography. Pursuing this path, therefore, allows us to see how spatial secularization undergirded what many people have understood as political secularization. From this perspective, Hobbes was also a child of the Spatial Reformation—even if, as we will see, he turned out to be a juvenile of a particularly recalcitrant sort.

To this point, no one has juxtaposed Hobbes with Münster, since the first is obviously a political thinker, while the second is a geographic one. I propose, however, to read Hobbes against the disciplinary grain and thereby to understand his political thought as a product of his encounters with Euclidean geometry. Such a reading shows, first, that both Hobbes and Münster produced their respective visions of humanity's original state, only after having confronted Euclid's geometry. Second, this reading also reveals an important difference in their thought that is essential for explaining the divergence in their respective anthropologies: Münster accepted Euclidean geometry's doctrines, while Hobbes resisted them to the bitter end.[4] The history of space opens, in this way, a new vista onto not only secular political thought's origins but also modern politics' relationship to developments in European anthropology.

Scholars have largely overlooked the significance of Hobbes's response to Euclidean space for the shaping of his politics. This is due partly to the mistaken conventional wisdom that Hobbes partook of a diffuse geometric spirit (*more geometrico*) that ostensibly pervaded both seventeenth-century and eighteenth-century thought.[5] The misunderstanding dates to 1693, when the biographer John Aubrey wrote: "[Hobbes] was forty years old before he looked on geometry; which happened accidentally. Being in a gentleman's library Euclid's *Elements* lay open, and 'twas the forty-seventh proposition in the first book. He read the proposition. 'By G——,' said he, 'this is impossible!' So he reads the demonstration of it, which referred him back to such a proof; which referred him back to another, which he also read. And so forth, that at last he was demonstratively convinced of the truth. This made him in love with geometry."[6] Influenced by Aubrey's story, many scholars have absorbed the notion that Hobbes's enthusiasm for Euclidean geometry undergirded his politics. This is misleading. As I explain below, the most that one can say is that Hobbes admired Euclid's use of logic. The *Elements*' actual manner of projecting space had no direct effect on Hobbes—and especially not on his State of Nature.

I do not dispute that the reception of Euclid's *Elements* sustained many and varied seventeenth-century discussions about the power of and utility of logic. I hold, however, that one must not forget that Euclid's logic was tied to an elaborate and rigorously argued system of spatial projection. Thus, any given thinker's views on logic were shaped by the effects that he perceived Euclidean homogeneous space to have on other constructs, including the Western triad. One classic example, to which I return in this book's conclusion, is the gap that separated René Descartes from Blaise Pascal on the matter of space's proper application. Consider, along these lines, that the former rejected infinite geometric space's full application to the physical cosmos, because, among other things, the emphasis on spatial infinity definitively elbowed aside God. The latter, in contrast, although also devoutly Christian, embraced the mathematical projection of infinity onto

All and, concomitantly, rejected both Descartes's theology and his physics. If we look to their contemporary, Hobbes, we see that he responded, in effect, to the same underlying problem—space's corrosive effect on the relationship between humanity and God. However, rather than contemplate humanity's relationship to God through homogeneous space, Hobbes appropriated geometry in a way that allowed him to discipline both God and humanity.

The misreading of Euclid's influence on Hobbes has yielded a contorted view of not only the Hobbesian State of Nature but also early modern political thought's development. Scholars routinely associate the *Leviathan* with John Locke's *Two Treatises of Government* (1690) and Jean-Jacques Rousseau's *The Social Contract* (1762) and thus have transmuted the three into a secular trinity of modern politics.[7] Chronology justifies some of this, since Locke had read Hobbes, while Rousseau knew of both his predecessors. Moreover, the three tomes align nicely in their positing of a State of Nature as a precursor to a social contract. The biggest reason, however, for scholars' fixation on the Hobbes-Locke-Rousseau trinity is that each of them was associated with revolutionary unrest—and if we moderns love anything, it is a good revolution.[8] The contemporary fixation on modern politics' rise, however, is distorting when applied to the early modern, precisely because our spatial sense exists in a different intellectual context. If we wish, therefore, to understand the rise of Hobbes's politics, we must confront the spatial backdrop against which this rise occurred.

The desire among contemporary academics to excavate (for good or ill) modern democracy's origins has driven homogeneous space to the existing literature's margins.[9] If, however, we consider secular politics' rise via homogeneous space, a previously undetected schism appears: both Locke and Rousseau studied geometry (in Locke's case quite intensely) and, moreover, both of them accepted unequivocally Euclid's chief assumptions about space. I return to Locke and Rousseau in the conclusion. Here, I foreground two commonalities that they share with Münster—but not with Hobbes. First, Locke and Rousseau's respective States of Nature were based on a traditional reading of Euclidean space. Second, and not coincidentally, these States of Nature were as politically and socially tranquil as Münster's earlier version. Thus, in addition to suggesting a possible alternative trinity for secular politics' rise, taking note of spatial continuity highlights an overlooked interrelationship between anthropology and the *Elements*. If a thinker accepted homogeneous space, then Man was a rational being that merited a government; if he rejected that space, however, then Man was an animal that needed governing.

From Geometry to Philosophy

Against this backdrop, I return to Aubrey's vignette. Contrary to what scholars have assumed, Aubrey's recounting highlights not the depth of Euclidean geometry's influence on Hobbes, but the discipline's absence from his classroom education. He was born in 1588 and entered Magdalen Hall, Oxford, in 1603 with a solid foundation in Greek, Latin, and arithmetic, but with little training in geometry. Hobbes's ignorance of geometry's core doctrines was reinforced, moreover, during his undergraduate years, by the discipline's exclusion from Oxford's curriculum. In general, the university emphasized Aristotelian philosophy, which Aubrey claims that Hobbes detested. This is an overstatement, however, as Hobbes's philosophical views were quite differentiated.[10] Although he had reservations about Aristotelian natural philosophy, he also adopted many of the Stagirite's views on rhetoric and logic.[11] This aspect of Hobbes's thought is significant, given what I have argued previously about Aristotle, because it suggests that Hobbes may have absorbed a bias against Euclidean visions of continuity, well before he turned to the study of geometry.

The nonspatial trajectory of Hobbes's education and training extended through his service as both a tutor and a personal secretary. Immediately after graduating from Oxford, he joined the household of William Cavendish, Baron of Hardwick (later Earl of Devonshire) as tutor to Cavendish's son, who was also named William. Between 1608 and 1628, Hobbes worked continually for the Cavendish family, entering into the son's service after the father's death. There is no doubt that this job sustained Hobbes's postgraduate education.[12] His employer possessed one of the age's great private libraries and Hobbes took advantage of the glittering collection of works to engage in intense study of the classics. Indeed, this private research yielded a historically significant result, the publication in 1629 of an English translation of Thucydides' *History of the Peloponnesian War*.[13] In essence, after leaving Oxford, Hobbes spent two decades becoming a first-rate philologist. It is not, therefore, coincidental that he remained unaware of critical developments in geometry.

Geometry's absence from Hobbes's education was sustained by the growing intensity of his philosophical studies. As Hobbes's ward, the younger Cavendish, came to require fewer services, Hobbes sought additional employment opportunities. Beginning probably in 1619, he served as personal secretary to Francis Bacon, a caustic anti-Aristotelian and a key figure in the rise of modern science.[14] Bacon's critical approach to older natural philosophical traditions suited Hobbes. It is thus significant that although Bacon used logic remorselessly in attacking Aristotelianism, he was no mathematician either.[15] Whereas Hobbes's negative assessment of Aristotle was probably sharpened by his exposure to Bacon's practice of flinging calumnies boldly, none of the resulting insults incorporated an understanding of idealized space.[16] In 1626, Bacon died and Cavendish the younger followed two years later, with the latter's death costing Hobbes both his friend and his remaining job. He found another position, however, with Sir Gervase Clinton, serving as tutor to Clinton's son, whom he also accompanied in

1629 and 1630 on a Continental tour. According to Aubrey, this was when Hobbes became smitten with geometry.

The Grand Tour with the young Clinton marked the beginning of a new stage in Hobbes's intellectual development, as it introduced him to atomist thought.[17] While traveling around the Continent, he encountered thinkers who embraced ancient Atomism, which was then experiencing a revival. Hobbes's introduction to this philosophical school is critical for understanding his relationship to both Aristotelianism and Euclidism. The Atomists held that the cosmos consisted only of corpuscles in motion—and this doctrine was inherently incompatible with Aristotle's blend of hierarchical space and pervasive logic. Moreover (and the intellectual historical scholarship often overlooks this point), Atomism was equally incompatible with the *Elements*' cultivation of continuity, given that Atomists believed reality to be discontinuous. It was therefore critical that, after Hobbes returned to England, he joined an intellectual circle that gathered around another William Cavendish (the future Duke of Newcastle and cousin to Hobbes's recently deceased employer). This circle was expressly Atomist and succored Hobbes's interest in cultivating discontinuity *philosophically*.

After acquiring new atomist weapons, Hobbes returned repeatedly to Paris. Between 1634 and 1637, he traveled there while in the service of the deceased William Cavendish's widow. (The Atomist cousin was still alive.) This trip marked another waypoint in Hobbes's development, because it allowed him to become acquainted with the Parisian scientific community, which was then searching for ways to combine Atomism with recent advances in mathematics. (Paris constituted another important center of mathematical thought whose participants, including especially Descartes and Pascal, were in regular contact with other centers in England and the Netherlands.) In 1640, Hobbes returned to Paris once again, this time as a refugee from the Puritan Revolution. He remained for over a decade, participating vigorously in the circle of Marin Mersenne, the learned friar and respected geometer.[18] Under Mersenne's influence and that of other French thinkers, such as Pierre Gassendi, Hobbes studied corpuscular physics and mathematics jointly. This mixture had profound implications, insofar as Hobbes's awkwardly atomist view of geometric space became his political anthropology's foundation.

The Elements of Hobbes's Space

Hobbes's Parisian studies yielded both his idiosyncratic geometry and the politics for which he is celebrated. Given the contemporary literature's political tendencies, it is important to underscore that Hobbes's geometry preceded his politics. We can see this, I argue, in a trilogy that Hobbes began to publish before the *Leviathan* appeared and under the general title *The Elements of Philosophy*.

Hobbes's inclusion of the word "Elements" was deliberate, as he meant to invoke classical geometry's exemplar, even if his own sense of space bore little connection to Euclid's elegant system of projection.[19] Hobbes's *Elements* suggests, in this respect, a different way to read Hobbesian thought's development, one that begins with his space and ends with politics.[20]

I begin my alternative reading of Hobbes by outlining *The Elements of Philosophy*'s structure. Each of the three volumes appeared separately. *On the Citizen*, which covered politics, was issued in 1642.[21] *On Body*, the geometric volume, followed in 1655.[22] *On Man*, an anthropological work, went to press in 1658.[23] Only *On the Citizen* appeared during Hobbes's period in exile, but there is no doubt that the trilogy emerged from his experiences in Paris. Significantly, the volumes appeared out of order, as *On Body* was intended to be first, while *On the Citizen* was to be last.[24] It is important to foreground Hobbes's intentions, because the actual publication history has justified the overemphasis on the *political* arc of Hobbes's thought.[25] It is common for commentators to move directly from *On the Citizen* to *Leviathan* to *Behemoth* (which was a history of the English Civil War) and to imply, thereby, that Hobbes's thought was *essentially* political.[26] In contrast, Hobbes's intended progression—geometry, anthropology, and politics—reveals how both his ideas on humanity and his politics emerged from his space.

Before exploring Hobbes's geometric rebellion, I recapitulate the main parts of Euclid's *Elements*, since this text's structure serves as the analytical backdrop to my reading of Hobbes. In this, his greatest work, Euclid began with plane geometry, propounding in Books 1–6 his core geometric doctrines (points, lines, and planes) while also presenting two of his most important arguments, the parallel postulate and the Pythagorean theorem. (I return to these below.) Euclid turned to number in Books 7–10, in which he tried to explain the relationship between number and the rest of his geometry by analyzing numbers as line segments. (As I argued in Chapter 3, in the fifteenth and sixteenth centuries these discussions dignified human attempts to measure both things and space within the real world.) In the final books, Books 11–13, Euclid covered geometric solids, including especially the Platonic solids, the tetrahedron, the cube, the octahedron, the icosahedron, and the dodecahedron, all of which, as I have noted, were important to Johannes Kepler's early work. Thus, in general, I would underscore that Euclid moved from two-dimensional to three-dimensional space, or from points and planes into spheres and geometric solids.

With this summary in mind, we can see immediately that Hobbes's *On Body* is not very geometrical, as it is light on analysis of spatial concepts and heavy on philosophical disputation. Part 1, for instance, which has the title "Logic," mentions neither points, lines, planes, nor geometric figures, but reduces all knowledge to complicated two-dimensional schemata that are presented as logic trees.

Part 2, which has the title "First Philosophy," takes this two-dimensional impulse further and explicates concepts such as time, place, quantity, and cause and effect, before discussing in chapter 14 (finally!) two-dimensional geometric notions, such as lines, curves, angles and figures—and then only briefly. This overview of the other *Elements* reveals a profound irony: Hobbes's use of spatial terms is more reminiscent of the medieval fractured space that I outlined it in the introduction than of the homogeneity that suffuses Euclid's *Elements*.

Although Hobbes was influenced by Euclidean geometry, he deliberately emptied the system of its most significant contents. Consider, in this vein, part 3. It is the most "geometric" section of the book in the sense that it actually troubles to define points and lines. Nevertheless, its presentation of geometry is actually nothing more than Atomism in disguise, as it concentrates on justifying and analyzing bodies that are in motion rather than constructing a vision of motionless space. Part 4, meanwhile, has nothing to do with geometry at all, as it describes the physical bodies of humans and animals and details how each type of being experiences the world. (Hobbes probably got this idea from René Descartes.) Neither of these two subsections had anything to do with classical geometry, at least if this discipline's subject matter is understood with reference to idealized space.

Hobbes's *On Body* comprises, thus, a vigorously contrarian attempt to read Euclid's *Elements* with respect to its logic alone. Two inevitable breaks resulted: one was with Euclidean geometry's construction of *nothing*, and another with its projection of three-dimensional space. With respect to the former rupture, consider that part 1 has the title "Computation, or Logic," while its first chapter bears the subheading "On Philosophy."[27] Space is missing in action here, as this section of the text made no mention of points and how they have no parts. Logic's obvious primacy for Hobbes is further underscored by his reliance on binaries for the mapping of knowledge. This is important, because the cultivation of binaries and the exclusion of Euclidean homogeneity went hand in hand. As an Atomist, Hobbes was interested in the existence (or nonexistence) of bodies. As he saw it, only bodies could be experienced and only for two reasons: (1) because the body had substance, and (2) because that substance was in motion. Hobbes could not, therefore, base his own geometry on Euclidean approaches to space, since "nothing" could not be the subject of experience. Geometry was, in this sense, a *philosophical* nonstarter.

With respect to the second break, Hobbes's reading of Euclid was completely "flat," as it failed to engage with the final three books of Euclid's *Elements*. For instance, Hobbes excluded geometric solids—and especially the Platonic ones—from his own geometry, because he wished to avoid the baleful political consequences of Plato's idealism. Here, I harken back to Chapter 3, in which I argued that melancholy was transformed by homogeneous space's permeation of

European culture. In this context, I held that Plato's own embrace of geometry animated, in part, Renaissance appropriations of Euclidean space—and I made this point with reference to Plato's cosmological work the *Timaeus*.[28] The *Timaeus* had, of course, occupied a critical position within European thought for a millennium, because it was one of the few works by either Plato or Aristotle to have been translated into Latin before the twelfth century.[29]

A summary of Platonism's foundations illustrates what Hobbes's geometric thought signified against a broader philosophical background. In general Plato held that there were two worlds, a perfect world of ideas and the physical world that human beings inhabited. Since geometry concerned itself with space in the abstract sense, this discipline promised a structured path to the world of ideas. Yet geometry also promised more. First, in the *Timaeus*, Plato argued that the four elements (earth, air, fire, and water) were composed, respectively, of cubes, octahedra, tetrahedra, and the icosahedra—and this allowed him to subject all physical bodies to geometry's underlying regularity. Second, some of Plato's associates used geometry to explain the Heavens' functioning, which opened a path to understanding the foundations of celestial motion. In these ways, Plato's geometry allowed the human mind to associate the physical world with the ideal one and, moreover, to impose rationality onto the whole.

Hobbes contested the ancient notion that rationality was diffused throughout the cosmos—or, at least, that human beings were capable of accessing its potential wisdom. As a result, to elicit exactly how homogenous geometric space could be applied to any part of the cosmos was never part of his program. Instead, Hobbes was interested only in justifying logic via the human ability to experience the presence and/or motion of things, since this philosophical underpinning limited human knowledge to direct experience. Even more illustrative, perhaps, of this "nonspatial" approach to geometry is Hobbes's exclusion of spherical triangles from his discussions of geometry. As I have noted, such triangles do not appear in the *Elements* but figured prominently within early modern discussion of mathematics. In the course of the sixteenth century, due especially to developments in navigation and astronomy, these triangles became a fundamental issue, since the manipulation of curved triangles made it possible for the human mind to orient itself on both a terrestrial and celestial sphere.[30]

Spherical triangles' absence from Hobbes's geometric thought is an example of how far outside the mathematical mainstream his atomist commitments took him. I deepen our understanding of this issue via the history of astronomy. In 1543, Nicolaus Copernicus's epochal *On the Revolutions of the Heavenly Spheres* was published, inaugurating Europe's heliocentric revolution. This moment has been well analyzed,[31] so instead I look to the prior year when Copernicus published a less discussed book, *On the Sides and Angles of Triangles, both Rectilinear Planar and Spherical*.[32] This work explicated the differences between the two kinds of triangles, including how spherical triangles differ from planar ones in that the sum of their inner angles exceeds 180 degrees. (This detail will be important later.) By publishing this geometric work, Copernicus established himself as one of Europe's foremost geometers and also prepared the way for the reception of his next (and greatest) book.

Copernicanism and Euclidism are, as I have intimated previously, closely related, on account of their common sense of space. From an expressly spatial perspective, the significance of Copernicus's *On the Revolutions of the Heavenly Spheres* lies not just in how it liberated the earth from the cosmos's center, but in how it incorporated the uniform geometric space of *On the Sides and Angles of Triangles* into both Heaven and Earth.[33] Put another way, this great work served both astronomy and navigation, because it situated each discipline inside a fully *human* sense for space. Moreover, in subjecting the cosmos to such a projected whole, Copernicus also made an important anthropological leap, insofar as he assumed that mere human beings could subject both celestial and terrestrial realms to the human mind.[34] Hobbes, as always, begged to differ.

However important heliocentrism was for the history of science's development, its rise should be understood against the backdrop of geometric space's older diffusion. Every natural philosopher from Copernicus to Kepler to Isaac Newton studied and applied ancient geometry's lessons to the cosmos. Much the same was true, as I have noted throughout, for early modern cosmographers and geographers, all of whom projected a *space* onto the cosmos's unseen places in a way that was congenial to the Renaissance eye—not to mention deadly for the medieval divine one. It is, therefore, more than a bit ironic that Hobbes, who was a militant heliocentrist, rejected the spatial sense that had made it possible for humanity to imagine an extraterrestrial position. The cause behind Hobbes's break with this aspect of early modernity's rise lurks precisely in homogenous space's disquieting anthropological implications.

Hobbes's views on geometry allowed him to deny to Natural Man the ability to acquire knowledge in the independent manner that was also heliocentrism's *sine qua non*. Thus, he linked knowledge's very existence to political power, rather than to an autonomous human imagination that also had Euclid's *Elements* on hand. Here, it is especially revealing to consider the lines that immediately precede Hobbes's characterization of the State of Nature in the *Leviathan*. He wrote: "In such condition [the State of War], there is no place for Industry; because the fruit thereof is uncertain: and consequently no Culture of the Earth; no Navigation, nor use of the commodities that may be imported by Sea; no commodious Building; no Instruments of moving, and removing such things as require much force; no Knowledge of the face of the earth; no account of Time; no Arts; no Letters; no Society; and which is worst of all, continuall feare, and danger of violent death."[35] Trapped in a world of aimless atoms and denied the regularities of space, human beings could not learn anything independently amid the chaos that surrounded them. In the absence of a leviathan, neither social peace, nor even the most basic forms of knowledge existed. Neither God nor Euclid offered any hope; only the leviathan was humanity's savior.

From Philosophy to Geometry

Hobbes's geometry belongs to the history of philosophy rather than the history of mathematics.[36] When understood from a philosophical perspective, *On Body* becomes an attempt to revive, for political-anthropological reasons, ancient Atomism's rejection of geometric homogeneity. In order to explain what Hobbes achieved by cultivating a nonspatial approach to geometry, I return to the classical world. Atomism had a distinguished ancient career. It originated in the fifth century BC with the Greek philosopher Leucippus and his student Democritus and extended up through Epicurus in the later fourth century, before being embraced by the Roman poet Lucretius in the first century BC. As a group, Atomists rejected geometry's infinite divisibility of space, because it suggested that reality was fundamentally continuous.[37] For Atomists, in contrast, reality comprised a myriad of indivisible atoms, each of which bobbed and weaved of its own accord.[38] Thus, where the geometers saw order, the Atomists saw chaos. For centuries, however, conflict between the two systems was put off, as ancient Atomism sank beneath antiquity's rubble, while geometry limped along in inferior translations and superficial compendia.

Atomism's fifteenth-century return was an important factor in the rise of modern physics, as has often been documented.[39] Less emphasized, however, is its complicated and productive interaction with the early modern reception of a reinvigorated Euclidean tradition. In the seventeenth century, many thinkers combined Atomism with geometry in ways that would have appalled Leucippus. Two examples will suffice. First, there is no doubt that Isaac Newton, the greatest of all Atomists, was also a Euclidean, given especially his connection to Isaac Barrow.[40] Second, Marin Mersenne, who was Hobbes's friend and mentor in Paris, maintained an odd balance between Euclidism and Atomism.[41] A great defender of mathematics, Mersenne not only published an edition of Euclid's *Elements*, but also attacked skeptics and deists on mathematical grounds of exactly the sort that appear in this chapter's epigraph from Bernard Lamy's work.[42] Within Mersenne's mind, religious belief and geometry had to cooperate in order to save God from Man—even if, as it turned out, the project was in vain.

Hobbes had a different take on all these spatial-philosophical issues. His thoughts on space's relationship to physical reality appear in chapter 7 of *On Body*, which has the title "Of Place and Time." As this chapter title suggests, Hobbes did not actually use the term "space" but preferred "place." In order to explain what this preference meant historically, I return to Hobbes's Atomistic commitments. Following Atomism's strictures, Hobbes held that reality comprised nothing more than bodies in motion. Thus, a given place was not defined by its relationship to a larger, tenebrous whole, but solely by whether a body occupied it. This emphasis on bodies, as opposed to points, imbued "place" with a binary quality that functioned in a manner that I would compare to "bits" in modern computers. A place either contained a body or it did not; it was either "on" or it was "off."

This "bit space" anchored Hobbes's geometry to his anthropology. By denying any epistemological warrant to human experience, beyond the difference between existence and nonexistence, Hobbes isolated the individual viewer from other human beings, who were also seeking to know things. Thus, the philosophical marginalization of space rendered impossible the cultivation of a collective knowledge base, or, as Hobbes put it in *Leviathan*, there was "no account of Time; no Arts; no Letters; no Society."[43] Hobbes's emphasis on place, as opposed to space, carved his break with Euclid in stone, as his "bit space" bears no relationship to geometry's core assumptions but is fundamentally Atomist in spirit. As Hobbes wrote in *On Body*: "Therefore, no one says, in fact, that space is that which is occupied at present, but that which can be occupied [by something]; otherwise, bodies would arrange to carry their own space with them."[44] With these words, Hobbes distinguished between the space that our imagination produces and a world that surely exists outside the human imagination but to which humans have no direct access. He added: *"Space is an apparition of a visible thing, while visible,* that is with none of its other accidents considered, than what appears outside the imagination."[45] Extending ancient Atomism's critique of geometry, Hobbes reduced Euclid's space to a remnant of experience that was corrupted and thus potentially corrupting. The latter aspect of Hobbes's geometry was particularly important, precisely because knowledge's "atomization" justified the leviathan's claims to sovereignty.

I draw out the broader implications of Hobbes's space through a return to sixteenth-century cosmography. Here I call the reader's attention back to Peter Apian's Eye (see figure 8). This image is pregnant with assumptions that were unacceptable to Hobbes, as in it Apian made clear not only that geometry's space could permeate the cosmos, but also that the human mind *commanded* geometry. This Renaissance-era image illustrates precisely why space's early modern diffusion shook the relationship between anthropology, cosmology, and theology so brusquely. On the one hand, Apian juxtaposed a disembodied God-like eye with terrestrial and celestial spheres that are, in turn, positioned outside of their "actual" arrangement. This *resituating* suggested, moreover, that the Heavens could be known via a space that was wholly imagined. On the other hand, alongside this humanly projected perspective onto the physical cosmos is an actual terrestrial globe, which was both a product of homogeneous space and was also visible to a real human eye. In this sense, Apian's Eye incorporated both the real world and an ideal one within early modern material culture.

In the early modern period, knowledge of the physical world was projected onto an essentially imagined spatial context. Apian's juxtaposition of real with ideal both inside this context and with reference to material culture suggested, in addition, the possibility of an expansive way of *seeing*. This form of seeing is exemplified, as I have suggested, in the winged eye that Matteo de'Pasti included in his medallion of Leon Battista Alberti. (I discussed this medallion in Chapter 1. See figure 5.) For Hobbes, however, an approach to seeing that was built on projected space was yet another nonstarter, precisely because of its baleful political implications. Thus, he denied any connection between his own geometry's logic and the three-dimensional projected space that had liberated the human eye, because to do otherwise would have legitimized the individual's knowledge and experience. It is therefore not surprising that *On Body* contains only drawings of two-dimensional geometric forms, while excluding images of three-dimensional ones. Any concession to three-dimensionality risked authorizing a generic viewer.

We are now in a position to understand the relationship between the "flatness" of Hobbes's geometry and his Atomism. Since Hobbes believed that reality consisted of bodies in motion, his geometry was based on bodies in motion.[46] This proposition is, however, not viable for a Euclidean, because classical geometry assumed space to be immobile. Put most simply, there is no motion in geometry—nor can there be. Hobbes took another tack, of course, holding: "By point is not meant that which has no quantity, or what cannot be divided in any fashion (for nothing of this sort exists in the order of things), but a thing whose quantity is not considered, that is, neither whose quantity nor whose part is reckoned in demonstration."[47] By making geometric points substantial and putting them in motion, Hobbes not only limited geometry to the physical realm, but also isolated each viewer within his own experience. Put another way, Hobbes rejected the possibility of an idealized mental world that suggested human reason extended to things that could not be seen.

Hobbes's commitment to a *physical* geometry had broad implications. First, his emphasis on points in motion undermined the Renaissance's multivalent approach to "seeing." If discontinuity underlay the individual's experience, then what any individual saw could not be aggregated into a common perspective that authorized both independent reasoning and the material culture that aided it. Second, by denying an independent foundation for knowledge, Hobbes underscored the need for a *political* leviathan that united humanity's malevolent atoms within the existing political and social order. The political nature of the core problem was, in turn, reflected in Hobbes's marginalization of spherical geometry. Spherical triangles have the peculiar mathematical property that the sum of

Figure 27. Illustration from the frontispiece to Hobbes's *Leviathan,* 1651.
Courtesy of Bayerische Staatsbibliothek München.

their angles exceeds 180 degrees. Within Hobbes's mental world this geometrical peculiarity was *politically* important—and we can see this if we reflect on Aubrey's description of Hobbes's "ah-ha" moment in Switzerland. Spherical triangles are incompatible with the Pythagorean theorem, because the planar triangles on which the theorem is based comprise, by definition, only 180 degrees. Hobbes's mind could thus never enter into spherical space, since doing so would have undermined the foundations of the two-dimensional logic with which he justified the leviathan's power.

With this backdrop in mind, I call attention to figure 27, which includes the image of the leviathan that appeared in the frontispiece to Hobbes's great work. If we compare this image to Apian's Eye, we note that "seeing" has been *resituated,* as a once disembodied eye has been placed entirely within a "political" body.[48] This body, in turn, unites the perspectives of each individual, insofar as all the real people are reduced to looking up at the leviathan's head. There is, therefore, no "independent" view on anything, whether it pertain to humanity, the cosmos, or God. Political authority (and not Euclid) legitimized the knowledge of unseen things. Most crucially, perhaps, in this sense the *Elements* was as politically dangerous as was the Bible, because both works encouraged reason's independent use—and that would not do.

The Great Schism

Hobbes's unique views on mathematics did not go uncontested from the mathematical side. In 1655, the renowned mathematician John Wallis published a harsh response to Hobbes's *On Body* under the title *Elenchus geometriae Hobbianae.*[49] The title was already an affront, since "elenchus" means trinket in Latin, while the whole translates as *A Trinket, [Appended] to Hobbes's Geometry.* Wallis essentially began by patting Hobbes on the head before adding still more demeaning comments in the text. In the introduction, for instance, he mocked the *Leviathan* for its impiety and equated Hobbes's views on spiritual matters with the juvenile belief in hobgoblins.[50] It may appear odd that Wallis not only mentioned a political work in the course of launching a geometric attack but also tied the debate to a theological issue. This was, however, normal practice among seventeenth-century thinkers, and especially among the English. The Puritan Revolution that ravaged England between 1640 and 1660 came as quite a shock to the elite. Indeed, Wallis, as a university don, could not but hope that training in geometry would bring discipline to the masses.

Here I return to a point that I made in Chapter 1, namely that geometry lacks *for us* many of the resonances that it held before the Spatial Reformation's end. Consider that by associating Hobbes's geometry with the *Leviathan,* Wallis smeared Hobbes with the charge that incoherent spatial thought legitimized childish approaches to the supernatural. Contemporary thinkers would probably not tread this path. Nevertheless, many among of the seventeenth-century elite did so, because they saw the revolution as an uprising among people whose susceptibility to religious enthusiasm impeded the pursuit of rational thought.[51] (Enthusiasm was understood, in this time, as a religious disease that induced the common sort to reject good sense.) As a mathematician and a moderate Presbyterian who supported Parliament in the 1640s and opposed Charles I's execution in 1649, Wallis saw both geometry and the educated elite that cultivated it as bulwarks against extremism. And in Wallis's view, Hobbes's geometry was no help, since it opened the door to ghost stories.

Even under ordinary circumstances, Hobbes could only have taken Wallis's *A Trinket* as an insult. The circumstances were not, however, ordinary, as Wallis was one of England's most prominent mathematicians.[52] In 1643, he became Parliament's chief cryptographer and held that position until 1689.[53] In 1655, moreover, he published an important treatise on conic sections that introduced the infinity sign into mathematics.[54] Equally fundamental publications would follow, including contributions to the still emerging calculus. Most significant, however, was that Wallis had been, since 1649, the Savilian Professor of Geometry at Oxford, which put a good deal of institutional heft behind his arguments. Given what I argued above, it seems clear that Hobbes's philosophical commitments prevented his entering into those realms where the continent's best mathematicians were clearly heading.

Wallis was too august a figure for Hobbes to ignore—and his response was furious. In 1656, he published in reply *Six Lessons to the Professors of the Mathematiques, One of Geometry the Other of Astronomy, in the Chaires Set Up by the Noble and Learned Sir Henry Savile in the University of Oxford.*[55] In this work Hobbes attacked both of the Savilian professors, although his chief target was Wallis, of whom he was righteously dismissive, writing in the dedication: "For first, from the seventh chapter of my book [*On Body*], to the thirteenth, I have rectified and explained the principles of the science; *id est,* I have done that business for which Dr. Wallis receives the wages."[56] And then he flung another calumny: "And I verily believe that since the beginning of the world, there has not been, nor ever shall be, so much absurdity written in geometry, as is to be found in those books of his."[57] The *Six Lessons* brought to a conclusion the first round in a twenty-three-year exchange of insults that terminated only with Hobbes's death in 1679. In the course of these often pungent exchanges, Wallis highlighted Hobbes's mathematical errors, whereas Hobbes ignored the ungentle corrections and hurled back new insults. The debate's sheer nastiness is apparent in the title of Wallis's rapidly published reply, *Due Correction for Mr. Hobbes: Or Schoole Discipline, For Not Saying His Lessons Right.*[58] Things went downhill from there. The contemporary scholarship judges Wallis to have won the debate (which he did, and by a lot) while granting that Hobbes acted more the gentleman (which he did, but not by much).[59] Neither aspect of this assessment helps us, however, to appreciate the debate's significance, when it is read through the history of spatial thought.

The Hobbes-Wallis debate marks the intersection of two important currents, the mid-seventeenth century move away from a more intuitive understanding of space and the consequent intensification of space's *ideological* significance.[60] With respect to the former, Hobbes and Wallis tussled during one of the great mathematical efflorescences of all time. This was not just a matter of a few intellectual giants improving the discipline, but of a renaissance that combined advanced research with improvements in teaching. The infrastructure began to emerge at the end of the sixteenth century with the founding of Gresham College in 1597 in London as a public lecture society and ran through the establishment of the Savilian professorships at Oxford before reaching maturity in the second half of the seventeenth century, with the founding in 1663 of the Lucasian chair in mathematics at the University of Cambridge. Mathematics, in addition to being a "public" discipline, was fully ensconced in English higher education's two jewels, Oxford and Cambridge.[61] And thanks to this infrastructure's emergence, England's culture became the most "mathematical" in Europe.

Mathematics' English dawn had many and profound intellectual effects. The Lucasian chair's second occupant was, of course, none other than Isaac Newton, who was both the chief figure in the seventeenth century's invention of the calculus and also the editor of Bernhard Varenius's seminal geographic work, which I discussed in the previous chapter. The calculus's invention highlights a critical shift in the Spatial Reformation's approach to space. In comparison to geometry, the calculus is less intuitively visual and requires more detailed instruction, which meant that only people who had acquired expertise in independent institutions could gain command of the subject. And as mathematics ceased being intuitive, it also became intellectually and socially more exclusive.

Hobbes was not enthusiastic about any of these developments. First, the move away from the visual was hostile to his cramped reading of space's philosophical implications. Hobbes's "bit-space" geometry remained visual in the sense that his geometric figures (circles and triangles) were rooted exclusively in viewers' experiences of bodies in motion. Were mathematics to take leave of this approach to the visual, then Hobbes's geometry would dissolve, with fatal consequences for both his anthropology and his politics. Second, the mathematical discipline's retreat into institutions had additional ideological implications. A mathematics that was based on the cultivation of a philosophically liberating continuity and, moreover, that resided in Oxford and Cambridge could serve as an alternative source of political authority. (And one can never predict what mischief academics will get up to.) For Hobbes, therefore, only a leviathan should determine knowledge's boundaries—and not some club of academic peacocks.

Significantly, Hobbes's resistance to alternative centers of mathematical authority matched exactly his assessment of the church's role in public politics. In the third part of the *Leviathan,* Hobbes detailed his argument that in a "Christian Commonwealth," only the sovereign can determine whether (and how) divine laws will be applied to the civil realm.[62] In short, for Hobbes it was the leviathan that established what Scripture meant for daily life. Moreover, this manner of integrating Christianity's sacred texts with the civil realm was valid for the *Elements,* too, insofar as its meaning could be determined only by the leviathan's views on space. Thus, regardless of the text to be interpreted, whether sacred or mathematical, Hobbes opposed the formation of alternative power centers that arrogated to themselves the right to interpret potentially dangerous works.

The foundation for Hobbes's critique of independent authorities was, as always, his atomistic philosophy. In the *Leviathan,* he denied permanence to human knowledge and described the imagination as "decaying sense": "The decay of Sense in men waking, is not the decay of the motion made in sense; but an obscuring of it, in such manner, as the light of the Sun obscureth the light of the Starres; which starrs do no less exercise their vertue by which they are visible, in the day, than in the night. But because amongst many stroaks, which our eyes, eares, and other organs receive from externall bodies, the predominant onely is sensible; therefore the light of the Sun being predominant, we are not affected with the action of the starrs."[63] Hobbes's philosophy not only ascribed chaos to the cosmos but also denied to humanity the ability to remember it clearly, that is, without external guidance. Nor was this a new theme. In 1640, Hobbes completed the *Elements of Law Moral and Politick*, a political treatise in which he used Atomism as a club with which to combat the notion that experience secured knowledge.[64] Euclid's space thus had no *place* within Hobbes's thought.

Keeping Hobbes's philosophical commitments in mind, I turn to his responses to Wallis. Rather than explicate the whole debate, which lasted for decades, I concentrate on only Hobbes's initial retort, the *Six Lessons*. I do so, first, in order to highlight the philosophical issues that motivated Hobbes and, second, to trace how spatial themes bled into both theological and anthropological discussions. I begin with an overview of the text. The *Six Lessons* was divided into six sections, with the first two relating to, as Hobbes put it, "the Principles of Geometry, &c," and "Of the Faults that Occurre in Demonstration."[65] The subsequent four lessons then responded to Wallis's retorts, although this constituted little more than Hobbes's fixating on peripheral issues, such as the perceived logical contradictions in Wallis's (and Euclid's) use of terms and Wallis's (in Hobbes's opinion) tenuous grasp of Latin.

I concentrate on the first lesson, since it contains the clearest statements of Hobbes's positions. Not coincidentally, it follows the same conceptual progression that characterized *On Body*, moving from Hobbesian definitions of geometry to politics and then to theology. In order to illustrate this progression, I reach back to the dedicatory epistle in which Hobbes wrote: "So that it is not only my own defence that I here bring before you, but also a positive doctrine concerning the true grounds, or rather atoms of geometry, which I dare only say are very singular, but whether they be very good or not, I submit to your Lordship's judgment."[66] Hobbes could not have drawn this distinction any more clearly from a philosophical perspective—or, more contradictorily from a geometric one. Euclid believed that points have no part and Leucippus held that only atoms have substance, which meant that there was no philosophical mid-point between Euclidean geometry and ancient Atomism. One accepted either spatial continuity or physical discontinuity—not both.

After running up his philosophical colors, Hobbes turned to detailing in the first lesson the foundations of his "atoms of geometry." On the first page, he defined the subject matter thus: "Geometry is the science of determining the quantity of anything, not measured, by comparing it with some other quantity or quantities measured."[67] Mirroring his approach in *On Body*, Hobbes substituted "things" for space and thus separated geometry from its Euclidean roots. He did this, of course, largely to avoid the murkiness that, in his view, had characterized projected space ever since Plato's *Timaeus*. Consistent with his desire to expel metaphysics from geometry, Hobbes continued: "And the science of geometry, so far forth as it contemplateth bodies only, is no more but by measuring the length of one or more lines, and by the position of others known in one and the same figure, to determine by ratiocination, how much is the superficies; and by measuring length, breadth, and thickness, to determine the quantity of the whole body. Of this kind of magnitudes and quantities the subject is body."[68] The reduction of geometry to bodies in motion was an act of philosophical discipline, as it not only kept the metaphysical realm at arm's length, but also diminished the viewer's epistemological authority. A student of geometry was authorized to discuss *only* the dimensions of a body and to compare them *only* to another body's dimensions. Thus, the boundary between real and ideal was placed outside everyone's portfolio, while anything that could not be measured was determined not to exist.

The philosophical radicalism that underlay Hobbes's break with Euclid is even more apparent in his attempts to redefine homogeneity. Euclid never bothered to justify the homogeneity of space but assumed it as a necessary condition. For example, the parallel postulate makes no sense without the adscription of homogeneity; but the *Elements* never justifies this assumption and asserts simply that points have no substance and that parallel lines run to infinity. Hobbes used this gap in Euclid's reasoning to "despatialize" geometry by emphasizing quantity, which he said derived from the measurement of things: "*Homogeneous* quantities are those which may be compared by (*etharmosis*) application of their measures to one another; so that solids and superficies are heterogeneous quantities, because there is no coincidence or application of those two dimensions."[69] Hobbes had thus disciplined homogeneity by denying that it had any broader applicability beyond the human mind. Only things that could be counted could, in turn, be measured (and vice versa), with the result that only concrete things partook of homogeneity. He continued: "No more is there of line and superficies, nor of line and solid, which are therefore heterogeneous. But lines and lines, superficies and superficies, solids and solids, are homogeneous."[70] With homogeneity now limited only to substance, whatever lay outside a given object—whether spatial or spiritual—was excluded from philosophy.

Hobbes's views on homogeneity also offer a glimpse of the theological backdrop to the Hobbes-Wallis debate. Here, I return to Wallis's *A Trinket*. Beyond his mathematical disputes with Hobbes, Wallis could not accept the theological implications of Hobbes's geometry. I noted earlier that Wallis's initial response criticized the *Leviathan*'s understanding of the spiritual realm. At first blush, it may seem odd that Wallis commented on a political work, while debating geometry's foundations. This decision makes sense, however, if we consider that Wallis saw Hobbes's rejection of spatial continuity as injecting gaps into reality that could be filled by specters, or as Wallis called them in Latin, *lemures*.[71] Wallis then added that only children and the mentally ill believed in such things, before expressing in the final line of *A Trinkets'* introduction the fervent hope "that you [Hobbes] may learn in the future, either to know rightly, or to rave modestly."[72]

Hobbes, for his part, did not dispute homogeneous space's theological implications; he just believed Wallis to be the one purveying fables. Consider that when Hobbes defended his argument that points were substantial, he referred to theological issues:

> Theologers say the soul hath no part, and that an angel hath no part, yet do not think that soul or angel is a point. A mark or as some put instead of it (*sigma*), which is a mark with a hot iron, is visible; if visible, then it hath quantity, and consequently may be divided into parts innumerable. That which is indivisible is no quantity; and if a point be not quantity, seeing it is neither substance nor quality, it is nothing. And if Euclid had meant it so in his definition, as you pretend he did, he might have defined it more briefly, but ridiculously, thus, *a point is nothing*.[73]

This juxtaposition of souls and angels with Euclid's spatial doctrines makes Hobbes's position clear: an ill-considered geometry can lead people to discuss things that they can neither see, nor count—and such untutored activity was politically problematic.

Although Hobbes's relationship to traditional geometry was skewed, his thought's internal development highlights two broader themes that suffused the entire early modern period. First, Hobbes and Wallis's serial tiffs reveal that every debate about space always had powerful anthropological, cosmological, and theological resonances, as the issue that split this irascible duo was not whether geometry was relevant to our understanding of humanity, the cosmos, and God, but only *how it was relevant*. Second, as a mathematical outlier Hobbes offers another perspective on the breadth of the shift within the Spatial Reformation that occurred around 1650. Much like the individual contributions of Kepler, Galileo, Descartes, Leibniz, and Pascal to European thought, Hobbes's philosophical system reflected a profound change in geometry's intellectual *position*. Points, lines, circles, and spheres no longer inhered *in the cosmos* but were becoming artifacts of the human mind—and this put both the cosmos and God into a wholly new philosophical light. Hobbes's geometry thus emerged from one of the seventeenth century's main preoccupations: how best to understand the relationship between humanity and its God.

Conclusion

As I noted at the outset, scholars' commitment to secular Trinitarianism has obscured the rupture in space that separated Hobbes's thought from that of Locke and Rousseau. Hobbes rejected the *Elements'* most important doctrines, while Locke and Rousseau never questioned any of them. This difference in attitude, moreover, was not a product of ignorance, as both Locke and Rousseau knew geometry quite well. In his *Confessions*, for example, Rousseau attested that in addition to plodding through works of algebra and the calculus, he studied Bernard Lamy's *Elements of Geometry*, a quote from which I took for this chapter's epigraph.[74] Locke, meanwhile, acquired advanced mathematical knowledge at Oxford, while studying under his favorite teacher, the Savilian Professor of Geometry, John Wallis.[75] Whatever else one may say about the secular Trinity's utility (and it still has some), it is clear that we cannot continue to ignore the spatial backdrop to early modern politics' dawn.

At this point, I step back and sketch out the broadest conclusions that my reading of Hobbes suggests, before I look to eighteenth-century debates. First, by 1650, Euclid's victory was complete, as there was no longer any debate about the need for educated people to study geometry, nor was there any doubt that educated readers could derive something of value from it. Second, Hobbes's rebellion against Euclid's geometry, although quixotic from a mathematical perspective, confirms from an intellectual-historical one that homogeneous space had implications for every aspect of the Western triad, including especially its anthropological component. To ponder idealized homogeneous space was to ponder humanity's relationship to both the cosmos and God. Finally, Hobbes's psychologizing of geometry reflected a broader shift, as European thinkers in multiple fields of intellectual endeavor were emphasizing the significance of geometry for understanding the human mind's bounds. In the next chapter I pursue one effect of this reorganization, when I explore early modern discussions of extraterrestrial life.

While keeping geometry's contested intellectual position in mind, I turn now to an alternative trinity of Münster, Locke, and Rousseau. Although Sebastian Münster was no political thinker, in his *Cosmography* he envisioned a State of Nature in which human beings lived in peace. Münster's anthropological assumptions with respect to this stage of human history do not differ significantly from those that Locke and Rousseau articulated in, respectively, the *Two Treatises on Civil Government* (1690) and *On the Social Contract* (1762).[76] Although the latter two works appeared much later, they also associated humanity's original state with a basic sociability and rationality.[77] This anthropological similarity is rooted, I hold, in the regularities of Euclidean space. Münster studied mathematics diligently and applied its methods to both celestial and terrestrial realms with his arboreal State of Nature being one result. Concomitantly, Locke and Rousseau were not only students of mathematics but also (and especially in the former's case) conversant in the latest advances in physics.[78] Thus, by linking Münster to Locke and Rousseau, we can understand how homogeneous space sustained a positive anthropology and, by extension, a liberal secular politics.

When read against the traditional secular trinity, the alternative trinity opens another perspective on how spatial homogeneity torqued the relationship between humanity and the divine. In returning to this issue, I concentrate on John Locke and examine two of his "non-political" works. In 1690, the same year that his *Two Treatises* appeared, Locke published *An Essay Concerning Human Understanding.*[79] A fundamental work in the history of philosophy, it was central to the rise of philosophical anthropology and thus is particularly important for reflecting on space's anthropological resonances. In the *Essay,* while discussing the relationship between mathematics and human ideas, Locke wrote:

> It is *not* only Mathematicks, or the *Ideas alone of Number, Extensions, and Figure,* that are *capable of Demonstration,* no more than it is these *Ideas* alone, and their Modes, that are capable of Intuition: For whatever *Ideas* we have, wherein the Mind can perceive the immediate Agreement or Disagreement that is between them, there the Mind is capable of intuitive Knowledge; and where it can perceive the Agreement or Disagreement of any two *Ideas,* by an intuitive Perception of the Agreement or Disagreement they have with any intermediate *Ideas,* there the Mind is capable of Demonstration, which is not limited to *Ideas* of Extension, or Figure, or Number, or their Modes.[80]

Locke's meditation on mathematics reveals that, although he never doubted that knowledge originated in experience, he did not restrict humanity's mental world *solely* to bodies in motion either. Not coincidentally, Locke never called for a leviathan that could "rescue" humanity from its natural state of blind ferocity. Instead, the notion that human intuition justified knowledge's demonstration allowed Locke to ascribe an independent (and pacific) rationality to all human beings. Humanity could thus progress toward civilization rather than be herded into it, since individuals were never trapped inside conceptual chaos but could identify truth all by themselves. In this sense, Locke's approach to geometric knowledge imbued his anthropology with a liberal tinge.

The links between regular space and the stability of human reason also undergirded an additional pivot within Locke's thought from anthropology to theology. Locke was more than simply aware of political instability's dangers. Born in 1632, he had lived them. The religiously inspired Puritan Revolution concluded only in 1660, while the Glorious Revolution, in which Locke was personally implicated, terminated the year before the *Essay* was published.[81] As a member of the educated elite (and a former student of Wallis), Locke naturally associated the discipline that emerged from mathematical study with the maintenance of domestic tranquility. We can see how profoundly Locke wove mathematics into this worldview through the second text, *Some Thoughts Concerning Education,* published in 1693.[82] Based on a letter that Locke wrote in 1686 to Edward Clarke, a businessman who sought advice on how to educate his son, this text includes an array of suggestions for not ruining young people's minds.[83] Locke's most significant recommendation is that children begin with practical topics, such as geography and globes, before tackling rarified ones, such as geometry and the Bible.[84]

Locke's main concern in structuring young people's educations was to dampen the youthful tendency to traffic in beliefs about incorporeal beings, such as goblins. Locke actually used this term—and in doing so demonstrated his allegiance to principles that were consistent with those that his teacher Wallis had articulated so savagely against Hobbes.[85] Concomitantly, Locke also turned to the problem of enthusiasm, arguing that children not be allowed to read the entire Bible, since not everything therein was appropriate for young brains: "And what an odd jumble of Thoughts must a Child have in his Head, if he have any at all such as he should have concerning Religion, who in his tender Age, reads all the Parts of the *Bible* indifferently, as the Word of God without any other distinction. I am apt to think, that this in some Men has been the very Reason, why they never had clear and distinct Thoughts of it all their Life-time."[86] People needed, in short, the discipline that came from the study of spatial topics (progressively arranged, in accord with students' advancing maturity) before any of them could be expected to read Christianity's sacred texts. Thus, Locke imagined a series of stages, holding that young people should be drilled, initially, in basic sciences such as geography, since these required only the use of their memories. Later, upon entering into maturity, they could study theoretical subjects, such as geometry and only thereafter consider the most difficult (and dangerous) truths of religion.

After outlining his preferred curriculum, Locke turned to what he believed was the most important of all subjects, Natural Philosophy, or, as we say, physics. This final step throws into relief homogenous space's permeation of not only Locke's State of Nature, but also the Western triad. Locke constructed a hierarchy of knowledge whose primary characteristics retraced homogeneity's progress through European thought. As I have argued, geometry's second return kicked off a transformation in the regnant vision of Heaven and Earth and thus resituated the relationship between humanity and God. Consistent with this point, Locke held:

> The Reason why I would have premised to the *study of Bodies;* and the Doctrine of the Scriptures well imbibed, before young Men be entered in

> *Natural Philosophy*, is, because Matter being a thing, that all our Senses are constantly conversant with, it is so apt to possess the Mind, and exclude all other Beings, but Matter, that prejudice grounded on such Principles often leaves no room for the admittance of Spirits, or the allowing of any such things as *immaterial Beings, in rerum natura*, when yet it is evident that by mere Matter and Motion, none of the Phaenomena of Nature can be resolved, to instance but in that common one of Gravity, which I think impossible to be explained by any natural Operation of Matter or any other Law of Motion, but the positive Will of a Superiour Being, so ordering it.[87]

Locke's anthropology was premised on a system whose strictures prepared the human mind to ascend from the Earth to the Heavens, while erasing all incorporeal beings—except, of course, for the Creator, who was now saved by His being located "not nowhere."[88] Seen from this perspective, the State of Nature, which was as essential to Locke's anthropology as it was to his politics, becomes an episode in space's broader reformation of Western thought.

Looking ahead to the next chapter, I now trace space's echoes into eighteenth-century anthropological and theological discussions through an example from Jean-Jacques Rousseau's correspondence. In 1778, a Polish count, József Teleki, explained in a letter to Rousseau how one could justify the belief in Christianity via geometry. He used this example:

> I only know of the existence of America through the reports of several geographers, who themselves never having been there, base their reports only on those of several explorers, and I have only learned that there is currently a war [ongoing] in America between England and the colonists through the reports of several pamphleteers, who themselves are not well-known for their truthfulness. And yet I would not be able to say that I am just as assured of America's existence and even of the present war as [I am] of any mathematical truth demonstrated by Euclid, or at least that I feel no difference between the degree of conviction that I have in one and the other.[89]

Teleki's ranking of unseens—God, geometry, and the New World—reflects exactly the manner in which Euclid's long rise had reorganized European ways of knowing. Moreover, this approach fit exactly with Locke's hierarchy of education, given that Locke expressly situated geometry between geographic and theological knowledge. In the end, to read Euclid allowed the mind both to understand the terrestrial globe in detail and to contemplate God in a rational manner.

Figure 28. Cover illustration from Müller, *Instruction in the Knowledge and Use of Man-Made Celestial and Terrestrial Globes,* 1792. Courtesy of Staatsbibliothek zu Berlin.

We can understand the significance of Teleki's emphasis on knowledge of global events by returning to the history of globes. In 1792, Johann Wolfgang Müller published the second volume of his *Instruction in the Knowledge and Use of Man-Made Celestial and Terrestrial Globes.* (I analyzed this text's contents in Chapter 2.) Here I concentrate on its frontispiece, which I have included as figure 28. In the image, we see a terrestrial globe placed amid contemporary events in global exploration. The ship on the left may be French, as it seems to be flying the Tricolor. The map in the center of the image depicts Australia, which is identified as "Hollandia." Finally, to the right is the Island of Hawaii (based on the German words "Ins[el] Owaihi"), along with an attending depiction of the events that precipitated the Englishman James Cook's death there in 1779.[90] If we reflect on Teleki's arguments with Müller's frontispiece in mind, we see again how early modern minds continually moved between print and material culture, using the two in order to situate both themselves and others on a great unseen sphere. Europe's culture of space gave a *place* to all things, including peoples and their politics.

Taken together, Teleki and Müller suggest that, by the eighteenth century, probably the majority of educated people had some knowledge of both projected space's theory and its attending material culture. We know that Teleki's correspondent, Jean-Jacques Rousseau, was hardly uninitiated, as the combative Genevan confessed to having studied numerous geometric texts. In his *Confessions,* which were published posthumously in 1782, he summarized his experiences:

> I went from there to elementary geometry; because I had never gotten much further, and persisting in my desire to overcome my poor memory, I was forced to retrace my steps hundreds of times and to recommence incessantly the same journey. I did not like [the geometry] of Euclid who seeks the chain of demonstrations, rather the connection of ideas; I prefer the geometry of Father Lamy, who from that point became one of my favorite authors, and I still reread his works with pleasure. Algebra followed, and I always took Father Lamy as my guide. When I was more advanced, I occupied myself with the science of calculation of Father Reynaud, then with his demonstration of analysis, which I have scarcely touched. I have never gotten far enough to have a feeling for the application of algebra to geometry. I do not at all like this way of operating without seeing what we do, and it seemed [to me] that to solve a problem of geometric equations, was akin to playing a tune by turning a crank.[91]

Although not educated to quite Locke's level of expertise, Rousseau had been exposed to the mathematical fundamentals of the period. And this background calls to mind two issues. First, Rousseau seems to have rejected none of his lessons but continued to live within a mathematical world. Thus, when he constructed his State of Nature, the geometry that he had studied led him to see the natural state as a *regular* one. Natural Man was, in this context, a child of homogenous space and, for that reason, was essentially nonviolent. Second, Rousseau also calls attention to the intellectual significance of mathematics' progress beyond the *Elements*, with the discipline leaving many observers behind when it jettisoned Euclid's intuitive space.

Rousseau's vision of a secular politics, like those of Locke and Hobbes, must be situated against deeper changes in the understanding of space that came to fruition in the early nineteenth century. In a sense that I pursue in the next chapter, and to which I also return in the book's conclusion, changes in spatial thought ushered Natural Man to European thought's margins, much as an earlier innovation in spatial thought had produced both Adam and Eve's marginalization and their ultimate dismissal from humanity's world. Profound changes in mathematics were coming and, as we have seen in Dostoevsky's modest ravings, the newer currents in spatial thought would shake European thought to its foundations yet again. Space had thrust the human mind upward, into God's realm—and God would pay the price.

Chapter 6

Strangers to the World

> Thy Travels dost thou boast o'er foreign Realms?
> Thou *Stranger* to the *World!* Thy Tour *begin*;
> Thy Tour through *Nature's* universal Orb.
>
> —Edward Young, *The Complaint: Or, Night Thoughts on Life, Death, and Immortality* (1750)

During the Spatial Reformation's five centuries, homogeneous space suffused and *resituated* every aspect of the ancient Western triad. It was not merely that tension developed among traditional anthropological, cosmological, and theological perspectives, but of each of the triad's parts becoming something different in response to space's advance. Homogeneity's rise required a new humanity, a new cosmos, and a new God, given how long these concepts had been nestled within fractured space. Equally important, the relative ranking among the three changed, as humanity came to promote itself over both the cosmos and God. I pursued one aspect of this shift in the previous chapter, where I explained how space shaped the anthropology on which modern secular politics is built. In this chapter, I pursue a related line and consider how heliocentrism's use of space produced another anthropological resituating, in which early modern thinkers dared to envision true "others" who gazed *back* from extraterrestrial realms. Early modern anthropology was transformed when Europeans embraced an open universe that hosted not only empty space but also, as we will see, multiple heliocentrisms.[1]

For at least the past forty years, scholars of the history of anthropology have emphasized the empirical encounter with difference, as opposed to the rise of projected space.[2] As the story goes, between 1500 and 1800 the direct experience of physical and cultural diversity by explorers and travelers prompted Europeans, first, to shed classical traditions that evaluated variety in terms of the monstrous and, second, to abstract a common humanity from the observation of difference. Opinions vary on when the transformation occurred. Some scholars look to the seventeenth century and emphasize philosophy, holding that the invention of the philosophical subject—especially through René Descartes's labors—allowed the processing of diversity in new ways.[3] The seventeenth-century reading, however, has been overshadowed by two traditions that pinpoint the change in either the sixteenth century (the Renaissance approach) or the eighteenth century (the Enlightenment approach).

The Renaissance and Enlightenment approaches both emphasize that anthropology emerged from theoretical reflections on the nature of difference and its meaning for diversity.[4] Thus, they see anthropology as originating either in the late Renaissance or in the Enlightenment and, moreover, as germinating specifically on non-European soil. Both approaches have distinguished pedigrees. With respect to the sixteenth century, the contributors include Edmundo O'Gorman, Margaret Hodgen, Anthony Pagden, Walter Mignolo, and Karl Butzer, all of whom argue that early explorers' confrontations with the "other" produced a new vision of unity.[5] Among these scholars, Pagden has made the most influential argument, which holds that sixteenth-century anthropologists, such as the Spanish missionary José de Acosta, learned to extract unity from diversity and, in doing so, founded an anthropological tradition that extended into the thought of the *philosophe* Denis Diderot and the *Aufklärer* Johann Gottfried Herder.[6]

The partisans of the Enlightenment, meanwhile, do not want for intellectual firepower. Contributors to this side include Claude Blanckaert, Michèle Duchet, E. E. Evans-Pritchard, Larry Wolff, and John Zammito.[7] Heavily influenced by the history of philosophy—and especially the work of Wilhelm Dilthey—this tradition emphasizes how the European philosophical subject discovered itself through confrontations with "others."[8] The profoundly introspective aspect of the Enlightenment's encounter with difference suggested, in turn, that both sides of the anthropological coin lived inside distinct conceptual worlds. Put another way, the Enlightenment's encounter with "self" and "other" produced the discovery of cultures as autonomous wholes, each of which could be studied with respect to its own unity. Humanity was thus justified as a concrete object of study precisely by its production of cultural difference.[9]

In this chapter I am not concerned with deciding which of the interpretations is correct, but with superimposing the history of space onto the whole debate. In this respect, it is significant that although both interpretations place the rise of anthropology within the early modern period, their explanations are conceptually incompatible. Consider that the Renaissance reading begins with an event (the landing in the New World) that produced a shock but whose anthropological theories were ultimately superseded, while the Enlightenment reading highlights a shift in theory (the discovery of culture) that has no clear starting point but whose effects are, apparently, still felt. The interpretive problems that emerge from this situation are evident in an admirably broad essay that the historian Harry Liebersohn published under the title "Anthropology

Before Anthropology."[10] Liebersohn organized the process into two phases: a Renaissance phase, which began with Acosta's views on difference and comprised the entire sixteenth century and most of the seventeenth, and an Enlightenment phase, which began with the invention of culture and comprised the late seventeenth century and all of the eighteenth.[11]

Liebersohn's excellent synthesis, however, harbors two problems. First, the attribution of distinctiveness to the Enlightenment renders problematic the traditional periodization, because it suggests that a break occurred in mid-seventeenth century, even as it foregoes any analysis of this break's foundations. Second, along these lines, this interpretation also undermines its own assessment of Acosta as anthropology's founder. If Enlightenment anthropology was truly distinctive, then it is difficult to see how Acosta can have a claim to the status of pioneer, especially if his advances are no longer relevant to the contemporary discipline's practices, whereas the Enlightenment's theorizing remains current. In this sense, it is unclear what the contemporary discipline actually sees as a theoretical foundation—and this is particularly problematic when trying to organize a historical study of a venerable discipline's origins.

Another important issue arises if we consider the connections between anthropology's origins and the contemporary discipline's interest in culture. Regardless of their respective starting points, both sides in the debate assume two things that undermine the possibility of a complete early modern narrative for anthropology. The first assumption is the belief that anthropology arose in response to a single phenomenon, the encounter with difference. Thus, the discipline was born either when Renaissance missionaries landed in the New World, or when Enlightenment theorists began to think more clearly about the cultures of non-Western European peoples.[12] This approach, however, leaves us without a criterion for judging which moment constituted a true beginning. And the problem looms larger when we consider that scholarship on the ancient world insists that difference has been important for anthropologists since the time of Herodotus.[13] Contemporary scholarship, therefore, needs to find a way to contextualize the history of difference if it is to establish why any given encounter with "others" transformed anthropology—or did not. I argue that the history of space provides exactly the crucial context.

To superimpose spatial thought affords us an analytical backdrop that can accommodate multiple changes in European thought. Partisans of the Renaissance approach may argue that the New World landing's historical uniqueness justifies a sixteenth-century claim to originality. This is partly right, but in a way that reveals the need for a broader approach. Here, I turn to the second assumption, namely that anthropology emerged *solely* from direct experience. A cursory glance at the disciplinary canon reveals such a reading to be incomplete, at best. José de Acosta, René Descartes, and Johann Gottfried Herder—canonical anthropologists whose works span the early modern period—all appended cosmologies to their anthropologies and, in so doing, expressly juxtaposed extraterrestrial spaces with the terrestrial ones.[14] Put another way, the status of unseen spaces and places was a central issue for all early modern anthropologists. Moreover, the cosmology of each of the thinkers that I just mentioned differed as radically from that of the other two as did their respective anthropologies from each other, all of which suggests the possibility that developments in cosmology occurred in conjunction with changes in anthropology. In this respect, I propose that in order to understand anthropology's emergence *down here,* we need to account for the evolution of Western views of the *out there.*

The literature, however, has failed utterly to emphasize both the Heavens and the spatial senses that shaped humanity's experience of them. As a result, a rupture within early modern anthropological discussion has gone unnoticed: Renaissance anthropologists were geocentrists, whereas their successors were heliocentrists.[15] This shift did not emerge from within anthropology itself but was due to developments in astronomy, which was already one of the Spatial Reformation's most significant disciplines, even before it resurrected ancient heliocentrism.[16] The honor roll includes names such as Nicolaus Copernicus, Johannes Kepler, Galileo Galilei, Christiaan Huygens, and Isaac Newton, and its period of glory extends from 1514, the year that Copernicus first put his ideas in manuscript form, up to 1687, the year that Newton published his *Mathematical Principles of Natural Philosophy.*[17] Rather than retrace heliocentrism's rise, I reiterate my central point: heliocentrism participated in a broader change in spatial thought that allowed human beings to impose an extraterrestrial perspective on the terrestrial. All the astronomers involved thus understood that the only way for the mind to "see" the earth was to take leave of direct experience. And given how the Western world interwove humanity, the cosmos, and God with fractured space, the effects of this change in perspective were profound.

I illustrate my point with a detour through Johannes Kepler's early work. In 1596, Kepler published *Cosmographic Mystery,* which is often celebrated as an important step in heliocentrism's rise. There is, of course, nothing wrong with this interpretation. I would, however, add that taking account of the *Cosmographic Mystery*'s relationship to geometry's space enriches this reading of Kepler's significance. In this context, I suggest that we see Kepler's great text as delineating Euclidean space's high-water mark, as in it Kepler propounded a doctrine that he would come to reject, namely that the planets traced circular orbits.[18] (Kepler was the first to suggest these orbits were, in fact, elliptical, which was an important step in geometry's passage into the human mind.) The underlying Euclidean nature of Kepler's space is on display in the text's cosmological illustration (figure 29). In it, we see that Kepler explained the distance between the planetary orbits

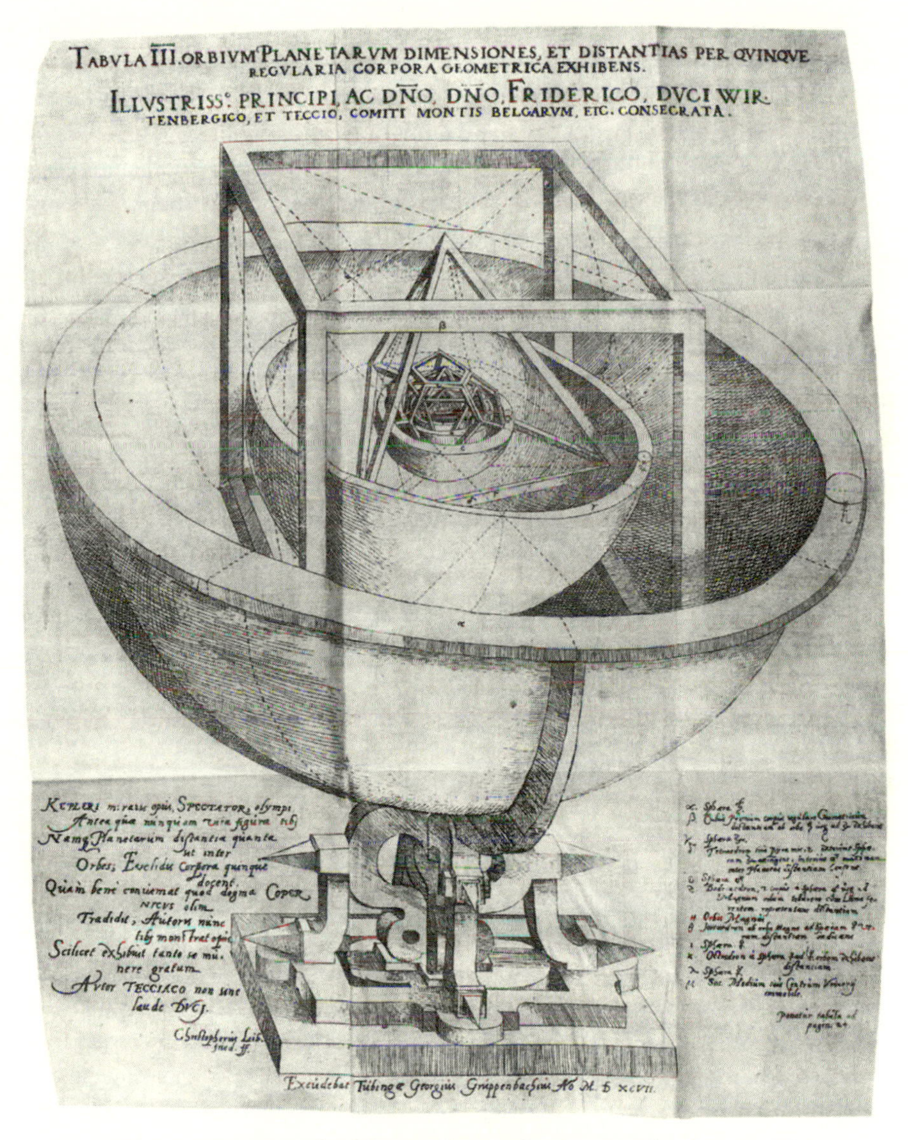

Figure 29. Cosmographical drawing from Kepler, *Cosmographic Mystery,* 1596. Courtesy of Herzog-August-Bibliothek Wolfenbüttel.

through the application of the Platonic solids to the celestial realm. (And he was inspired, of course, to do this by Plato's *Timaeus.*)

Rather than go into greater detail, I touch upon three themes that relate to the history of space as I have outlined it. First, Kepler's heliocentric cosmos is pervaded by geometric space from top to bottom, that is, including both Heaven and

Earth expressly. Second, the image's visual perspective is clearly extraterrestrial, as the viewer *looks back* to our solar system in a way that fatally undermined the tradition of fractured space. Finally, this representation of the cosmos incorporates prominent characteristics of early modern material culture, insofar as the whole is resting on a pedestal in the manner of a celestial globe. The cosmological whole has thus been *resituated,* given that homogenous space pervades both Heaven and Earth in a way that liberates the human imagination by allowing the viewer to deploy his own sense of space. Kepler's cosmological vision, I suggest, was anchored to the alliance between Euclidean space and material culture.

When viewed against this backdrop, the history of astronomy reveals an alternative way to *situate* the history of anthropology. From the ancient world into the early modern one, the experience of Heaven and Earth was geocentric, as every human being looked *up* to the Heavens. This incontestable aspect of daily life dominated Western thought, as many of its intellectual traditions—in philosophy, theology, cosmology, and anthropology—anchored humanity to a place that was understood as being both "central" and "below." Heliocentrism repudiated this perspective, however, by not only projecting an extraterrestrial position that was out of reach, but also valuing the information gathered *out there* as knowledge that was applicable down here. In the early modern period, to be a heliocentrist was to deny experience and, correspondingly, to privilege the imagination. The anthropological consequences of this leap into nothingness echo into our own day.

Geocentric Anthropology

I now pursue the Renaissance-versus-Enlightenment debate by understanding anthropology's history with respect to two contemporaneous shifts. I agree, following the Renaissance school, that the sixteenth century wrought changes that were crucial to anthropological thought's course, but I identify a different cause and suggest a different way of understanding the resulting changes' significance. The Renaissance school typically looks to the New World and its Christian missionaries when arguing for anthropology's Renaissance origins. One popular figure is José de Acosta, who in 1590 published *Natural and Moral History of the Indias,* the first anthropological study to break with classical approaches to difference.[19] Acosta is not the only possible choice; one could make a good case for other figures, such as Bernardino de Sahagún.[20] Still, Acosta is a good option because his written works diffused in multiple translations, some of which were cited approvingly into the nineteenth century.[21] Hence, if a tradition can be traced to the sixteenth century, Acosta is a plausible founder.

Against this backdrop, it is significant that the contemporary literature on Acosta has not examined either his geocentrism or its underlying sense of space. If we take these characteristics into account, however, the great Jesuit traveler appears not as the founder of a new anthropological tradition, but as the finest example of a Christian Aristotelian approach to knowledge.[22] Such a reading does not signify that Acosta was unimportant to the history of anthropology, but that his contributions must be evaluated with respect to the intellectual system that his education enshrined in him. Thus, against the backdrop of Kepler's *Cosmographic Mystery*, I would suggest that the scholars have profoundly misread Acosta, as they have failed to frame his contributions against the Euclidean space whose diffusion was already producing, during his own lifetime, radically different perspectives on humanity's physical position within (and conceptual relationship to) the cosmos.[23]

My emphasis on revealing the conceptual roots of Acosta's cosmology dovetails with attempts by scholars to redress the distortions caused by the sometimes celebratory tone of the anthropological literature. As Laura Ammon has noted, scholars' generally positive assessments of Acosta's contributions to anthropology obscure how his thought, especially his theology, served the Spanish Empire's ends.[24] In a similar vein, Joan-Pau Rubiés has highlighted the significance of Acosta's theological commitments, noting how his opposition to idolatry was as important to his anthropology as any encounter with difference.[25] These critiques reinforce a trend toward greater contextualization that is apparent in the work of Fermín del Pino and Walter Mignolo. Del Pino has insisted (rightly) that Acosta must be read as a sixteenth-century Christian Aristotelian, while Mignolo has underscored that the Old World's colonial system had a dark side that cannot be excluded from our memory of the early modern.[26] In sum, today's scholarship argues that Acosta must be understood as a Christian who served the Old World rather than the New.

Taking my cues from specialists in sixteenth-century anthropology, I understand Acosta as a traditional Christian Aristotelian theologian rather than an incipient anthropologist. For that reason, I underscore that his mental cosmos was fundamentally incompatible with both the burgeoning tradition of heliocentrism and the homogeneous space that undergirded it.[27] Based on what I have argued in previous chapters, it is significant that Acosta's education made theology and cosmology coterminous. In 1552, at the age of twelve, he joined the Society of Jesus and entered the College of Medina del Campo, which, like all Jesuit institutions, emphasized the study of the classics, including works by Cicero, Terence, Virgil, Aristophanes, and Homer. After completing this phase, Acosta taught at posts around the Iberian Peninsula before pursuing an advanced degree, between 1559 and 1567, at the legendary College of Alcalá de Henares. This postgraduate period is critical to understanding the philosophical commitments that underlay Acosta's anthropology. The curriculum at Alcalá de Henares prescribed four years of philosophy, which consisted of readings in the Aristotelian corpus. Next came four years of theology, which meant imbibing the thought of St. Thomas Aquinas, the jewel in Christian Aristotelianism's crown.[28] Acosta's mental cosmos was thus soaked in Aristotelianism and Thomism, both of which cleaved, as I have explained, to not only geocentrism but also hierarchical space.[29]

With Acosta's intellectual formation in the background, I turn to his professional career. In 1572, Acosta arrived in the New World and spent the next fifteen years working as an administrator and missionary. This posting allowed him to travel widely in Peru and Mexico, where he encountered an array of indigenous cultural practices. In 1587, he returned to Europe and, three years later, published his *Natural and Moral History*.[30] This work comprises seven books that can be organized into three parts. The first two books make up the first part and evaluate both classical and biblical traditions in cosmology and geography against the reality of the unforeseen New World.[31] Books 3 and 4 form the second part and describe the New World's environments, including its waters, soils, winds, flora, and fauna.[32] Books 5–7 constitute the third part and are dedicated to discussions of indigenous peoples, with Book 5 examining their religious practices and Book 6 explicating their time reckoning, language, governance, and burial practices.[33] Book 7 is historical and covers events in Mexico, beginning with original indigenous settlements and extending up to Spain's uninvited arrival.[34]

The literature on Renaissance anthropology is quite impressed with how Acosta broke with the classical anthropology that he had learned in Jesuit lecture halls.[35] In particular, the literature notes how Acosta discarded Herodotus, Pliny the Elder, and Strabo, all of whom had populated the known world's boundaries with monstrous beings—which was an approach that medieval thinkers also adopted.[36] Based on his own direct experience, however, Acosta argued that the indigenous peoples of the New World were not monsters but human beings, and he suggested that any differences between Old World and New should be understood as historical developments within a divinely ordained cosmos rather than as deviations from a perceived ancient norm.[37]

The contemporary literature has not emphasized, however, that Acosta's anthropological break occurred within traditional continuities in both cosmology and epistemology. I examine the former immediately and return to the latter below. The *Natural and Moral History* envelops both the New World and its peoples in geocentricity. In Book 1, for instance, Acosta defended the Aristotelian cosmos against alternative Christian and classical approaches, melding the traditions wherever possible. The first non-Aristotelian cosmologies came from John Chrysostom and the Bible, both of which Acosta refuted, before turning to Augustine, Plutarch, Lactantius, Seneca, Dioscorides, and Plato, among others.[38] He concluded: "There is no doubt but that Aristotle and the other Peripatetics, together with the Stoics, felt that the entire figure [of the universal orb was] rounded and rotated in circular fashion and, in turn, this is precisely correct, as we have seen with our own eyes in Peru."[39] Acosta thus began his work by fixing humanity in a particular kind of space. Following thereon, in the first two books he also endorsed Aristotle's chief cosmological doctrines, including that the cosmos was round, that the earth sat immobile at its center, and that the Heavens (unlike the earth) were incorruptible.[40] Acosta, in short, traveled to the other side of the Atlantic and looked up.

Acosta's Aristotelianism brought with it a commitment that we today would call empiricism, although of a particular sort. As I have already mentioned, geocentrism comported with humanity's direct experience, which meant that its core doctrines appeared to align with nature. Moreover, Aristotle had emphasized that knowledge of the natural world should come from observation, such as observation of the sun and other heavenly bodies as they passed *overhead*. Although Aristotle's commitments were never wholly empirical (no system of inquiry is), the commitment to the chronicling of experience over the elaboration of theory meant that Aristotelianism could be both vibrant and (partially) self-correcting, as long as its cosmology remained untouched. (No one could *see* from above, of course, which meant that this imagined perspective was excluded from Aristotelian science.)

The significance of Acosta's absorption of both Aristotle's cosmology and his emphasis on interpretive flexibility is on display in Book 2 of the *Natural and Moral History*. Here, Acosta broke with Aristotle's view that the area around the equator, the "Zona Torrida," was uninhabitable due to its extreme heat.[41] This was not truly an Aristotelian doctrine, as it came from an even older tradition in Greek geographic thought, in which the earth's surface was divided into five parallel regions known as "climata."[42] The five bands ranged from extremely hot in the equatorial region to extremely cold in the two polar ones. Only the two strips between the extremes were habitable: the one above the equator that contained the Mediterranean and another unknown realm below it. Pursuant to my discussion of terrestrial globes, it is significant that this mode of organizing terrestrial space was basically two-dimensional, as the earth's climates were arranged on a conceptual disc that answered to latitude (up and down) but that lacked any reference to longitude.

Against the backdrop of ancient Greek geography, it is particularly important that Acosta traveled to the so-called "Zona Torrida," where he encountered peoples who should not have been where they were. Acosta's unexpected encounter with flourishing populations suggests that, far from marking a break with ancient thought, his views partook of profounder continuities.[43] Acosta went to Peru, where he observed Heaven and Earth and confirmed traditional visions of the former, while "correcting" older approaches to the latter. Given the nature of his education and training, it is therefore simply incorrect to separate Acosta's geocentrism from his empiricism, for it was the latter that allowed him to integrate the information that he had gathered on the newly encountered populations with traditional doctrines.

With this hermeneutical issue in mind, it is imperative to note that Acosta's thought was completely isolated from the new astronomical traditions that Johannes Kepler's work exemplifies. Acosta's astronomy was rooted in classical works—and then almost purely in Aristotelian ones.[44] He never mentioned any fifteenth- or sixteenth-century astronomers, regardless of their position on geocentricity, failing to cite even important geocentrists, such as Georg Peurbach and Johannes Regiomontanus.[45] (The latter, of course, was particularly important for his contributions to the mathematical circle in Rome that produced the first known paired globes.) Nor did Acosta cite Tycho Brahe, who, although not a doctrinaire geocentrist, had gained renown already in the 1570s.[46] Given how Acosta excluded geocentrist astronomers, it is also not surprising that he ignored Copernicus, even though *On the Revolutions* had been circulating around Europe for decades, including in ostensibly backward Spain.[47] Cutting-edge astronomy and its methods of projection played no role in Acosta's renovation of Christian Aristotelian anthropology.

The gap that separated Acosta from other rising intellectual currents within sixteenth-century thought looms even larger if we take into account Claudius Ptolemy's curiously low profile in the *Natural and Moral History*. As I explained in Chapter 2's analysis of the history of globes, Ptolemy's astronomical magnum opus, the *Almagest*, accepted some Aristotelian cosmological doctrines even as it maintained a critical distance from other ones.[48] Although he was a geocentrist, like Aristotle, he was more interested in using mathematics to trace the movement of heavenly objects than in defining philosophically what those things were.[49] The *Almagest* was fifteenth-century astronomy's cornerstone, which makes it significant that Acosta mentioned this work only once and without reference to Ptolemy's mathematics.[50] Acosta's geocentrism was philosophical—Aristotelian—which meant that the Spaniard remained isolated from not only heliocentrism but also the mathematical debates that undergirded the new astronomy.[51]

The nonspatial mode in which Acosta presented terrestrial space underscores the significance of Ptolemy's absence from the Spaniard's thought. Acosta's single mention of Ptolemy was couched in a formulaic reference to his contributions to geography. As I have also explained, Ptolemy was the chief source not merely for early modern astronomy, but also for its attending geographic thought, which had become manifest in changes in globe making, cartography, and especially in the rise of cosmography. Ptolemy's geographical thought was crucial to multiple aspects of European thought in the fifteenth and sixteenth centuries, because its mathematics encouraged new departures in the material culture of space.[52] Acosta did not, however, apply any of Ptolemy's mathematical techniques to his depiction of the New World, even though globe makers, cartographers, and cosmographers had been using them for over a century. Instead, when Acosta

esence. First, Acos-
Vorld, because Hell
already discussed,
fferent direction. A
Hell's dimensions,
trist suggested that
space.[59] Pursuant,
ion of melancholy,
onal geocentrism's
us, whatever mod-
Acosta, from a spa-

ology that, in turn,
ow from this point.
e break with either
ptual apparatus not
them. Acosta was
of a narrow set of
embedded inside a
ovel had less to do
ained Aristotelian,
d when it was still
of new information
s to see more con-
ained in the main
n absolutely neces-
nce for subsequent
ce to the heliocen-

ent anthropology.
s issue, Enlighten-
parated them from
entist and anthro-
Acosta not as an
oldt asserted:

y—all mathe-
Acosta's *Nat-*
rk of Gonzalo

ce to the spatial tools that were

nts in early modern geography
o relief what truly characterized
oples. Here, I turn to Book 3 of
he New World's environments.
rs in Books 5–7 follows below.)
aneous astronomical work, the
e New World is the absence of
s that allowed unseen things to
essed the efflorescence of globe
and places he had seen via non-
ster some fifty years before, he
pect to the four elements—Air,
to the virtue of studying God's
d chapter of Book 3 and consid-
ect of moving air.[53] In chapter 10
e the oceans touched the New
springs, and rivers in nine addi-
he soil before examining fire in
's use of the four elements in his
g sense of space.
e from the Spatial Reformation's
story's anthropological sections.
lian natural philosophical catego-
d of people one would expect a
t needed saving. Backstopped by
e diversity of indigenous cultures
s before God. As he wrote in the
say of [the indigenous peoples],
us], and since the goal of this
in the Indias, but also to put this
e derived from the knowledge of
ir salvation and to [give] glory to
of the slippery shadows of their
ight of His Gospel."[56] Within this
Vorld's ostensible idolatry, in the
the diversity that he encountered
[7] For example, Acosta portrayed
onic torments that had the (sole)
ity.[58]

It is critical to underscore the spatial logic of the Devil's p ta's geocentrism justified this fallen angel's role in the New had a place, too, beneath the earth's surface. Second, as I hav the main currents of European thought were heading in a d prime example of this trend is Galileo Galilei's 1588 talk on which I have previously discussed, and in which the heliocen neither Hell nor the Devil could exist within a fully geometri moreover, to the arguments that I made in Chapter 3's discus it seems clear that the Devil was as embedded within tradit epistemological frame as were the four physical elements. Th ernizing virtues contemporary anthropology may identify in tial perspective, he remained primarily a medieval thinker.

Acosta's anthropology was rooted in a geocentric cosm reached back through the medieval period. Two implications f First, Acosta's innovations cannot be represented as a comple classical or medieval traditions, given how much of his conce only originated with these traditions, but also remained tied t a Christian Aristotelian, whose knowledge base consisted ancient texts and whose cosmology was both geocentric and fractured space. Second, what made Acosta's anthropology n with its specific natural scientific doctrines, most of which ren and more with his arrival in the New World having occurre "new." The brute fact of Acosta's having to assimilate masses into the Aristotelian system of thought has induced historian ceptual change than was really there.[60] Acosta's thought ren Aristotelian, however, and he broke with the Stagirite only whe sary, that is, when direct experience mandated it.[61] His significa anthropological thought thus cannot be judged without referen tric cosmology that superseded his geocentrism.

Heliocentric Anthropology

Heliocentrism stands between Renaissance and Enlightenm Although the contemporary literature has not emphasized th ment anthropologists understood that something important se their Renaissance forebears. The late Enlightenment natural sc pologist Alexander von Humboldt, for example, characterize anthropologist but as a geographer.[62] Writing in *Cosmos*, Humb

> The foundation for what we presently call physical geograp matical considerations aside—is contained in the Jesuit Jos *ural and Moral History of the Indias*, as well as in the wo

> Fernández de Oviedo, which appeared barely twenty years after the death of Columbus [in 1506]. At no other moment since the emergence of civilization had the realm of thought expanded so suddenly and wonderfully with respect to the external world and its spatial circumstances, nor had the desire to observe nature at different latitudes and at different elevations above sea level, and to multiply the means by which phenomena could be observed, been felt more keenly.[63]

Nineteenth-century physical geography unconsciously mixed Acosta's emphasis on empirical research with heliocentrism's abstract space, insofar as geographers put local environments within a mathematical whole that was foreign to Acosta.[64] Thus, for Humboldt, Acosta represented a first step in a conceptual transformation that had been unleashed by global expansion, but reached fruition inside the projected space that was the Spatial Reformation's singular achievement.[65] By specifically calling out mathematics, Humboldt indicated that idealized space had become anthropology's cornerstone.

Heliocentrism's victory in the second half of the seventeenth century brought profound conceptual changes to anthropology. This was not a simple process, just a matter of Copernicus's ideas unfolding through the efforts of the great thinkers mentioned above. Instead, heliocentric anthropology was built on a culture of projection that had been produced by scholars of many kinds, including astronomers, cosmographers, cosmologists, geographers, and natural philosophers.[66] Among this culture's manifestations were the construction of observatories and the founding of university chairs in astronomy, as well as the publication of maps, books, pamphlets, calendars, and journal articles, in addition to the production of instruments such as telescopes, quadrants, and globes.[67] Taken together, these things made it possible for the entire continent to imagine an unseen whole. Not all the contributors in this context were heliocentrists, as in the examples of Johannes Regiomontanus and Tycho Brahe, nor even necessarily astronomers, as in the case of Peter Apian. Nonetheless, all of them *situated* the unseen with mathematical tools that were absent from Acosta's work.

One important aspect of heliocentrism's rise, therefore, is the way that knowledge of extraterrestrial space was diffused in word, image, and object.[68] In explicating how astronomy and its space diffused, I consider four examples, two from the seventeenth century and two from the eighteenth. The first is simply the title to an astronomical compendium that appeared in 1684 in London:

> Astronomy's Advancement, or, News for the Curious; Being a Treatise of Telescopes: and an Account of the Marvelous Astronomical Discoveries of late years made throughout Europe; With the Figures of the Sun, Moon, and Planets; with Copernicus his System, in twelve Copper Plates.

Figure 30. Cover illustration from Fontenelle, *Conversations on the Plurality of Worlds,* 1686. Courtesy of William Andrews Clark Memorial Library, UCLA.

> Also an Abstract touching the Distance, Faces, Bulks, and Orbs of the Heavenly Bodies, the best way of using Instruments for satisfaction, &c. out of the best Astronomers, Ancient and Modern, viz. Mr. Hooke, Mr. Boilleau [*sic*], Mr. Hevelius, Father Kircher, &c.[69]

Such compendia appeared regularly in the seventeenth century, keeping the public informed about the latest advances in astronomy and other related disciplines. As the title indicates, the advances were presented along many lines, including the celebration of individual astronomers, the printing of images, and the description of instruments. In sum, the universe was constructed by specialists in a variety of fields, with news of their achievements being diffused to a reading public.[70]

The next example is an image from Bernard de Fontenelle's *Conversations on the Plurality of Worlds,* which appeared in 1686 in Paris[71] (see figure 30). The text itself comprises a delightful series of nighttime colloquies between an educated man (who is Fontenelle) and a noble lady, concerning both the cosmos's heliocentric physical structures and the possibility of life on other planets.[72] I return to the Plurality of Worlds debate below. Here, I concentrate on the significance of the image's spatial perspective. There is nothing inherently heliocentric about the image itself, as it simply depicts a man and woman looking the night sky. To the extent that the image betokens heliocentrism, this sense only derives from the text itself, which goes into great detail on the cosmos's functioning. I have already touched upon the relationship between text and image at various points, and especially with my discussion of the image of Creation in Martin Luther's Bible. In this case, given the text's contents, it is clear that the man and the woman in the image know that they were peering *out* into the Heavens, rather than *up* into the cosmos—even though their physical experience should have indicated the reverse. Against this backdrop, the image's full meaning becomes clear: thanks to the Spatial Reformation's progress, viewers now brought a different sense of space to representations of the unseen cosmos.

Scholars' continual illustrations of the cosmos for a consuming public made extraterrestrial realms *present* within European thought.[73] Another powerful example that incorporates both the spatial perspective that we have just seen with depictions of material culture comes from the cover image of Johannes Rost's *Portable Atlas of the Heavens,* which appeared in 1723 in Nuremberg[74] (figure 31). In this image, we see an astronomy lesson for children that is set against the Heavens in the background. A woman, the personification of knowledge, teaches the youngsters and also gestures toward the Heavens, while the surrounding globes, maps, and the other objects incorporate the mental apparatus that the children were to acquire. In this respect the work's title, *Portable Atlas of the Heavens,* is particularly apt, since it accentuates how both the book itself and the space that it cultivated were to accompany readers for the rest of their lives.

The portability of the new spatial sense is an important theme for understanding European culture after 1650. Another important example that illustrates how the command of space structured the European imagination comes from *Cosmological Letters on the Arrangement of the World Edifice,* a popular text published in 1761 by the Alsatian mathematician and astronomer Johann Heinrich Lambert. Hurling both himself and his readers into the cosmos, Lambert wrote: "To each fixed star, I gave a similar host of such bodies, which receive from it their light and warmth, and on each I imagined innumerable occupants of every shape and stripe. I have, thus, extended my imagination along with the universe and expend no effort now in taking as a yardstick the distance between our sun and a fiftieth-magnitude star and, by extending it a million times, setting it up as a measure against the limits of those star systems that we can see with telescopes—and even those that lie beyond."[75] The centuries since Copernicus's death in 1543 had witnessed a transformation in the early modern world's view of the Heavens, as

Figure 31. Cover illustration from Rost, *Portable Atlas of the Heavens*, 1723. Courtesy of Staats- und Universitätsbibliothek Göttingen.

those who looked "up" now saw suns rather than mere points of light. Two significant themes emerge here. First, the Spatial Reformation's progress up through heliocentrism's victory shattered the geocentric hierarchy that had once *situated* Acosta's anthropology. Second, the mind had become so free to flutter about the universe that its eye now dared to imagine a cosmos in which multiple heliocentrisms were possible.

Heliocentrism transformed anthropology by requiring that it assess humanity with reference to spaces and places that no human eyes had ever seen, but that the mind's eye could put into space. Developing most fully in the eighteenth century, this anthropology demonstrated three characteristics. First, it assumed a homogeneous space in which the earth was suspended as one planet among many rather than as a cosmological center. Second, it projected a position out in space and "returned" the mind to the terrestrial realm, sometimes in the course of only a few pages. Finally, as the Lambert quote implies, this anthropology made a leap more profound than anything that Acosta could have imagined: it embraced extraterrestrial life. Against the backdrop of outer space, anthropology could no longer simply be about *human* beings but had to include the possibility of nonhuman ones.

The epigraph at the beginning of this chapter provides an excellent point of entry into heliocentric anthropology. Taken from *Night Thoughts,* a popular anthropological poem by the English writer Edward Young that appeared between 1742 and 1745, it contextualizes the earth against outer space.[76] I consider the lines again:

> Thy Travels dost thou boast o'er foreign Realms?
> Thou *Stranger* to the *World!* Thy Tour *begin;*
> Thy Tour through *Nature's* universal Orb.[77]

Young deemed terrestrial travel to be insufficient for a true anthropology, and in subsequent lines he attributed blindness to those whose minds had not floated through open space:

> And *Man* how purblind, if unknown the whole!
> Who circles spacious *Earth,* Then travels *here,*
> Shall own, He was never from *Home* before![78]

In contrast to the scholarly literature, which emphasizes the direct encounter with exotic terrestrial peoples, Young points to an anthropology that emerged only after a tour of extraterrestrial spaces and places had *resituated* the earth itself.

Young's work is but one example of a continent-wide trend toward using imagined space to redefine the meaning of humanity's spaces. A decade earlier, between 1732 and 1734, Alexander Pope, master poet and enthusiastic Newtonian, wrapped humanity within celestial space, writing in the poem "Essay on Man":

> He, who through vast immensity can pierce,
> See worlds on worlds compose one universe,
> Observe how system into system runs,
> What other planets circle other suns,
> What varied peoples circle every star,
> May tell us why Heaven has made us as we are.[79]

Like all heliocentric anthropologists, Pope used the celestial to situate the terrestrial, beginning not with the stars but out in the stars. Pope's celestial journey, in turn, structured the work itself. The first epistle is titled "Of the Nature and State of Man with Respect to the Universe." Thereon follow "Of the Nature and State of Man with Respect to Himself as an Individual" and "Of the Nature and State of Man with Respect to Society."[80] Pope's anthropology was, in effect, anchored to the imagined Heavens rather than the lived Earth.

The English were not alone in applying heliocentrism to anthropology. In France, the great Voltaire, a canonical literary anthropologist, was steeped in astronomy, applying the discipline's findings liberally to his writings.[81] Like Pope, Voltaire was a convinced Newtonian, which placed his mind squarely amid heliocentric space. In 1738, for example, he published *Elements of the Philosophy of Newton,* which became a landmark in France's acquisition of Newtonianism.[82] In 1752, in a different vein, he added *Micromégas,* a mordant tale in which a traveler from the Sirius system visits a friend on Saturn, after which they embark for Earth and discover that humanity is pathetic.[83]

The effects of this spatial change are also apparent in the work of the anthropologist and natural scientist the Comte de Buffon.[84] Buffon published numerous astronomical tracts that backstopped his anthropological works.[85] His *Natural History,* for instance—the first volume of which appeared in 1749—began with two methodological discourses. The first was the famous "On How to Study and Treat of Natural History," which laid the foundation for eighteenth-century research into the natural world. The second, "History and Theory of the Earth," is less renowned but extremely significant for heliocentric anthropology, because it constructed the world from a position so fully spatial that our planet did not yet exist.[86] For Buffon, the entire solar system began as a diffuse cloud of matter that coalesced into a system with a star and attending planetary bodies.

Astronomy's vast space figured as prominently in German anthropological debates of the eighteenth century. Johann Wolfgang Goethe, another literary anthropologist of note, was thoroughly versed in the celestial science, not only serving as the director of the observatory in Jena but also personally ordering (and reading) much of its library.[87] His verdict on the discipline: "Astronomy is, for me, so valuable, because it is the only science of them all that rests on generally accepted, uncontested foundations, with which and in complete security [it] progresses ever further into infinity."[88] Goethe merely echoed general sentiments. In 1747, the year before Goethe's birth, the astronomer and mathematician Abraham Gotthelf Kästner exclaimed in the periodical *Hamburg Magazine* (*Hamburgisches Magazin*): "Do you wish to know how far the powers of human understanding extend? You must study astronomy!"[89] Astronomy was now the key to revealing new perspectives on humanity's remarkable abilities. The little God had ceased to tremble.

Thanks to homogeneity's early modern diffusion, extraterrestrial space became anthropology's constant companion—and vice versa. Indeed, in 1787, Immanuel Kant, a giant of philosophical anthropology, folded celestial space into his sense of self, writing in *Critique of Practical Reason*: "Two things fill the mind with ever new and increasing admiration and awe, the more often and steadily reflection is occupied with them: the starry Heaven above me and the moral law within me. Neither of them need I seek and merely suspect as if shrouded in obscurity or rapture beyond my own horizon; I see them before me and connect them immediately with my existence."[90] Kant could put the starry Heavens before his eyes, because he knew astronomy well, having not only taught classes on the subject at the University of Königsberg but also published in 1755 a cosmological work, *General Natural History and Theory of the Heavens*, in which he described the new universe and also credited the astronomers Tycho Brahe, John Flamsteed, Christiaan Huygens, and Edmond Halley with having elaborated it.[91] It was astronomers who produced the context for what Kant "saw" when he looked out.

Against this backdrop, I turn to Johann Gottfried Herder, Kant's greatest student and one of the early modern period's great anthropologists. Born in 1744, he reached intellectual maturity in the second half of the eighteenth century, which gave him the advantage of retrospection, insofar as he digested a vast heliocentric literature before penning his own ideas on humanity. His main contribution to anthropology is *Ideas on the Philosophy of the History of Humankind*, an unfinished, four-volume work that appeared between 1784 and 1791 and, according to Goethe, was widely influential.[92]

The literature on Herder's anthropological thought is expansive, comprising studies by historians, theologians, philosophers, and literary scholars.[93] It is curious, however, that not one major work on Herder's anthropology mentions astronomy, in spite of numerous references in the *Ideas* to astronomical realms and themes. Indeed, the subtitle of the work's first chapter is "Our Earth Is a Star Among Stars," while the first line reads, "Our philosophy of the history of humanity must begin with the Heavens, if it is to be considered worthy of the name."[94] And with respect to our planet, Herder added (still within the first paragraph), "so must one behold it, in the first place, not as alone and lonely, but as part of a choir of worlds, among which it is placed."[95] Finally, at the end of this paragraph Herder presented a genealogy of his cosmology, citing Nicolaus Copernicus, Johannes Kepler, Isaac Newton, Christiaan Huygens, and Immanuel Kant, before adding in a footnote Johann Heinrich Lambert and an astronomer named Johann Elert Bode, whose astronomical contributions inspired Hegel.[96] (Returning to a point that I made in the introduction, I note that it was Bode who elaborated the theories that Hegel would use for his *Philosophical Dissertation on the Orbits of the Planets*.) Herder, like many thinkers of his era, relied on (and credited) astronomical thinkers when constructing his anthropology's space. Thus, his tour began.

I now present an overview of Herder's text before analyzing its main themes. Herder produced four volumes, each of which was divided into multiple parts and, in turn, into chapters. The first volume constructs the universe before moving to the earth, its terrestrial web of life, and its inhabitants.[97] Volume 2 discusses the earth's topographic features and offers a geographic tour of human diversity via descriptions of the physical appearances of the peoples from different parts of the globe.[98] Volume 3 presents histories of the ancient world, including sections on Asia, the Middle East, and the Mediterranean.[99] Volume 4 covers the history of Europe, including the migration of Germanic tribes, the rise of Christianity, the growth of European cities and economy, and, finally, the first hints of European exploration.[100] In sum, Herder began with a universe and considered the earth and its inhabitants only after he was done with the stars.

In order to underscore how radically Herder's perspective on the universe differed from Acosta's, I examine volume 1 of the *Ideas* in more detail. Herder divided this volume into four parts, with the first being dedicated to discussing the rise of heliocentrism and to detailing our solar system. He wrote:[101] "The earth has below it two planets, Mercury and Venus, with Mars (and perhaps another hidden above it), Jupiter, Saturn, and Uranus above—and whatever others may still be above, until the regular zone of the Sun trails off and the eccentric orbit of the final planets jumps off into the wild ellipses of the comets' trajectories."[102] In the third chapter Herder described how the earth began as cloud of dust, citing (not surprisingly) Buffon, Kepler, and Newton.[103] In the next he constructed the earth as a satellite under the subtitle "Our Earth is a sphere that rotates around itself and, with respect to the Sun [is] askew," which is, of course, exactly what Buffon had done.[104] Herder then continued his voyage home in the fifth chapter by detailing our planet's atmosphere and climates before explaining in the sixth how its varied topography supported biological diversity.[105] In the second and third parts of the initial volume Herder analyzed the diversity of plant and animal life, before discussing, at the very end, humanity's position within this terrestrial web. Herder, in sum, began out in space and looked back to his Earth. Indeed, in volume 2 he recognized having done so: "To this point, we have considered the earth in general, as the home of the human species, thereafter seeking to identify the position that the human being occupies in [the terrestrial] chain of living creatures. Let us now, after having fixed an idea of [humanity's] nature in general, consider the many guises in which [our] species appears on this rounded stage."[106] For Herder, humanity belonged to a planet, whose physical unity as an object (within a larger whole, of course) undergirded human diversity. Here I reach an important point. The Enlightenment's sense for diversity was *situated* by the Spatial Reformation's production of the space inside which the earth became a planet.[107] Thus, although humans may not look or act alike, they are all terrestrial creatures whose *location* can only be understood inside a human projection of extraterrestrial space.

As was the case in with Pope's "Essay on Man," Herder's extraterrestrial perspective penetrated the structure of his text. Having constructed our planet in the first volume, Herder examined in later ones the terrestrial realm's diversity. Far from anchoring anthropological unity to his theological concerns, as Acosta had done, Herder saw our planet as an entity that, in turn, made the human species' diversity possible. Along these lines, it is no accident that the Devil fails to appear in Herder's work, since his realm had been crushed under homogeneous space's heel. These religious points are particularly significant, because Herder was, like Acosta, a trained cleric, albeit a Protestant one.[108] Moreover, it is also notable that Renaissance anthropology plays no role in Herder's text. Throughout the *Ideas,* Herder cited eighteenth-century travel literature and world histories, including works by Georg Forster, James Cook, Carsten Niebuhr, and William Robertson.[109] And even in the *Ideas'* section on Mexico and Peru, he relied exclusively on eighteenth-century works, leaving Acosta and his cohorts unmentioned.[110] Herder's footnotes suggest, in short, that to the extent that his anthropology had Renaissance roots, these traced back to Copernicus rather than to Acosta.

Herder's anthropology was suffused with a spatial ethos that was alien to Renaissance anthropology. I illustrate the difference by returning to Herder's genealogy for anthropology, which runs from Copernicus to Kant. By beginning with Copernicus, Herder not only announced his allegiance to heliocentrism, but also made this cosmological system into his anthropology's cornerstone. He did not fully document his own heliocentrism's textual foundations but cited one text directly, Kant's *General Natural History.* Unlike Copernicus's *On the Revolutions of the Heavenly Spheres* or Newton's *Mathematical Principles of Natural Philosophy,* this text has long been considered a minor work—and rightly so, as even in the eighteenth century almost no one paid attention to it.[111] Given, however, that the *General Natural History* is one of only three cosmological works that Herder cited in all of volume 1, it obviously occupied an important position within his own anthropology and merits closer examination.

Although it is not considered an anthropological work, Kant's *General Natural History* is most definitely a heliocentric one.[112] It is divided into three parts and follows the same progression that was apparent in the other works that I discussed above. In part 1, Kant freed his mind from the earth's confines, discussing the universe as a system of stars, with each star serving as the center for a system of planets. In part 2 he explained how each system began as a nebula before condensing into a group of bodies, in accord with Newton's law of gravitation. (Known as the nebular hypothesis, it was first proposed in 1734 by the Swedish mystic and amateur astronomer Emmanuel Swedenborg.[113] Buffon, in turn, took up this idea for his *Natural History* and Herder, later, cited Buffon.) Having framed a universal spatial context for all solar systems, Kant turned to each system's parts, including suns, planets, moons, and comets, explaining how each fit into the whole. Thus, by the end of the second part of the text Kant had not just built our world, *but all possible worlds*—and this philosophical generality is crucial for understanding Kant's next step in part 3, the discussion of extraterrestrial life.

As I noted in this chapter's introduction, heliocentricity's anthropological implications emerged from the new cosmology's requirement that the mind leave its home planet. The human imagination's liberation from the tyranny of the terrestrial had, however, an additional effect: the growing certainty of extraterrestrial life's existence. Beginning in the late sixteenth century, Europeans began to ask whether other planets hosted intelligent life and, if so, what such life would be like. Known today as the "Plurality of Worlds" debate, the discussion raged through the seventeenth and eighteenth centuries and still has not ended.[114] This debate emerged directly from heliocentrism and cannot, therefore, be separated from its conception of space.[115] For example, one of the earliest "pluralists" was Giordano Bruno, who was both a heliocentrist and something of a geometer.[116] Early modern thinkers, in turn, thought the connection between heliocentrism and the plurality of being was obvious. Christiaan Huygens, the physicist and mathematician whom Kant and Herder cited, held: "It is scarcely possible that he . . . who believes with Copernicus that the earth we dwell upon is one of a number of planets that orbit the sun and receive all of their light from it, would not occasionally ponder that the remainder [of the planets] are, perhaps, not without their accouterments and ornaments, in the same manner as our globe."[117] Huygens was a well-known natural philosopher, and the work from which I have taken this quote, *Kosmotheoros,* occupied a prominent position within the "plurality" debate. It is, therefore, significant that in part 3 of the *General Natural History,* Kant tread exactly the same path as Huygens, moving rapidly from contemplating the stars to understanding them as centers for systems of planets.

The specific details of Kant's thought on this matter are not important. Suffice it to say that Kant believed it likely that life existed on other planets. Nevertheless, in the course of his life, he became increasingly reticent to consider what such life would be like, since humanity had no experience of it. Extraterrestrial life was, in essence, a formal possibility within the human projection of space—but that was all it could be, since geometric space itself inhered in the mind rather than in the cosmos. (In taking this position, Kant disagreed expressly with Huygens, who believed that all sentient beings cultivated geometry.) The most important aspect of part 3 is, however, its epigraph, which comprised a German translation of famous poetic lines:

> He, who through vast immensity can pierce,
> See worlds on worlds compose one universe,
> Observe how system into system runs,
> What other planets circle other suns,

What varied peoples circle every star,
May tell us why Heaven has made us as we are.[118]

Kant's quote from Pope's "Essay on Man" suggests, from another angle, the amalgamating power of heliocentrism's space. Whereas Pope framed his own anthropology with heliocentrism's extraterrestrials, Kant framed extraterrestrials with Pope's anthropology. Their minds' eyes floating through space, these eighteenth-century anthropologists looked *out* and *back*—and the continual exploration of this human space had profound anthropological consequences. As Leon Battista Alberti asked, "What now?"

Conclusion

I have argued that across the entire early modern period, anthropology, cosmology, and theology were interrelated, with changes in any one profoundly affecting the other two. Nothing I have said in this respect is new, but it builds on the arguments that Hans Blumenberg made decades ago about the interrelationship between anthropology and cosmology.[119] Following Blumenberg's lead, in this chapter I have overlaid the rich and lively literature on the rise of anthropology in the early modern period with the profound changes in cosmological thought that accompanied these changes. It was not simply that new questions and protocols arose within anthropology, as Harry Liebersohn has argued, but that the rise of a new view of the universe altered how humanity *knew things*—including itself, its own location, and, of course, its relationship to the divine.[120]

In this context I turn to a curious essay, *Dreams of a Spirit-Seer, Elucidated via Metaphysical Dreams*, which Kant published in 1766.[121] It is a review of a Latin text that Emmanuel Swedenborg contributed in 1758 to the Plurality of Worlds debate.[122] Although Swedenborg's work is quite a fun read, I mention only its title: *On the Earthly Planets in Our Solar System, And on the Earths in the Starry Heaven and their Inhabitants; Thereupon, On the Spirits and Angels from There Heard and Seen*. Since this title speaks for itself, I note only that Kant argued in his review that the doctrines of geometric space undermined Swedenborg's notion that the cosmos was filled with angels and spirits, since space had no place for such beings.[123] He then added a striking comment that illustrates the breadth of the Spatial Reformation's effect on the divine: "When one speaks of Heaven as the seat of the blessed, the popular imagination fancies to put it above itself, high in boundless universe. One does not consider, that our Earth, [when] seen from these regions, also appears as one of the heavenly stars, and that the inhabitants of other worlds could point to us, and with equally good reason, say: See there, the home of eternal pleasures, a Heavenly residence that is prepared to receive us one day."[124] Space had opened the possibility not only that true "others" were looking

Figure 32. Cover card to Andreae, *Newly Invented, Educational and Entertaining Astronomical Card Game,* 1719. Courtesy of Herzog-August-Bibliothek Wolfenbüttel.

back at us, but also that these beings had exported their own vision of the divine to the starry Heavens above and, in doing so, transformed our Earth into a divine place. Only a radically different view of the relationship between humanity, the cosmos, and God could sustain such a profoundly anthropological insight.

I now frame Kant's mature views on space against two eighteenth-century images that, together, illustrate space's centrality to emerging European traditions in pedagogy. As I noted in the previous chapter, John Locke was quite specific about the need for young people to pursue (in rough sequence) geography, astronomy, and geometry, if they were to become rational adults. In this context, I call attention to figure 32, which contains the cover card to "Astronomical Card Game," which Johann Philipp Andreae published in 1719 in Nuremberg.[125] The full deck contains forty-eight cards, each of which depicts a sign of the zodiac or some other heavenly body along a line that demarcates the ecliptic. The young mind was thus drawn into space via the games that it played. Moreover, the cover

Figure 33. Cover illustration to *New Children's Atlas*, 1776. Courtesy of UCLA Library Special Collections, Children's Book Collection.

card illustrates precisely the mixture of spatial thought with material culture that drove the reformation in space. One child holds a compass (the like of which we have seen before) and applies it to a celestial sphere, while the other child uses a telescope to observe the Heavens. Given the completeness of heliocentrism's victory in the previous century, the young observer could only have been looking *out* as opposed to *up*.

Homogeneous space's presence within (and effect on) pedagogy is even more obvious in the second image (figure 33), which contains the cover illustration to *New Children's Atlas*.[126] It was published in Amsterdam in 1776 as a work of geography. Thus, the image includes all the things that one would expect to see, including the express mixture of spatial knowledge with material culture. Rather than analyze the image in detail, I juxtapose it with the anthropological implications that the text itself ascribes to geography. It held: "This science, necessary and agreeable in whatever condition one be, is brought today to such a point of perfection that to our eyes the entire world is scarcely a large city and all the people that occupy it together a great family."[127] In this context, it is crucial to note that the cherubic Atlas in the image does not shoulder a terrestrial globe—which one would expect, given the work's topic—but a celestial globe. On the one hand, this is consistent with the traditional story of Atlas. On the other hand, it represents exactly how knowledge of both the earth's surface and of its inhabitants was now enveloped by a larger, nebulous sense of space.

Against this eighteenth-century backdrop, I turn to Alexander von Humboldt, the last and the greatest of the heliocentric anthropologists.[128] Born in 1769 in Berlin, he was a child of the Enlightenment and, concomitantly, absorbed a heliocentric intellectual apparatus from both his tutors at home and his studies at a variety of institutions, including the University of Göttingen, where he studied under the anthropologist Johann Friedrich Blumenbach.[129] After ending his formal studies, Humboldt headed to the New World, where he traveled, observed, and catalogued the flora, fauna, and peoples that he encountered. In 1804, after returning to Europe, he and his traveling companion, Aimé Bonpland, published reports of their experiences under the summary title *Voyage to the Equinoctial Regions of the New Continent,* a work that reached 30 volumes.[130] All the tomes appeared in Paris, where Humboldt had settled and would remain until 1827, continually interacting with luminaries in astronomy, anthropology, mathematics, and geography.[131] In short, in addition to everything else that he already was, Humboldt became the most broadly educated anthropologist of his age.

While keeping the richness of Humboldt's intellectual background in mind, I turn to his final natural scientific work, *Cosmos.*[132] Written after he left Paris for Berlin, it was published in five volumes between 1845 and 1862 and became wildly popular, landing on coffee tables on both sides of the Atlantic in multiple translations, including one in Spanish that appeared in Mexico in 1851.[133] Humboldt's text expressly incorporated both the history of European expansion and the rise of the new astronomy within its vision of the cosmos.[134] For that reason, although it is often celebrated as a work of natural science, it should also be read as a contribution to heliocentric anthropology. In the first volume Humboldt wrote:

> We begin with the depths of outer space and the regions of the remotest nebulae, descending step-by-step through the star cluster to which our solar system belongs, [and down] to the terrestrial spheroid around which flow air and oceans, [considering] its form, temperature, and magnetic tension, [as well as] the fullness of life that flourishes on its surface, vivified by light. Thus, with only a few strokes, is a portrait of the world painted that [includes] both the immeasurable regions of space and the tiny microscopic organisms of the animal and plant kingdoms that inhabit our inland waters and the weathered surfaces of rocks.[135]

This is heliocentric anthropology distilled, with boundless and bounded, massive and miniscule being incorporated into one sweeping view that is *situated* by Humboldt's absolute command of unseen space.

Like Pope and Herder before him, Humboldt contextualized terrestrial knowledge against his capacious vision of outer space. It is worth noting, in this context, that Humboldt mentioned humanity only at the end of the first volume—and then in twenty pages that cover (roughly) these topics: "Universality of Animal Life," "Geography of Plants and Animals," "Floras of Different Countries," "Man," "Races," and "Language," before tracing the geographic discipline's history in the second volume.[136] Astronomy's great, unseen spaces were both textually and logically prior to the geographic space in which Humboldt imagined human beings to flourish. The celestial-to-terrestrial progression is also apparent in volumes 3 and 4, which are dedicated, respectively, to Humboldt's observations on, as he put it, "regions of cosmic phenomena" and "regions of telluric phenomena."[137] (The fifth volume, which was published posthumously, is an extension of the fourth.) Humboldt, like all his heliocentrist predecessors, constructed humanity's *place* by projecting his mind outward and looking back.

Humboldt's *Cosmos* offers an ideal perch from which to reflect on the mismatch between the Renaissance and Enlightenment readings of anthropology. The root incompatibility between them is a product of the literature's failure to grapple with anthropology's profoundly retrospective intellectual currents. All eighteenth-century anthropologists understood that they had predecessors and identified them via the themes that dominated their own educations and experience—that is, they tromped into the past looking for traces of who they imagined themselves to be. For heliocentric anthropologists, this could mean praising Copernicus and his heirs while circumscribing or even ignoring Acosta.[138] (Retrospection cut both ways, of course. In 1724, the French Jesuit Joseph-François Lafitau published an anthropological work in which he defended religion against pesky French atheists by pointing to New World peoples' rationality and also citing Acosta's views. Copernicus went unmentioned.)[139] Thus, a flaw similar to the one that bedevils contemporary approaches to anthropology also characterized Enlightenment perspectives on the sixteenth century. The enlightened culled the geocentric context from earlier anthropology—just as modern interpreters have cast aside the heliocentrism that the Enlightenment celebrated. Herder's genealogy for anthropology was, in this sense, as unfair to Acosta as contemporary readings have been unfair to Copernicus.

There is, however, a way out of this cul-de-sac: scholars must put both the Renaissance and the Enlightenment approaches into the broader history of space. Between 1500 and 1800, a variety of works on spatial issues that have nothing to do with contemporary anthropology's image of itself influenced early anthropological debates as profoundly as the texts that scholars have enshrined in the contemporary canon. Thus, attempts to read anthropology by moving along a daisy chain of approved *anthropological* texts excludes currents that may no longer be relevant to the discipline but once had powerful effects. This is not to say that canons are bad—writing the history of a discipline without one is impossible—but that they must be used with sensitivity to how canon building can narrow our historical vision. Perhaps the clearest example of the problem is that the vast literature on Herder's anthropology makes nary a mention of astronomy, cosmology, or heliocentrism, in spite of the *Ideas*' multiple references to all of them.[140] In a sense, both the starry Heavens and their assumed space have been filtered out of anthropology's history.

In this chapter I have traced how the new astronomy brought the Spatial Reformation to early modern anthropology. My emphasis on the Heavens should not, however, be taken as exclusive (or even conclusive) given the many contexts that I have covered throughout this book.[141] Nevertheless, the recovery of astronomy for the history of anthropology suggests a way to reframe the separate Renaissance and Enlightenment readings in a way that respects each era's emphases, even as it situates both of them within a great reformation in space. If the historical discipline is to take seriously the idea that the years between 1500 and 1800 formed an anthropological whole, then it makes sense to understand anthropology's rise in terms of the spatial themes that cut across the entire discussion.

With this in mind, I suggest that the history of early anthropology should be understood with respect to two shocks that took centuries for the early modern mind to process. One was the unexpected landing in the New World, which was a product of early modern spatial thought's command of space, while the other was the rise of a heliocentrism that emerged from the same spatial currents. Although both shocks were crucial to early anthropology, the contemporary literature has emphasized the empirical aspects of the New World landing to the detriment of the spatial thought that had made the entire enterprise possible. Just how complete space's exclusion of from anthropology has been is exemplified in Claude Lévi-Strauss's canonical *Structural Anthropology*, which appeared in 1958.[142] While summarizing his own views on the discipline, he noted: "*The anthropologist is the astronomer of the social sciences*: he is charged with finding a meaning in the extremely different configurations—according to their size and their distance—of those things that are immediately before the viewer."[143] This sentence highlights exactly what separates modern anthropology from its early modern ancestors—a near total spacelessness. Lévi-Strauss not only put astronomy at the end of his work rather than the beginning, but his vision of the celestial realm's relationship to humanity was devoid of the space inside which humanity had once found meaning for itself. Anthropology's mental cosmos has, in short, gotten much smaller.

Along these Lévi-Straussian lines, we can begin to understand why contemporary scholars have concentrated so intensely (and exclusively) on culture. If anthropology is about understanding what is before our eyes, then the Heavens and their potential "others" are of little consequence. In view, however, of what I have argued throughout this book, a true history of anthropology must consider how discussions of humanity were implicated in the collective vision of unseen spaces and places, even if the contemporary discipline has expelled the cosmos from its purview. The shift toward contemporary cultural anthropology is a product of changes in Western thought that produced, as I have explained, Euclid's demotion within modern mathematics. I turn to the broader historical implications of this final shift in this book's conclusion. At this point, however, I suggest that anyone who would explore anthropology's beginnings should glance, first, to the night sky, since the early modern vision of humanity here was rooted in what people once thought they saw out there.

Conclusion

Prosaic Reflections

> It is, therefore, quite clear in the results that we are going to enumerate, that it is not a question of the concrete notion of our space, but of a purely logical notion [that is] defined in accordance with certain conventions.
>
> —Paul Tannery, "La Géometrie imaginaire et la notion d'espace" (1877)

Between 1350 and 1850, European thinkers applied Euclidean space so thoroughly to the relationship between humanity, the cosmos, and God that they expelled the latter two from their mental cosmos, only to leave the former floating amid *nothing*. Friedrich Nietzsche (as always) was on the case, writing in 1882:

> "Where has God gone?" he cried. "I shall tell you. We have killed him—you and I. We are his murderers. But how have we done this? How were we able to drink up the sea? Who gave us the sponge to wipe away the entire horizon? What did we do when we unchained the earth from its sun? Whither is it moving now? Whither are we moving now? Away from all suns? Are we not perpetually falling? Backward, sideward, forward, in all directions? Is there any up or down left? Are we not straying as through an infinite nothing? Do we not feel the breath of empty space?"[1]

After noting the disorientation that humanity felt within what was now its very own space, the hammer-wielding philosopher added, "God is dead. God remains dead. And we have killed him."[2]

Nietzsche offers an ideal perspective from which to evaluate the profundity of homogeneous space's effects on the fraying Western triad. His reference, in the quote above, to the oceans and the horizon is directly related to humanity's having learned to see Earth from *above*. Concomitantly, the mention of our planet's repositioning with respect to other bodies illustrates how space's rise forged a new cosmological whole—the full embrace of which was possible only for those people who had already been drilled in homogeneity's fundamentals. Finally, Nietzsche's reference to up and down's dissolution into an infinite nothing echoes Friedrich Schiller in that it brings humanity nearer to the nothingness that, according to the Bible, had preceded Creation itself. Still, where Schiller concluded, "And in the abyss lives Truth," Nietzsche sensed a restlessly inventive (and mere) humanity to be lurking.[3] Space's renovation of anthropology was complete.

If we juxtapose Nietzsche's vision of space to this book's epigram, in which Fyodor Dostoevsky has Ivan Karamazov point to a perceived gap between God and Euclidean space, we can see how early modern thought's profoundest problem never disappeared. No one, from Cusa to Newton and up through Kant, could relate the divine to empty space in a way that made any sense—and in the course of the nineteenth century people elaborated a variety of responses to this problem. For example, whereas Nietzsche used space's humanness to dispatch God, Dostoevsky saved God by limiting humanity exclusively to its own space. Thus spoke Karamazov: "I have a Euclidean earthly mind, and how could I solve problems that are not of this world? And I advise you never to think about it either, my dear Alyosha, especially about God, whether he exists or not. All such questions are utterly inappropriate for a mind created with an idea of only three dimensions."[4] For Dostoevsky, space's merciless three-dimensionality required that God's existence be elevated to a different plane, where the most divine of beings could, perhaps, breathe freely again. Dostoevsky intuited, in short, that God needed to remain ensconced within the "not nowhere," in order to survive space's depredations. Of course, given the nature of early modern thought's development, this "not nowhere" could not long survive Euclid's final demotion.

Both Dostoevsky and Nietzsche were children of the Spatial Reformation. Nevertheless, as this chapter's epigram from Paul Tannery makes clear, they lived in a time when homogeneous space's underpinnings were giving way, before the onrush of a new geometry. Dostoevsky and Nietzsche are thus best seen as transitional figures who lived just as early modernity's space was supplanted by a more mathematically sophisticated successor. Along these lines, we can characterize their historical significance in a way that is similar to my earlier assessment of Nicholas of Cusa. As I noted, the Renaissance Cardinal's vision of space did not quite *fit* with traditional medieval yearnings for the divine, with the result that we moderns see him in Janus-faced terms. For Dostoevsky and Nietzsche, meanwhile, the underlying intellectual situation, although subtly reversed in content, remained formally similar, insofar as any attempt to comprehend the divine's relationship to space no longer *fit* with the ever more expansive projection of a human space onto everything. Geometric homogeneity supplanted the divine, in the end, but this triumph came only just before earthly Euclidean minds confronted even more radical conceptions of space.

The history of space's absence from contemporary studies of modern intellectual history has largely precluded the cultivation of "early modernist" readings of Nietzsche and Dostoevsky. I suggest, therefore, that if we pursue space forward, we will be able to reflect on both of these thinkers in a new way. Given, however, the normal page limitations, I examine only Nietzsche here and leave a full reexamination of Dostoevsky to others.[5] According to regnant scholarly interpretations, Nietzsche is a prophet of postmodernity, one of the first thinkers to liberate European thought from the linearity that had undergirded the West's fondness for metaphysics. There is some truth in this belief, insofar as postmodernists such as Jacques Derrida have laid claim to Nietzsche as a forerunner. At issue here, however, is whether postmodernity's claims are compatible with Nietzsche's participation in a culture of space from which contemporary theorists are, at least, partly insulated. Based on what I have argued previously, I insist now that postmodern theorists have failed to recognize how they look back to Nietzsche from a peculiar moment, in which Euclid's *Elements* is completely accessible, while this work's spatial doctrines, along with the intellectual problems that once attended them, have become innocuous.

Against this alternative historical backdrop, Nietzsche's atheism is especially significant, since scholars often see it as pointing the way beyond modernity. The dyspeptic Prussian's insistence on understanding the divine as a cultural phenomenon, however, was not an incipient postmodern break with modernity, but represented the final stage of early modern space's marginalization of the divine. In this vein, it is significant that Nietzsche's connection to Euclidean space has been overshadowed by his tendency to substitute snark for argument. The greatest of the immoralists adopted an oblique, oracular voice, preferring to issue nonlinear blasts rather than to construct logical explications. This rhetorical tick has led contemporary commentators to see Nietzsche as a destructor of Christianity's logically infused meta-narratives. The contemporary interpretation is not wholly wrong, but it is woefully incomplete. Nietzsche's view of the divine as a disruptive *human* force, whose periodic eruptions infused the mundane with almost godlike creative energy, is consistent with precisely the spatial secularization that I have explored throughout this book. Put another way, Nietzsche's humanizing of the divine was an outgrowth of the spatial trends whose origins I have traced back to the fourteenth century.

Contemporary theory reads Nietzsche, however, in an effectively nonspatial way. If we combine this chapter's epigraph from Paul Tannery with Claude Lévi-Strauss's characterization of anthropology as the astronomical study of what lies before our lies, with which I concluded the previous chapter, we see that the problem of *spatium absconditum* is a relatively recent phenomenon. Contemporary spatial sense is simply bifurcated in a way that Nietzsche's Euclidism was not. (There is no evidence that Nietzsche was conversant in the latest mathematics, or even that he cared in the slightest about its innovations.) Moreover, this contextual difference with respect to geometry suggests an alternative approach: rather than think in terms of modern versus postmodern, with Nietzsche pegged to the latter, we may wish to privilege the shift from a Euclidean space to a post-Euclidean one, with Nietzsche's religious barbs tying him firmly to traditional geometry.[6] Thus, if we read Nietzsche through the history of space, he becomes an early modern thinker who reveled in the cultural wreckage that five centuries of Euclidean dominance had left behind.

Revaluing Space

Having drawn a line the interpretive sand, I turn to the problem of postmodernism. A product of multiple intellectual currents that began to emerge in the fin-de-siècle, before cutting across the twentieth century's world wars, postmodern thought is exceedingly complicated. I cannot offer a complete history but refer the reader to the literature that has emerged over the past three decades.[7] My working definition comes from the postmodernist Jean-François Lyotard, who in 1979 wrote:

> I define postmodern as incredulity toward metanarratives. This incredulity is undoubtedly a product of progress in the sciences: but that progress in turn presupposes it. To the obsolescence of the metanarrative apparatus of legitimation corresponds, most notably, the crisis of metaphysical philosophy and of the university institution which in the past relied on it. The narrative function is losing its functors, its great hero, its great dangers, its great voyages, its great goal. It is being dispersed in clouds of narrative language elements—narrative, but also denotative, prescriptive, descriptive, and so on. Conveyed within each cloud are pragmatic valencies specific to its kind. Each of us lives at the intersection of many of these. However, we do not necessarily establish stable language combinations, and the properties of the ones we do establish are not necessarily communicable.[8]

For Lyotard, postmodernity was built on a linguistic foundation that hosted multiple possible platforms—including some mathematical ones, such as algebra, while excluding traditional geometry's three-dimensionality.[9] Rather than delve into the multiple traditions in linguistics and semiotics that also sustained Lyotard's perspective, I underscore that Lyotard's emphasis on language was, to no small degree, made possible by the deliberate marginalization of three-dimensional space.

Against this modern backdrop, it is crucial to understand that postmodernity's rejection of Euclidean space originated in profound changes within nineteenth-century mathematics. Here, I pick up the mathematical thread that I have pursued previously. In the early seventeenth century, geometry ambled toward abstraction, as mathematicians such as René Descartes developed ways to subject Euclid's space to equations. This shift was crucial for ushering traditional spatial analysis beyond the visibly intuitive, with the search for new, more abstract methods leading, ultimately, to Isaac Newton and Gottfried Leibniz's independent inventions of the calculus after 1660.[10] Hence, by the middle of the seventeenth century the Spatial Reformation's initial phase was over, with Newton and Leibniz confirming the change.[11] Thus, even if Euclid's great work was still studied, many and varied innovations in mathematics slowly shifted this text's intellectual position, making it into a prefatory, pedagogical work, rather than an essential mathematical one.

Euclid's post-1650 *resituating* was, however, not the end of the story. In the early nineteenth century, pioneering mathematicians such as Nikolai Lobachevski and János Bolyai began to critique the assumptions that underlay Euclidean space in essays that appeared mostly in the 1830s.[12] Thus began another great renovation, as mathematicians ruthlessly undermined Euclidean space's chief assumptions. The crowning result was Bernhard Riemann's oral presentation in 1854, at the University of Göttingen, of his second doctoral dissertation (the *Habilitationsschrift*), in which he elaborated the first complete and logically consistent non-Euclidean geometry.[13] This text would ultimately be published in 1867, after which point there was no going back.[14] Having presided regally for five centuries, Euclid's reign was over.

Both the time and place of Riemann's triumph are important to our understanding of postmodernism's dependence on the rise of new visions of space. Riemann's talk inaugurated another viciously corrosive process that continued up to 1899 and was centered on the University of Göttingen. This particular story begins in the early nineteenth century, when the young institution (it was founded in 1737) incorporated in its ranks one of the greatest mathematicians of all time, Carl Friedrich Gauss (1777–1855).[15] Gauss is a pivotal figure in the history of mathematics, as he made fundamental contributions to multiple areas, including especially the study of geometric solids. Moreover, with the Spatial Reformation in mind, Gauss becomes particularly important because he not only oversaw Riemann's studies but also suggested the topic for the epochal *Habilitationsschrift*.[16]

In the second half of the nineteenth century, the University of Göttingen became the center of the mathematical universe, as many of Europe's finest mathematicians either studied or taught there. Riemann eventually became the head of the mathematics department, but he fled the university when war broke out in

1866 between Prussia and Hanover. That year also brought the end of Riemann's life, as he died of tuberculosis while traveling through Italy. Riemann's home institution nevertheless continued to attract top-tier mathematical talent, as exemplified by its hiring in 1886 of Felix Klein, who was poached from the University of Leipzig, and whose mathematical contributions were as substantial as those of Riemann and Gauss.[17] The greatest mathematician that Göttingen lured to town, however, was David Hilbert, whom Klein recruited in 1895 from the University of Königsberg. Among Hilbert's seminal contributions is what is now known as the axiomatization of geometry.[18]

David Hilbert put paid to the Spatial Reformation, as he finally and completely robbed geometry of the intuitive sense for space that had shaped European thought and culture since the late fourteenth century. In 1899, he published a crucial work, *Foundations of Geometry*, which established that geometric axioms cannot be understood as self-evident truths but must be viewed as linguistic descriptions of the human intuition of space.[19] Accordingly, Hilbert held that only the logical relationship between geometric arguments was significant—and not the content of any term, such as point, line, or plane. Indeed, these words could be replaced by any randomly chosen ones without undermining the system's rigor.[20] Hilbert's insight is particularly significant for understanding the Spatial Reformation's end, because with it he established that geometry does not deal with space but only with logic. Thus, the *Elements*' first line, "A point is that which has no part," was denuded of all philosophical significance, with the most crucial result being that early modern *space making* had become logically impossible.

I cannot trace the history of mathematics beyond Hilbert's insights into geometry's foundations. Instead, I note that the reevaluation of three-dimensional geometric space bore fruit in many areas, including additional advances in mathematics, the rise of a new physics, and, as I explain, the emergence of a new philosophy. With respect to mathematics, the twentieth century brought elaboration of the notion of Hilbert space, in which Euclidean space is generalized to include an infinite number of dimensions, by John von Neumann, a Hungarian who had studied in Germany.[21] Mathematics had come a long way from imagining points, lines, planes, and spheres within an idealized realm. Indeed, space was now so fully transformed that the great *nothing* in which Nietzsche had found himself only a few decades before could no longer even be imagined visually. As an example of the change's radical nature, consider this comment from Paul Tannery, from whose work this chapter's epigraph was taken: "Now, geometry, whether Euclidean or not, assumes completely implicitly that any figure can be transported through space in all possible manners without losing its shape. The supposition to the contrary had, at first sight, seemed so absurd that up to Riemann, perhaps no one even dreamed of it."[22]

Tannery's mathematical reflections bring two issues into focus. First, in a backward-looking sense, Riemann and company obliterated Platonism's remnants, insofar as even the most eternal of Plato's forms were now subject to change. Second, keeping in mind the other changes that were still to come, non-Euclidean geometry's rise breathed life into the next revolution in science, which came with Albert Einstein's reordering of traditional visions of space and time's relationship to motion. Thanks to the rise of modernist mathematics, the human mind constructed (and then ascended to) a different plane in which a once immobile and unchanging space itself seemed susceptible to change. Humanity's *All* had thus become quite a different thing.

Two-Dimensionality Returns

Three-dimensional space's precipitous decline between 1850 and 1900 impelled profound philosophical changes that must be taken into account if we are to understand postmodern thought's emergence. Here, I turn to Edmund Husserl, who was one of Europe's most important philosophers of the early twentieth century. Given that I have identified him as a philosopher, it may come as a surprise that Husserl was a trained mathematician, as he studied mathematics at the universities in Leipzig, Berlin, and Vienna. In 1883, he received a doctorate in mathematics from the University of Vienna before traveling to Berlin to work with the mathematician Karl Weierstrass. Within a year, however, he returned to Vienna, where he entered into postdoctoral studies in philosophy. In 1886, he moved to the department of philosophy at the University of Halle, where he completed his own *Habilitationsschrift.*

Husserl's education and background positioned him to transfer mathematics' most recent spatial insights to philosophy. An overview of the titles of his works illustrates what happened. Husserl's *Habilitationsschrift* bears the unmistakable imprint of mathematics, as its title reads, *On the Concept of Number, Psychological Analyses.*[23] Having studied in Vienna and Berlin, he was heavily influenced by the rise of psychology as an academic discipline. It was, therefore, natural for him to oscillate between European culture's longstanding nurturing of "inner" experience and mathematics' search for generality. That particular impulse is also clear in his next work, *Philosophy of Arithmetic, Psychological and Logical Investigations.*[24] Although Husserl continued to flirt with psychology, this text marked, on the whole, a move toward the greater emphasis on logic. This latter development was punctuated in the period 1900–1901, with the publication of Husserl's great two-volume work, *Logical Investigations, Part I: Prolegomena to a Pure Logic* and *Logical Investigations, Part II: Investigations in Phenomenology and Cognition.*[25] In this text Husserl elaborated a fully philosophical doctrine that he called phenomenology.

Over time, Husserl transformed his great work's most basic insights into a school that became the most powerful philosophical force on the Continent.

The fin-de-siècle publication of the *Logical Investigations* marked the watershed in Husserl's intellectual development. First, the text presented a mature vision of phenomenology, which Husserl would cultivate for almost the next four decades. Consistent with my discussion of Euclidean space's demise, this work's initial volume articulated a "deracinated" logic that was not dependent on an intuitively accessible three-dimensional space. The succeeding volume, in turn, used mathematics' independence from Euclidean space to part company with the dualities that had characterized European philosophy since René Descartes's construction of the philosophical subject. Phenomenology established (or so Husserl believed) the *objectivity* of the subject's experience by limiting philosophy to the study of phenomena. Put another way, phenomenology placed under a common rubric both the subject's experience of objects in the "external" world and its "internal" experience of itself as an apperceptive unity. Husserl would call this approach the "epoché," or a bracketing that unified philosophy by limiting it to the logical analysis of phenomena. As a consequence, there is no "room" within phenomenology for a humanly projected, three-dimensional space.

The turn to a new century also marked an important moment in another way for Husserl. After publishing his *Logical Investigations,* he left the University of Halle, where he had been an assistant professor of philosophy, and proceeded to the University of Göttingen, where he occupied a position as associate professor and also became personally acquainted with David Hilbert.[26] Living in this effervescent environment could only have pushed Husserl's thought more deeply into post-Euclidian territory, and he remained in Göttingen until 1916, when he became a full professor of philosophy at the University of Freiburg. While in Freiburg, Husserl continued to elaborate his system and also guided the studies of many important philosophers, including the only one whose significance for contemporary postmodern theory exceeds his own, Martin Heidegger. I cannot pursue that thread, however, but circle back to my main point: Husserl's arrival in Göttingen coincided with the moment when the city's university defined the mathematical world's pinnacle. It is thus not coincidental that much of what would become postmodernism's conceptual apparatus was borrowed from post-Euclidean mathematics.

Husserl's phenomenology was as logically formal and as *spaceless* as geometry had only just become. I illustrate my point by examining two of Husserl's later works, the *Cartesian Meditations,* which appeared originally in French in 1931, and *Origin of Geometry,* which was published in German in 1939 in the Belgian journal *Revue Internationale de Philosophie.*[27] My choices are deliberate. The first work allows us to understand how Husserl fit his phenomenology into modern philosophy, whose beginnings scholars traditionally ascribe to the works that

René Descartes wrote in the decades before 1650. The second work, meanwhile, offers a way to understand the interrelationship between postmodern views of space and postmodernism's preferred approach to periodization. The disregard for homogeneous space's *historical* significance traces back, in my view, to Husserl's attempts to justify his phenomenology without reference to the early modern embrace of three-dimensional space.

Husserl's *Cartesian Meditations* highlights a shift that was crucial to the postmodern rejection of Descartes. In the text, Husserl expressly noted geometry's significance for Descartes's thought, even as he himself reduced the seventeenth century's interest in geometry to its logic:

> Obviously, it was, for Descartes, a truism from the start that the all-embracing science must have the form of a deductive system, in which the whole structure rests, *ordine geometrico*, on an axiomatic foundation that grounds the deduction absolutely. For him a role similar to that of geometrical axioms in geometry is played in the all-embracing science by the ego's absolute certainty of himself, along with the axiomatic principles innate in the ego—*only this axiomatic foundation lies even deeper than that of geometry and is called on to participate in the ultimate grounding even of geometrical knowledge.*[28]

Husserl followed Hilbert in seeing geometry as a collection of axioms that were fundamentally devoid of *space*. He then imposed this reading onto a Cartesian system that had struggled with space's corrosive effects on traditional visions of God and the cosmos. (I have more on this below.) Indeed, Husserl went further, as he used a pre-geometric *logic* as an alternative foundation for the history of human thought.

Husserl reduced geometry's past to nothing more than a prologue to contemporary debates. I illustrate this point by considering Husserl's assessment of Descartes. Within Cartesian philosophy the subject had to be isolated from the experiential world, because this would otherwise have limited the subject's knowledge to the physical—and such a limitation made it difficult to argue that God existed and that the soul was immortal, since both of these ideas were *beyond* experience. Descartes's philosophical maneuver created the problem, however, that the subject's "internal" register of itself was not sufficiently general to justify the existence of the world that it was ostensibly experiencing. Husserl reframed the issue with his epoché:

> But perhaps, with the Cartesian discovery of the transcendental ego, *a new idea of the grounding of knowledge* also becomes disclosed: the idea of it as a transcendental grounding. And indeed, instead of attempting to use *ego*

> *cogito* as an apodictically evident premise for arguments supposedly implying a transcendent subjectivity, we shall direct our attention to the fact that phenomenological epoché lays open (to me, the meditating philosopher) *an infinite realm of being of a new kind,* as the sphere of a new kind of experience: transcendental experience.[29]

As I have explained, the epoché is a logical frame that reduces the subject's experience both of itself and of external stimuli to a single pool of phenomena. This approach represents a plausible extension of Cartesian thought, insofar as Descartes kept the world at bay, precisely so that (a) the subject could authorize itself and (b) God could be justified primarily in logical, rather than in physical terms. In this sense, therefore, Cartesianism was already implicitly two-dimensional. (Descartes could not rebel further against Euclidean space, since there was, as yet, no mathematical alternative.) The Husserlian approach to the subject is, in contrast, *explicitly* two-dimensional, insofar as the epoché functions according to the non-Euclidean logic that had only just conquered three-dimensional space.

Against this backdrop, I turn to the second text. In 1939, the *Origin of Geometry* was published posthumously (Husserl having died the year before). This short work was originally an appendix to a larger one, *The Crisis of the European Sciences and the Philosophy of Phenomenology,* which was completed in 1936 but appeared in its entirety only in 1954.[30] In the larger work, Husserl attempted to refound philosophy as a source of secure knowledge—and he did so expressly against the ominous backdrop of Nazism's rise. However significant are the interpretive issues that arise from Husserl's difficult situation, I must set aside the dark politics of the 1930s. Instead, I juxtapose his essay with my vision of the Spatial Reformation. Two issues are important here. First, I explain how the *Origin of Geometry* betrays modern mathematics' crucial effect on Husserl's (mis)reading of Descartes. Second, I explain how *Origin of Geometry* cast this reading in amber, given Husserlian thought's influence on key postmodern thinkers.

Postmodernism's assessment of its own historical position emerged from the suppression of geometry's historical development. Thus, Husserl projected his era's view of geometry's meaning onto the discipline's nebulous historical origins and, in so doing, erased the many and various debates about space that had occurred in between. I illustrate this point with a quote:

> The question of the origin of geometry (under which title here, for the sake of brevity, we include all disciplines that deal with shapes existing mathematically in pure space-time) shall not be considered here as the philological-historical question, i.e., as the search for the first geometers who actually uttered pure geometrical propositions, proofs, theories, or for the particular propositions they discovered, or the like. Rather than

> this, our interest shall be the inquiry back into the most original sense in which geometry once arose, was present as the tradition of millennia, is still present for us, and is still being worked on in a lively forward development; we inquire into that sense in which it appeared in history for the first time—in which it had to appear, even though we know nothing of the first creators and are not even asking after them. Starting from what we know, from our geometry, or rather from the older handed-down forms (such as Euclidean geometry), there is an inquiry back into the submerged original beginnings of geometry as they necessarily must have been in their "primally establishing" function.[31]

These words reveal two things about Husserl's historical imagination. First, his approach to the past is not "philological-historical," as he put it, which is to say that it takes little account of time's actual passage. Indeed, Husserl seems to reject time, with the result that geometry not only boasted no *historical* development but was also reduced to the contemporary world's idea of how things *ought to have been*.[32] Second, as Husserl made clear in the next sentence, he was well aware of geometry's "lively forward development," and by mentioning the intellectual ferment whose triumphs he had just witnessed, he exposed how his sense of *how things must have been* emerged from his understanding of *what geometry had just become*. There is, I suggest, a "Hilbertian" backflow at work, in which geometry's past was reorganized via mathematics' present.

Husserl's views on geometry's history then led him to the history of language. Here, the philosophical shift away from Euclidean space becomes palpable. Husserl's idea was that human beings constituted their world *only* linguistically, that is, without reference to space. He put his case thus:

> It is precisely to this horizon of civilization that common language belongs. One is conscious of civilization from the start as an immediate and mediate linguistic community. Clearly it is only through language and its far-reaching documentations, as possible communications, that the horizon of civilization can be an open and endless one, as it always is for men. What is privileged in consciousness as the horizon of civilization and as the linguistic community is mature normal civilization (taking away the abnormal and the world of children). In this sense civilization is, for every man whose we-horizon it is, a community of those who can reciprocally express themselves, normally, in a fully understandable fashion; and within this community everyone can talk about what is within the surrounding world of his civilization as objectively existing.[33]

Thus, for Husserl, nothing actually exists within space, but all is encoded exclusively in language. It is obviously true that language is essential to the construction of a world. As I have shown, however, with respect to the culture of the early modern period, this view is incomplete, at best, as Euclidean space permeated every aspect of early modern thought and culture.

Superficiality and the Spatial Turn

In this book's introduction, I situated the Spatial Reformation against regnant approaches to the early modern period. As I explained there, historians generally place this period between 1400 and 1800 and call everything that comes after Napoleon Bonaparte's coup in 1799 "modern."[34] In some contemporary circles, however, it is common to break up the time that separates us from Napoleon by pasting a postmodernity on the end of modernity. It is not clear when postmodernity began, but scholars look to late nineteenth century, when seeking its origins—and especially to Friedrich Nietzsche.[35] They do so, as my earlier quote from Lyotard illustrates, because they associate Nietzsche with their preference for nonlinear approaches to thought. I argued above, however, that although Nietzsche's view of the divine was nonlinear, his space remained essentially Euclidean. And since the divine was, for Nietzsche, all too human, any religious eruptions could only have occurred within this very human space. This latter point casts light on postmodernism's misreading of its relationship to not only Nietzsche but also the entire early modern world.

Postmodernity actually fits poorly with still-dominant (and profoundly historical) approaches to the early modern period that I have discussed previously. As I explained in the introduction, the contemporary vision of the early modern emerged, in part, from the Annales School's *longue durée* and thus combines a broad chronological sweep (roughly from 1500 to 1800) with an emphasis on deep historical rhythms. Postmodern theorizing, however, is based on a sense of rupture that not only breaks with all previous rhythms—be they early modern or modern—but also overwrites them as it directs its gaze into the past. Thus, the contemporary historical discipline's construction of an early modern period, in which older beliefs and practices endured, remains profoundly in tension with the postmodern preference for declaring the present moment to be philosophically unique. This tension is manifested, as I show, in three-dimensional space's conspicuous absence from postmodern theory.

Along these lines, I stress that postmodernity is not a historical-empirical category of analysis, but a philosophical-speculative one that emerged from modern mathematics. The University of Göttingen's contributions had the effect of, first, driving an accessible (and mathematically naïve) vision of space from European philosophy, and, second, of marginalizing the historical study of Europe's experience with projected space. Thus, Göttingen's own Edmund Husserl read geometry's past as nothing more than a preface to the conceptual revolutions that had been wrought by his own age. In short, from Husserl's perspective ancient geometry's modern endpoint was ordained at the discipline's inception, regardless of the two millennia's worth of stops and starts.

If we return to Lyotard's definition of postmodernism as being characterized by a profound skepticism toward meta-narratives, it becomes significant that the German philosophy of the period before 1933 inspired the postmodern thought that arose in France after 1945.[36] French students who studied at the universities in Göttingen and Freiburg, among other institutions, brought postmodernism's conceptual underpinnings back to France. One such student was Emmanuel Levinas, a crucial early figure in French postmodernism, who in 1928 not only studied in Freiburg with Edmund Husserl, but also came into contact with Martin Heidegger, whose thought he later promoted.[37] The connections between pre–World War II Germany and Levinas's own thought are legion. Levinas's doctoral thesis, which he submitted in 1929 to the University of Strasbourg, had the title *The Theory of Intuition in Husserl's Phenomenology*, and one of his most important books, which appeared in 1949, bears the title *Discovering Existence with Husserl and Heidegger*.[38] Finally, and most significantly, it was Levinas who translated the *Cartesian Meditations* into French, thus injecting Husserl's epoché into a culture that was determined to break with one of its greatest thinkers.[39]

German philosophy's role in shaping French postmodernism's agenda is most clear, however, in the work of a younger philosopher who owed a great debt to Levinas, Jacques Derrida. Derrida played, perhaps, the key role in the export of Husserl's views on geometry, since it was he who translated the *Origin of Geometry* into French and also appended to it an extensive introduction-commentary.[40] Appearing in 1962, Derrida's translation marked a key moment in postmodernity's erasure of Euclidean space as a *historical* phenomenon. In many ways, the translator's introduction is more important than the translation itself, as the former is longer and occupies the most prominent position within the text. Equally important, the entire book was, in turn, translated into English in 1978, which extended the reach of Husserl's thought, at exactly the moment when French postmodernism's influence was reaching its height in the United States.[41] (An intense reception of Derrida's thought was then under way, as the Frenchman's greatest work, *Of Grammatology*, had been translated into English just two years before.)[42]

Following Husserl's lead, Derrida used mathematics' dramatic pursuit of generality to sketch out an alternative vision of historical time. He explained how Husserl constructed a different notion of history, in which geometry's origins were permanently incorporated within all subsequent evolutions in the discipline's thought. From this perspective, the contemporary emphasis on two-dimensional logic was not an outcome of divergent intellectual processes but a fulfillment of geometry's original promise. In this sense, there was no actual development within geometry, and its history was reduced to the recapitulation of the discipline's nebulously prophetic beginnings. Derrida held:

> Provided the notion of history is conceived in a new sense, the question posed must be understood in its most historic resonance. It is a question of repeating an origin. In other words, reflection does not work upon or within geometry itself as "ready-made, handed-down." The attitude taken, then, is not that of a geometer: the latter has at his disposal an already given system of truths that he supposes or utilizes in his geometrizing activity; or, further, at his disposal are possibilities of new axiomatizations which (even with their problems and difficulties) already are announced as geometrical possibilities. The required attitude is no longer that of the classic epistemologist who, with in a kind of horizontal and ahistoric cut, would study the systematic structure of geometrical science or of various geometries.[43]

Against this backdrop, we see how the postmodern attitudes that emerged from Husserl's understanding of geometry undercut the sense of time that is the historical discipline's *sine qua non*. A research program that derives its primary inspiration from postmodern thought risks, as a result, doing little more than commemorating the past's fulfillment of the present.

At this point, I take leave of postmodernity's theoretical foundations in order to consider its practical implications for historical research. In doing so, I shift my attention from Jacques Derrida toward his countryman Michel Foucault. Although Derrida has influenced students of the early modern period, his effect has been felt more in literary theory and philosophy than in traditional historical research. Foucault, in contrast, has been extremely influential in the realm of historical practice, insofar as his analyses have shaped the research agenda for generations of historians of the both early modern period in general and early modern thought, in particular.[44] (And the latter category includes this historian.)

Ironically, Foucault's strongest effect has been felt through his so-called Spatial Turn. This turn is itself a narrower application of the Linguistic Turn, which postmodern theory, as a whole, has inspired among students of early modernity. Thus, the contemporary intellectual historical discipline, which has borrowed heavily from Foucault, is intensely interested in the problem of language and textuality, while showing less interest in projections of space, whether textual or otherwise. In this sense, the Spatial Turn constitutes the application to intellectual history of spatially infused words, such as field, web, and site. Indeed, it has become fashionable to expand on Foucault's terminological arsenal by deploying the word "interstices," that is, those tiny spaces that lie between (and within) larger (and flat) fields of thought. Nevertheless, amid all the spatial talk from both

Foucault and his followers, none of them has engaged directly with early modern conceptions of space.

The Spatial Turn is oddly spaceless. This element in the Foucauldian paradigm has gone unnoticed, however, because Foucault's own sense of space has not been situated historically. Born in 1926, he was, unlike Nietzsche, a product of the post-Euclidean revolution in geometry, which meant that he naturally absorbed a bifurcated vision of space, in which Euclidean geometry went uncontested precisely because it had been superseded.[45] In short, Euclid's intuitively visual space would not have appeared to Foucault to constitute a historical problem, because, as I have already explained, the questions that geometry raised between 1350 and 1850 were ancient history by 1950.

It is significant, therefore, that when compared to the Spatial Reformation's main currents, Foucault's analytical space is two-dimensional. This accusation requires some explaining. Foucault was interested primarily in understanding how, over the course of the early modern period, once unrelated concepts came to be associated logically. This is the subject of his (in my view) greatest book, *The Order of Things,* which appeared in 1966.[46] Foucault began this tome with a speculative story, which he borrowed from the Argentine writer Jorge Borges, and retold as follows:[47]

> This passage quotes a 'certain Chinese encyclopaedia' in which it is written that 'animals are divided into: (a) belonging to the Emperor, (b) embalmed, (c) tame, (d) suckling pigs, (e) sirens, (f) fabulous, (g) stray dogs, (h) included in the present classification, (i) frenzied, (j) innumerable, (k) drawn with a very fine camelhair brush, (l) *et cetera,* (m) having just broken the water pitcher, (n) that from a long way off look like flies'. In the wonderment of this taxonomy, the thing we apprehend in one great leap, the thing that, by means of the fable, is demonstrated as the exotic charm of another system of thought, is the limitation of our own, the stark impossibility of thinking *that.*[48]

Foucault highlighted this imagined taxonomy—there is no evidence that such an encyclopedia existed—in order to justify the notion that all knowledge systems are arbitrary and determined by power. Thus, when he turned to the early modern order of things, he pounced on the perceived arbitrariness of its tables. These were favorite tools of early modern scientists and scholars, who wished to summarize the era's growing (and increasingly unruly) body of knowledge. Tables were, moreover, also among the early modern state's favorite administrative tools, since what can be listed in a table can be taxed, conscripted, or thrown into prison.[49]

I set aside Foucault's narrower critique of the positivism that suffused tables and instead pursue three broader problems that loiter at the margins. First, simple tables are two-dimensional, as they arrange things on a flat surface. Second, the attending planes and grids represent merely one aspect of early modern space, given that Euclid's *Elements* proceeds deliberately from two-dimensional to three-dimensional geometry.[50] Finally, as I have argued, Euclid's early modern readers used his geometry expressly to *make space* via the production and/or contemplation of objects of material culture. It was, of course, David Hilbert who collapsed the distinction within the *Elements* between planar geometry and solid geometry that had long sustained the association of texts with objects. Yet early modern Europeans (Descartes included) knew nothing of this advance and, for that reason, assiduously incorporated many of the *Elements*' profoundest assumptions into their science, art, and philosophy. Thus, when Foucault's analytical two-dimensionality is applied to a period that veritably worshipped the *Elements*' three-dimensionality, an uncorrectable distortion results.

Foucault's arguments derive much of their force from postmodernity's narrow understanding of geometry's history. This weakness can be understood, in the first instance, against the backdrop of French academic culture of the post–World War II period, which chafed at the pretensions of early modern—and especially Cartesian—attempts to systematize knowledge. Here, I turn to two of this period's *bêtes noires*. First, there was the fierce reaction against encyclopedism (and Foucault was, by no means, alone in reacting against it).[51] Between 1751 and 1765, the *philosophes* Denis Diderot and Jean d'Alembert, two paragons of the Enlightenment, published their *Encyclopédie*, which, although it was not the first European encyclopedia, became the model for all future ones.[52] One thing that irked postwar, postmodern commentators, aside from this monumental work's obvious faith in reason, was the *Encyclopédie*'s alphabetical organization of topics—as if "alpha" and "beta" were significant beyond the Western world's linear arrangement of vowels and consonants. In this, Foucault and others had a point: the *Encyclopédie* did not just present knowledge but imposed an arbitrary structure on it.

The resistance to encyclopedism rested, in turn, on Foucault's even deeper critique of the second *bête noire*, Cartesian rationalism. As the quote above from Lyotard illustrates, the twentieth century brought a severe reaction in France against both the philosophical subject and the faith in reason that had attended it. Rather than enter into the broader debates, I concentrate on Foucault's perspective. Taking off from phenomenology's critique of the subject, Foucault opposed histories that celebrated reason's emergence in chronicles of progress—all of which rely, of course, on meta-narratives. Instead, in his *Order of Things* he substituted for the early modern period what he called the Classical Age, a rigidly defined, wholly separate era, in which reason not only failed to emancipate humanity, but also empowered systems of oppression that perdured into the modern day. Thus, on the one hand, Foucault held that around 1650—not coincidentally, the year of Descartes's death—a sudden shift occurred in what he called Europe's *episteme,* or its system of knowing. On the other hand, this shift was bookended by another one that appeared around 1800 and that, curiously, shunted the continent into modernity. It is, in my view, more than just happenstance that this was the period when Napoleon and his armies prowled the Continent. Foucault's most profoundly critical work was, in a very important sense, trapped between the history of philosophy and the rise of a modern politics.

Given that the rejection of the Cartesian subject played such a prominent role in Foucault's thought, I thus read Descartes in a way that highlights the Foucauldian approach's vulnerabilities to a fully spatial reading of the early modern period. In 1637, Descartes published *Discourse on Method,* a fundamental work that called for the application of a systematic logic to every aspect of European philosophy.[53] In essence, Descartes tried to discipline thought by applying to it the *Elements*' axiomatic methods, while also avoiding homogeneous space's potential pitfalls. As a materialist in physics, Descartes could not embrace Euclid's substanceless points, since doing so would have raised the question of how indivisible things related to substance. (Homogeneous space caused other problems for Descartes's physics, too. I cannot, however, pursue them here.) And then there was the problem of God. As we have seen in Isaac Newton's thought, three-dimensional space raised profound theological issues. Descartes, in my view, seemed not only to sense the looming contradictions, but he also attempted to sidestep them by taking the simple measure of not discussing them. Thus, he dealt with the small matter of God's location by announcing that the physical cosmos's extent was indeterminate rather than fully infinite.[54] In this respect, with its underlying space being constrained in this manner, it was now the cosmos itself that was "not nowhere."

When Descartes combined geometry with philosophy, the problems that space augured led him to emphasize the logically formal rather than the fully spatial. This is clear in his *Meditations on First Philosophy,* which appeared four years after the *Discourse on Method.*[55] In this text Descartes combined his axiomatic philosophical method with the application of radical doubt to individual experience; that is, he doubted absolutely everything except the fact that he was the one doing the doubting. Descartes's cultivation of doubt had plangent echoes, because it allowed the subject not only to authorize its own existence, but also to justify the existence of God from within itself, without reference to either the external world, or to a problematic spatial homogeneity. Moreover, and along similar lines, it is notable that the *Meditations'* subtitle is *in which the Existence of God and the Immortality of the Soul are Demonstrated,* because this illustrates how the relationship between humanity and God had been saved by Descartes's emphasis on logic over a creeping homogenous space. Within Cartesian thought, logical relationships were not connected to a sense of location but existed merely as they were—and this left God pristinely unmolested by space.[56] In effect, Descartes had *desituated* the medieval emphasis on rational ubiquity, in order to save the divine from space's ravages. Within the early modern world's Euclidean paradigm, however, this logical approach to the Western triad would not go uncontested.

I illustrate the significance of this last point by looking to another French thinker whom I have already mentioned, Blaise Pascal. While writing his *Pensées* in the years after 1654, Pascal had this to say about the human relationship to infinite space: "Let us, then, understand our condition: we are something and we are not everything. Such being as we have removes us from knowledge of first principles, which arise out of nothingness. And the smallness of our being conceals us from the sight of the infinite."[57] Significantly, these reflections came only a few pages before Pascal issued his famous cry into eternal silence. Pascal's approach to the relationship between humanity and space thus illuminates the depths of the philosophical-religious problems that space's march into the human mind caused for the triad. If geometric space was both infinite and applicable to the cosmos, then traditional views of the cosmos and God would have to be rethought, with a serious anthropological upgrade in the offing. Pascal, for his part, confronted the theological and anthropological issues squarely, while also wringing his hands furiously; Descartes, in contrast, looked away. Perhaps this was why the intensely inspired Pascal judged his brilliantly discursive contemporary so harshly, writing in the *Pensées*: "Descartes useless and uncertain."[58]

Nor is the tension between Descartes and Pascal the only example of a debate that ostensibly concerned the divine wading inexorably into empty space's depths (and vice versa). In the period 1715–1716, Gottfried Wilhelm Leibniz engaged in a written debate with Samuel Clarke, who was serving as Isaac Newton's faithful lieutenant, over the Newtonian notion of space and time.[59] The debates were extremely detailed, so I cannot go down its many avenues but note only that one of its central issues was whether homogeneous space could be aligned with extant visions of God. It is, therefore, especially important to underline that the problem over which Leibniz and Clarke wrestled never went away, as debates about space's potential theological implications reached well into the late eighteenth century, with the greatest of all enlightenment philosophers, Immanuel Kant, also taking up the issue. Thus, as Nietzsche and Dostoevsky's respective comments on God underscore, European thinkers never figured out how to relate empty space to the divine in a way that preserved God for Man.

Thanks, in part, to the influence of Foucault and others, however, homogeneous space's tense relationship with the Cartesian subject's inherent religiosity has been overshadowed by the literature's misleading overemphasis on Descartes's embrace of logic. Consider, along these lines, that Descartes's *Meditations on First Philosophy* introduced the immortal phrase: "Cogito, ergo sum" ("I think, therefore, I am").[60] Were one to concentrate exclusively on these words, one might conclude the following: (1) that modern European philosophy begins *solely* with invention of the subject, and (2) that this subject exists in a spaceless realm.[61] However, Descartes's *On Geometry*, which I have already mentioned as the key work in the history of analytic geometry, was originally written as an appendix to the *Meditations*. Based, thus, on a simple textual genealogy, it would seem that the subject's philosophical cultivation and the mental domination of space were linked closely enough to merit their association within a single work. Along these lines, it is important that the work's internal structure separated these two issues for largely theological reasons. As a result, Descartes's construction of the subject, to which so much of postmodern thought stands opposed for philosophical reasons, was actually an attempt to avoid space's enervating effects on the divine. Postmodernism's embrace of non-Euclidean space may be philosophically justified. Nevertheless, we miss a crucial element in Descartes's thought if we overlook that he turned to logic, precisely because he could not embrace the entirety of Euclidean space.

Homogeneity's Revenge

Postmodern theory's failure to confront homogenous space's rise and decline has impoverished contemporary views of the early modern world. In this context, I return to Foucault's Classical Age and do so with reference to my discussion of early modernity's underpinnings. Among those historians who criticize Foucault, it is common to cast aspersion onto his casual approach to evidence. This criticism is valid: Foucault did, sometimes, play fast-and-loose with the sort of details that concern historians. Nevertheless, the resulting deficiencies were not reflective of bad historical practice but of Foucault's tendency to project onto the past his ideological disagreements with his own present. In effect, Foucault channeled Husserl by holding the Classical Age responsible for the mess that modernity made.[62] None of this is particularly historical, which has led some historians to respond in an intemperately critical manner.[63] Foucault, however, was quite correct to look to the middle of the seventeenth century as a turning point in European thought, even if he overlooked the broader context in which the shift occurred. Thus, if we integrate Foucault's historical instinct with a spatial view of European thought, we can account for the significance of early modernity's encounter with three-dimensionality.

Foucault's error lay in his uncritical imposition of a two-dimensional sense of space onto a three-dimensional culture. The clearest evidence of this error is the *Order of Things'* most famous historical example. In the book's first chapter

Figure 34. Velázquez, *Las Meninas*, 1656. © Museo Nacional del Prado.

Foucault included a striking analysis of Diego Velázquez's painting *Las Meninas*[64] (see figure 34). Foucault highlighted this image, because he believed that it encapsulated that moment when reason not only presented itself publicly, but also announced the erasure of previous ways of knowing.[65] In particular, he was impressed with how the painting's cultivation of multiple visual perspectives within one work marginalized what he saw as an older respect for difference that had permeated Renaissance thought. For Foucault, therefore, Velázquez's image epitomized the Classical Age's marginalization of alternative ways of knowing.

One of Foucault's more astonishing arguments is that changes in the structure of reason *erased* the individual subject itself. Here, I turn directly to his reading of Velázquez's great canvas. For Foucault, the subject's decline is exemplified in the strange interplay within *Las Meninas* between viewers and viewed. Standing before *Las Meninas,* the viewer sees in the image a painter who is Velázquez himself and who appears to be looking back at the viewer. Nevertheless, inside the *painting's* space, the painter is peering at two figures who are assumed to be standing before him, but are not depicted, namely the king and queen of Spain.[66] (They appear dimly in the mirror that is suspended on the back wall.) Foucault thus sensed a tension within what he called this painting's representation of Classical representation.[67] Put another way, for Foucault the blurring of the boundary between the image's actual viewers and those imagined to be within its space revealed the philosophical subject's erasure. In my view, however, the opposite was true, as Velázquez's painting reveals that humanity was seizing control of its mental world via ever more elaborate projections of space. This was, after all, the world in which the terrestrial globe had become thinkable.

Nevertheless, Foucault saw *Las Meninas* as offering visual proof of a shift in the early modern *episteme* that subordinated all difference to a procrustean and peculiarly generalized philosophical subject. From this perspective, it was convenient that the work was completed in 1656, since this allowed Foucault, as the critic, to plant the painting as a signpost along his accusatory trundle into the past. He concluded:

> Perhaps there exists, in this painting by Velázquez, the representation as it were, of Classical representation, and the definition of space it opens up to us. And, indeed, representation undertakes to represent itself here in all its elements, with its images, the eyes to which it is offered, the faces it makes visible, the gestures that call it into being. But there, in the midst of this dispersion which it is simultaneously grouping together and spreading out before us, indicated compellingly from every side, is an essential void: the necessary disappearance of that which is its foundation—of the person it resembles and the person in whose eyes it is only a resemblance. This very subject—which is the same—has been elided. And representation, freed finally from the relation that was impeding it, can offer itself as representation in its pure form.[68]

Amid Foucault's abstruse theorizing, it is important to point out that he pursued only the *logic* that he sensed to be lurking within the image. As a result, he never actually dealt with how the painting *produced* space, which, in contrast to his own postmodern critical apparatus, was rooted in Euclidean three-dimensionality.

Foucault's reading can be contested on multiple grounds. Consider, first, the mature nature of the spatial knowledge that *Las Meninas* deploys. The image is executed in an extremely geometric manner—and Velázquez could not have done this without having mastered perspective, a method that requires knowledge of geometry's basics.[69] Moreover, *Las Meninas* also boasts the artful diffusion of light, which is significant, because light had been an important issue for early modern optics, a discipline that made extensive use of geometry.[70] It is therefore significant that Euclid also wrote a work on optics that had a prominent early modern career.[71] In addition, the painting also depicts a mirror, which connects Velázquez's work to catoptrics, another field that prospered in the sixteenth and seventeenth centuries, thanks to a Pseudo-Euclidean work on mirrors.[72] Thus, if we read *Las Meninas* with respect to spatial thought's history, it becomes apparent that we cannot separate the homogeneity of the painting's space from the painter's acquisition of geometry and its attending disciplines. Velázquez was a child of the Spatial Reformation, too.

With this point in mind, I consider Velazquez's relationship to homogeneous space more narrowly. If one follows Foucault, one would never know that Velázquez worked as an architect, a discipline whose practice was constructed on the prior study of geometry.[73] Nor would one know that, while serving as an apprentice in Seville, he studied geometry, which also happens to be what every major artist since the Renaissance did, including other luminaries such as Leonardo da Vinci and Albrecht Dürer. Along similar lines, Foucault mischaracterized Velázquez's master in Seville, Francisco Pacheco. Pacheco was more than just a competent artist; he also participated in a circle of humanist scholars that collected and discussed ancient written works, including those by Euclid.[74] There is, thus, a direct connection between the return of classical works on space and Velázquez's training.

Pacheco's position as Velázquez's teacher is particularly illustrative of the problem. Foucault mentioned Pacheco in the *Order of Things*' first chapter, when he described the painting as "a strangely literal, though inverted, application of the advice given, so it is said, to his pupil by the old Pachero [*sic*] when the former was working in his studio in Seville: 'The image should stand out from the frame.' "[75] The inclusion of the words "strangely literal" is key here, because this allowed Foucault to transmute Pacheco's artisanal wisdom into a pretext for doing away with (in his view) archaic terms, such as "the painter," "the characters," "models," "the spectators," and "the images." Foucault needed, in short, to separate Velázquez rhetorically from Pacheco, in order to be able to associate *Las Meninas* with a purported philosophical break.[76] In my view, however, Foucault used Pacheco as a rhetorical trampoline. Consider that Pacheco was born in 1564, which meant that he emerged from a profoundly humanist world that used space to reorient European views of humanity, the cosmos, and God. Pacheco, like his greatest student, lived inside a Euclidean world.[77]

We are now in a position to evaluate the teacher's advice to the student quite differently. In this context, I turn to Pacheco's manual *The Art of Painting*, because it highlights the chief problem with Foucault's reading of *Las Meninas*. In this textbook Pacheco associates painting with the seven liberal arts.[78] (As I have explained in previous chapters, these are divided into two groups: the Trivium, which comprises dialectic, grammar, and rhetoric, and the Quadrivium, which comprises arithmetic, astronomy, geometry, and music.) Pacheco was particularly careful to justify painting's relationship to geometry, which had become the most prominent of the arts: "Geometry is the best suited of all to serve the art of painting, because through drawing and the use of lines, the painting manifests itself for all to see: and (in a way) it refers back to Philosophy: and this is why it is a noble and liberal art."[79] Echoing sentiments that we have already encountered in Dürer, Pacheco elevated painting's status by subordinating geometry to it. Put another way, he held that painting's particular virtue was that it brought geometry's imagined lines and planes *into view* by transmuting geometric doctrines into material culture. Thus, much as we saw in de'Barbari's image of Luca Pacioli, Pacheco's vision of painting suggested that the painter remained situated between a copy of the *Elements* and the canvas itself.

If we take Pacheco's role in Velázquez's development seriously, then Foucault's quotation signifies the opposite of what his *Order of Things* argued. A painting that jumps beyond its frame represents the fulfillment of Euclidean space, with its homogeneity not only rushing into every conceptual corner, but also empowering the viewing subject to dominate its own space. Moreover, I cannot help but note that Pacheco's work was published in 1649, the year before the Classical Age ostensibly began, and the same year that Descartes's *Geometry* appeared as an independent text. It is thus difficult to justify Foucault's view of *Las Meninas* as a cutting-edge form of non-spatial representation. In fact, as Descartes's rather desperate isolation of geometric space from his own justifications for God's existence confirms, space was becoming all too human.

Conclusion

Let us redirect our own gaze to the Heavens. Figure 35 includes another famous image of Earth from NASA's heyday. It is called "Earth Rising" and captures what must surely have been a startling moment for the American astronauts who took the snapshot: their feet planted lightly on lunar soil, they observed their own Earth rising above an altogether different horizon. As I argued with respect to the Blue Marble, although the physical position of the astronauts who were viewing Earth was new, the spatial perspective itself was not. In 1634, for example, a literary work by Johannes Kepler appeared posthumously under the title *The Dream*.[80] In the course of this work's pages, Kepler imagined a lunar civilization

Figure 35. "Earth Rising," 1972. Courtesy of NASA.

whose denizens peered *up* at Earth—or so they thought. Kepler's tome was not the first early modern text (nor even the first Western one) to consider the possibility of life on the moon.[81] His work is unique, however, in that his interest was specifically spatial, as he wanted to show how deceptive appearances could be when trying to imagine the spatial relationship between exceptionally large bodies. In the case of the moon's fictitious residents, only a mathematically justified, extra-lunar perspective could reveal the truth of the moon's rotation around the earth. A student of both mathematics and astronomy, Kepler, in effect, used his sense for idealized space as a means to free European culture from the tyranny of having to look up. In short, the mental domination of empty space preceded humanity's conquest of physical space by centuries.[82]

With this lunar backdrop in mind, we can see how postmodernism's blind spot on space has distorted our historical perspective. Among other things, it has impeded the formation of new perspectives on the early modern world, on Nietzsche's relationship to that world, and especially on the contemporary tripartite approach to historical time. Foucault's Classical Age is, in this context, one example of a problem that plagues much of the early modernist literature, namely that postmodernism's two-dimensionality obscures three-dimensional space's once-awesome amalgamating power. Consider that once Foucault dispensed with *Las Meninas,* he went fully two-dimensional in the next chapter, "The Prose of the World," where he reduced all human knowledge to language.[83] Or as Foucault held in his discussion of *Las Meninas,* "The space where they achieve their splendor is not that deployed by our eyes but that defined by the sequential elements of syntax."[84] Foucault effectively closed off space to intellectual historical inquiry, although he was by no means alone in doing so.

The postmodern condition would not be a problem for the historical discipline had the early modern world not been soaked in spatial projections. But it was. The history of art would look quite different without the Renaissance rediscovery of ancient spatial thought. (Certainly, Pacheco's work would have been different.) And much the same can be said of the history of philosophy. The great Immanuel Kant knew Euclid well and plumbed geometric space's depths. In an unpublished text on space from 1768, "On the Primary Justification for the Difference of Areas within Space," Kant held:[85]

> For the positions of the parts of space [that are fixed] in relation to each other presuppose a region, according to which [positions] are arranged within this relationship, and the most abstract understanding of a region does not consist of the relationship of a [given] thing in space to another [thing,] which is actually the notion of location, but [lies] in the relationship of these positions to the absolute space of the cosmos. With respect to all extension, the positions of its parts with respect to one another are determined sufficiently out if [extension] itself. The area, however, toward which the order of the [various] parts is dedicated, relates itself to the space beyond itself and, indeed, not to its places, because this would signify nothing more than that the locations of its fragments are in relation externally, and [are not] related to the general unity of space, from which perspective any extension must be seen as a fragment.[86]

As Kant saw it, space is a unity (*Einheit*) of human thought and for that reason it *situated* all knowledge, including knowledge of those things that people did not see, but imagined. Kant's sense of space, in short, incorporated exactly the three-dimensionality that suffused European thought up through the mid-nineteenth century.

With the temporal expanse of this book's argument in mind, I complete the rhetorical circle by turning my attention forward. The most important work on modern spatial sense remains Stephen Kern's *The Culture of Time and Space, 1880–1918.*[87] In this path-breaking monograph Kern surveyed fin-de-siècle art and culture and held that modernity's chief characteristic was the fluidity of its sense for time and space. He wrote: "From around 1880 to around the outbreak of World War I a series of sweeping changes in technology and culture created distinctive new modes of thinking about and experiencing time and space. Technological innovations including the telephone, wireless telegraph, x-ray, cinema, bicycle, automobile, and airplane established the material foundation for this reorientation; independent cultural developments such as stream-of-consciousness novel, psychoanalysis, Cubism and the theory of relativity shaped consciousness directly."[88] Kern looked primarily to technology to explain the rise of the new sense. Taking into account the quotes from Nietzsche and Dostoyevsky that I highlighted above suggests, however, that technology's rise only intensified an intellectual shift that had been gathering momentum since the mid-nineteenth century. Although modern spatial sense *developed* in conjunction with technological advance—on this point, Kern was quite right—the true break with the early modern came before 1880 and was a direct result of post-Euclidean geometry's rise.

I conclude by reiterating that contemporary intellectual history should do more to integrate its vision of early modern thought with spatial homogeneity's rise and decline. I offer my notion of a Spatial Reformation and the concomitant "long" early modern period as one way to reorient our understanding of European thought's relationship to space. At this point, however, I remain content to note that postmodern theory's reduction of the early modern period to a collection of spatial *terms* erases precisely the richness that emerged from the intense cultivation (in text and object) of Euclidean space. However interesting postmodern analyses of texts are, postmodernism itself lacks the apparatus for understanding how the language of space united with material culture, in order to create a human *All.* We cannot, therefore, understand Nicholas of Cusa, Velázquez, or any thinker up through Nietzsche, if we do not confront the homogeneity that suffused five centuries of European constructions of humanity's relationship to itself, to the cosmos, and to God. It is time, in short, for contemporary historians of early modern thought to bring the space back in.

Notes

Preface

1. Michael J. Sauter, "Clock Watchers and Stargazers: Time Discipline in Early-Modern Berlin," *American Historical Review* 112, no. 3 (2007): 685–709.

2. For an example of the kind of book that I had in mind, see Alexis McCrossen, *Marking Modern Times: A History of Clocks, Watches, and Other Timekeepers in American Life* (Chicago: University of Chicago Press, 2013). On these four cities and their public clocks, see Gerhard Dohrn-van Rossum, *History of the Hour: Clocks and Modern Temporal Orders* (Chicago: University of Chicago Press, 1996), 346.

3. Clifford Geertz, *Local Knowledge: Further Essays in Interpretive Anthropology* (New York: Basic Books, 1983).

4. Georges-Louis Leclerc Buffon, *Histoire naturelle, générale et particuliére avec la description du cabinet du roy*, 29 vols. (Paris: L'Imprimerie Royale, 1749–1788); Alexander Pope, *An Essay on Man. Address'd to a Friend. Part I*, 2nd ed. (Dublin: Printed by S. Powell, for George Risk . . . George Ewing . . . and William Smith . . . 1734); Christian Wolff, *Vernünfftige Gedancken von der Würckungen der Natur* (Halle: Rengerischen Buchhandlung, 1725).

5. The working title of this project is *The Decline of Space*.

6. The most important influence on me has been the work of Sara Schechner. See Sara J. Schechner, *Comets, Popular Culture and the Birth of Modern Cosmology* (Princeton, N.J.: Princeton University Press, 1997); Sara J. Schechner, "The Material Culture of Astronomy in Daily Life: Sundials, Science, and Social Change," *Journal for the History of Astronomy* 32 (2001): 189–222.

Introduction

1. *Hamlet* 2.2.305–8.

2. On Cusa's pivotal role in European intellectual history, see Ernst Cassirer, *The Individual and the Cosmos in Renaissance Philosophy* (Oxford: Blackwell, 1963).

3. For a complementary perspective on space's intellectual effects that has influenced my own views, see Alexandre Koyré, *From the Closed World to the Infinite Universe* (New York: Harper & Brothers, 1957) (see also n. 8 below). My emphasis on homogeneity comes from my reading of the interesting essay by Branko Mitrović, "Leon Battista Alberti and the Homogeneity of Space," *Journal of the Society of Architectural Historians* 63, no. 4 (2004): 424–39.

4. In general, see Stephen Wagner, "Seven Liberal Arts and Classical Scholarship," in *The Seven Liberal Arts in the Middle Ages*, ed. Stephen Wagner (Bloomington: Indiana University Press, 1983), 12; Stephen K. Victor, *Practical Geometry in the High Middle Ages: Artis Cuiuslibet Consummatio and the Pratike de Geometrie* (Philadelphia: American Philosophical Society, 1979), 2–5; Edward Grant, *The Nature of Natural Philosophy in the Late Middle Ages* (Washington, D.C.: Catholic University of America Press, 2010), 21; O. A. W. Dilke, *The Roman Land Surveyors: An Introduction to the Agrimensores* (Newton Abbot: David and Charles, 1971). For an example, see Albert the Great, *The Commentary of Albertus Magnus on Book 1 of Euclid's Elements of Geometry*, trans. Anthony Lo Bello (Boston: Brill Academic Publishers, 2003). This work comments only on planar geometry—and this was standard practice. I discuss this issue further in Chapter 1.

5. Extending Stephen Greenblatt's arguments, I hold that Renaissance thinkers fashioned their world, in addition to themselves. Stephen Greenblatt, *Renaissance Self-Fashioning: From More to Shakespeare* (Chicago: University of Chicago Press, 1980).

6. John Donne, *Complete Poetry and Selected Prose of John Donne* (New York: Random House, 1952), 30.

7. Elly Dekker and P. C. J. van der Krogt, *Globes from the Western World* (London: Zwemmer, 1993), 16–17.

8. Alexandre Koyré, *Closed World; Newtonian Studies* (Cambridge, Mass.: Harvard University Press, 1965), and *Galileo Studies* (Atlantic Highlands, N.J.: Humanities Press, 1978).

9. On theology's relationship to science, the fundamental work is Amos Funkenstein, *Theology and the Scientific Imagination from the Middle Ages to the Seventeenth Century* (Princeton, N.J.: Princeton University Press, 1986).

10. Cusa, *Opera omnia*, 3, 144.

11. It is not surprising, in this context, that Cusa also speculated on how God saw things in his *De visione dei*. For an English translation, see *The Vision of God*, trans. Emma Gurney Salter (New York: Ungar Publishing Company, 1960).

12. A well-argued alternative view holds that heliocentrism's intellectual origins trace back to medieval Islam. See George Saliba, "The First Non-Ptolemaic Astronomy at the Maraghah School," *Isis* 70, no. 4 (1979): 571–76; "Arabic Versus Greek Astronomy: A Debate over the Foundations of Science," *Perspectives on Science* 8, no. 4 (2000): 328–41; *Islamic Science and the Making of the European Renaissance* (Cambridge, Mass.: MIT Press, 2007); F. Jamil Ragep, "Tusi and Copernicus: The Earth's Motion in Context," *Science in Context* 14, no. 1/2 (2001): 145–63.

13. As an example of NASA's continued cultural significance, see Becky Little, "NASA's 'Blue Marbles': Pictures of Earth from 1972 to Today: To Celebrate NASA's Newest Photo of Earth, Here's a Look at Some of the Ones That Came Before It," *National Geographic*, July 21, 2015, https://news.nationalgeographic.com/2015/07/150721-pictures-earth-nasa-dscovr-spacex-space-science/.

14. Donne, *Complete*, 246.

15. On the Judeo-Christian backdrop, see Clarence J. Glacken, *Traces on the Rhodian Shore: Nature and Culture in Western Thought from Ancient Times to the End of the Eighteenth Century* (Berkeley: University of California Press, 1976), 150–70.

16. My approach here can be understood as expressly combining the disciplinary perspectives of historians of science with those of historians of cartography. For examples of each current, see Denis E. Cosgrove, *Apollo's Eye: A Cartographic Genealogy of the Earth in the Western Imagination* (Baltimore: Johns Hopkins University Press, 2001); Owen Gingerich, *The Eye of Heaven: Ptolemy, Copernicus, Kepler* (New York: American Institute of Physics, 1993).

17. Funkenstein, *Theology and the Scientific Imagination.*

18. With respect to Cusa, see the editor's comments in Nicholas of Cusa, *De Ludo Globi = The Game of Spheres* (New York: Abaris Books, 1986), 18.

19. Space is an overlooked of topic within the contemporary intellectual historical literature, as there is no broadly conceived history of space for the early modern period, while general intellectual histories do not highlight the issue. Herewith, a sample of standard works: Marcia L. Colish, *Medieval Foundations of the Western Intellectual Tradition, 400–1400* (New Haven, Conn.: Yale University Press, 1997); Anthony Levi, *Renaissance and Reformation: The Intellectual Genesis* (New Haven, Conn.: Yale University Press, 2002); Steven E. Ozment, *The Age of Reform, 1250–1550: An Intellectual and Religious History of Late Medieval and Reformation Europe* (New Haven, Conn.: Yale University Press, 1980); John N. Stephens, *The Italian Renaissance: The Origins of Intellectual and Artistic Change Before the Reformation* (London: Longman, 1990). More surprisingly, perhaps, even a work as profoundly philosophical as Hans Blumenberg's *The Legitimacy of the Modern Age* does not mention Euclid once. See Hans Blumenberg, *The Legitimacy of the Modern Age* (Cambridge, Mass.: MIT Press, 1983).

20. The relevant works: Nicolaus Copernicus, *De revolutionibus orbium coelestium, libri VI* (Norimbergae: Ioh. Petreium, 1543); René Descartes, *Discours de la méthode pour bien conduire sa raison et chercher la vérité dans les sciences, plus la dioptrique, les météores et la géométrie qui sont des essais de cette*

méthode (Leyde: L'Imprimerie de Jan Maire, 1637); Isaac Newton, *Philosophiae naturalis principia mathematica* (Londini: J. Societatis Regiae ac Typis J. Streater, 1687); Thomas Hobbes, *Leviathan; or, the Matter, Form, and Power of a Common-Wealth Ecclesiastical and Civil* (London: Printed for Andrew Crooke, 1651).

21. I have cited from the English version. See Isaac Newton, *The Mathematical Principles of Natural Philosophy by Sir Isaac Newton*, trans. Andrew Motte, 2 vols. (London: Benjamin Motte, 1729), 2: 389–90.

22. On this point, see also Koyré, *Closed World*, 77–94.

23. Newtonian space *is* Euclidean space. Jeremy J. Gray, *Ideas of Space: Euclidean, Non-Euclidean, and Relativistic*, 2nd ed. (Oxford: Clarendon Press, 1989), 177.

24. Margaret C. Jacob has defined the literature on this topic: see her *The Newtonians and the English Revolution, 1689–1720* (Ithaca, N.Y.: Cornell University Press, 1976); *The Cultural Meaning of the Scientific Revolution* (New York: A. A. Knopf, 1988); "The Enlightenment Redefined: The Formation of Modern Civil Society," *Social Research* 58, no. 2 (1991): 475–95; *Strangers Nowhere in the World: The Rise of Cosmopolitanism in Early Modern Europe* (Philadelphia: University of Pennsylvania Press, 2006).

25. Immanuel Kant, *Critique of Pure Reason* (Cambridge: Cambridge University Press, 1998), 387.

26. Ibid., 117.

27. Jeremy Gray pointed out the significance of Dostoevsky's quote almost three decades ago. Gray, *Ideas of Space*, 175.

28. There was an older Latin translation by Boethius, but it was incomplete. Menso Folkerts, *Boethius Geometrie II: Ein mathematisches Lehrbuch des Mittelalters* (Wiesbaden: Steiner, 1970); "The Importance of the Pseudo-Boethian Geometria During the Middle Ages," in *Boethius and the Liberal Arts: A Collection of Essays*, ed. Michael Masi (Bern: Peter Lang, 1981).

29. For views from specialists that have heavily influenced my own, see Gillian R. Evans, "The 'Sub-Euclidean' Geometry of the Earlier Middle Ages, up to the Mid-Twelfth Century," *Archive for History of Exact Sciences* 16, no. 2 (1976): 105–18; Evgeny A. Zaitsev, "The Meaning of Early Medieval Geometry: From Euclid and Surveyors' Manuals to Christian Philosophy," *Isis* (1999): 522–33.

30. On the relationship between Euclid's decline in late antiquity and the history of cartography, see Lloyd A. Brown, *The Story of Maps* (Boston: Little, Brown, 1949), 83–85.

31. In general, see Rudolf Simek, *Heaven and Earth in the Middle Ages: The Physical World before Columbus*, trans. Angela Hall (Woodbridge: Boydell Press, 1996).

32. On the medieval cosmos's physical structures, see Edward Grant, *Planets, Stars, and Orbs: The Medieval Cosmos, 1200–1687* (Cambridge: Cambridge University Press, 1996); "The Medieval Cosmos: Its Structure and Operation," *Journal for the History of Astronomy* 28 (1997): 147–67; *Science and Religion, 400 B.C. to A.D. 1550: From Aristotle to Copernicus* (Westport, Conn.: Greenwood Press, 2004).

33. I am borrowing heavily from the following works, although their tendency is to emphasize the rise of science rather than of space more broadly understood: James Hannam, *God's Philosophers: How the Medieval World Laid the Foundations of Modern Science* (London: Icon, 2010); Edith Wilks Dolnikowski, *Thomas Bradwardine: A View of Time and a Vision of Eternity in Fourteenth-Century Thought* (Leiden: Brill, 1995); George Molland, *Mathematics and the Medieval Ancestry of Physics*, vol. 481 (Aldershot: Variorum Publishing, 1995); John Freely, *Before Galileo: The Birth of Modern Science in Medieval Europe* (New York: Overlook Duckworth, 2012).

34. This issue is very complex and needs to be explicated in a separate work. For discussions of what Nominalism was, how uniform it was (or not), and whether the term should be used, see Gordon Leff, *Medieval Thought: St. Augustine to Ockham* (Harmondsworth: Penguin Books, 1958); *William of Ockham: The Metamorphosis of Scholastic Discourse* (Manchester: Manchester University Press, 1975); *The Dissolution of the Medieval Outlook: An Essay on Intellectual and Spiritual Change in the Fourteenth Century* (New York: Harper & Row, 1976), 12–13.

35. I have found the following to be helpful: Leff, *William of Ockham*, 561–66; John Marenbon, *Medieval Philosophy: An Historical and Philosophical Introduction* (London: Routledge, 1998), 296–304; Colish, *Medieval Foundations*, 311–18.

36. William of Occam, *Quodlibetal Questions*, 2 vols. (New Haven, Conn.: Yale University Press, 1991), 52.

37. See André Goddu, *The Physics of William of Ockham* (Leiden: Brill, 1984).

38. Marshall Clagett, "The Use of Points in Medieval Natural Philosophy and Most Particularly in the 'Questiones de spera' of Nicole Oresme," in *Studies in Medieval Physics and Mathematics* (London: Variorum Reprints, 1979), 216–21.

39. Ibid., 218.

40. Occam, *Quodlibetal Questions*, 324.

41. This shift also changed the sense that some had of the church's mission, as it became more about the cultivation of moral lessons than understanding of the divine. See Heiko A. Oberman, *The Harvest of Medieval Theology: Gabriel Biel and Late Medieval Nominalism* (Cambridge, Mass.: Harvard University Press, 1963), 164–66.

42. Funkenstein, *Theology and the Scientific Imagination.*

43. George Molland, "The Geometrical Background to the 'Merton School,'" *British Journal for the History of Science* 4, no. 2 (1968); Molland, *Mathematics*, 481.

44. Edith Sylla, "The Oxford Calculators," in *The Cambridge History of Later Medieval Philosophy from the Rediscovery of Aristotle to the Disintegration of Scholasticism, 1100–1600*, ed. Norman Kretzmann, Anthony Kenny, and Jan Pinborg (Cambridge: Cambridge University Press, 1982), 540–63; "The Oxford Calculators in Context," *Science in Context* 1, no. 2 (2008): 257–79. On Ockham's philosophical influence over this group, Leff, *Dissolution of the Medieval Outlook*, 112–15.

45. André Goddu, "The Impact of Ockham's Reading of the Physics on the Mertonians and Parisian Terminists," *Early Science and Medicine* 6, no. 3 (2001): 204–36; L. W. B. Brockliss, *The University of Oxford: A History* (Oxford: Oxford University Press, 2016), 104–5; Gyula Klima, *John Buridan* (Oxford: Oxford University Press, 2009), 259–65. See also Marshall Clagett, "Nicole Oresme and Medieval Scientific Thought," *Proceedings of the American Philosophical Society* 108, no. 4 (1964): 298–309; *Nicole Oresme and the Medieval Geometry of Qualities and Motions: A Treatise on the Uniformity and Difformity of Intensities Known as Tractatus de configurationibus qualitatum et motuum* (Madison: University of Wisconsin Press, 1968); Dana B. Durand, "Nicole Oresme and the Medieval Origins of Modern Science," *Speculum* 16, no. 2 (1941): 167–85.

46. Edward Grant has published the best expository works on this complicated issue: Edward Grant, "Motion in the Void and the Principle of Inertia in the Middle Ages," *Isis* 55, no. 3 (1964): 265–92; "Medieval and Seventeenth-Century Conceptions of an Infinite Void Space Beyond the Cosmos," *Isis* 60, no. 1 (1969): 39–60; "Aristotelianism and the Longevity of the Medieval World View," *History of Science* 16, no. 2 (1978): 93–106; "Scientific Thought in Fourteenth-Century Paris: Jean Buridan and Nicole Oresme," *Annals of the New York Academy of Sciences* 314, no. 1 (1978): 394–423; *Much Ado About Nothing: Theories of Space and Vacuum from the Middle Ages to the Scientific Revolution* (Cambridge: Cambridge University Press, 1981); *Planets, Stars, and Orbs: The Medieval Cosmos, 1200–1687*; "The Medieval Cosmos"; "Scientific Imagination in the Middle Ages," *Perspectives on Science* 12, no. 4 (2004): 147–67.

47. On the discovery of the earth by the medieval "French Schools," see the fundamental work by Jean-Marc Besse, *Les Grandeurs de la Terre: Aspects du savoir géographique à la Renaissance* (Lyon: ENS, 2003), 91–96.

48. For general overviews, see Gerhard Bott and Johannes Karl Wilhelm Willers, *Focus Behaim Globus: Germanisches Nationalmuseum, Nürnberg, 2. Dezember 1992 Bis 28. Februar 1993*, 2 vols. (Nuremberg: Verlag des Germanischen Nationalmuseums, 1992); Oswald Muris, "Der Globus des Martin Behaim," *Mitteilungen der Geographischen Gesellschaft Wien* 97 (1955): 169–82.

49. See Chapter 2.

50. On this point, see the different view of Hannah Arendt, "The Conquest of Space and the Stature of Man," *New Atlantis*, no. 18 (2007): 43–55.

51. Johann Gabriel Doppelmayr, *Kurze Einleitung zur edlen Astronomie, in welcher zum Fundament derselben das Systema Solare & Planetarium nach des weitberühmten Herrn Chrisiani Hugenii Copernicanischen Grund-Sätzen erkläret/ Und in unterschiedlichen neu-verfertigten Homännischen Charten allen wahren Freunden der Sternen-Wissenschaft deutlich vor Augen gelegt Wird* (Nürnberg: Joh. Baptistae Homann, 1708).

52. Ibid., 32.

53. On this globe, see Dekker and Krogt, *Globes*, 98. An original is in the possession of the Utrecht University Museum, to which I am indebted for having provided the images. Carl Bauer, "Die Erde und ihre Bewohner" (Nuremberg: Homann, 1825).

54. On the slow "civilizing" of unseen space, see Larry Wolff, *Inventing Eastern Europe: The Map of Civilization on the Mind of the Enlightenment* (Stanford, Calif.: Stanford University Press, 1994); "Discovering Cultural Perspective: The Intellectual History of Anthropological Thought in the Age of the Enlightenment," in *The Anthropology of the Enlightenment*, ed. Larry Wolff and Marco Cipolloni (Stanford, Calif.: Stanford University Press, 2007).

55. Wilhelm Schickard, *Kurze Anweisung wie künstliche Land-Tafeln auss rechtem Grund zu machen/ Und die biss her begangne Irrthumb zu verbessern/ sampt etlich new erfundenen Vörtheln/ die Polus Höhin auffs leichtest/ und doch scharpff gnug zu forschen* (Tübingen: Johann Georg Cotta, 1669), preface (unpaginated).

56. Blaise Pascal, *Pensées*, trans. Roger Ariew (Indianapolis: Hackett Publishing Company, 2005), 64.

57. Ibid., 265.

58. See the rest of "Good Friday, 1613. Riding Westward," in Donne, *John Donne*, 246–47.

59. On this score, I have been heavily influenced by Max Scheler, *Die Stellung des Menschen im Kosmos* (Darmstadt: O. Reichl, 1928).

60. The most famous articulation of this "New World" approach to changes in anthropology appears in the work of Anthony Pagden. See Anthony Pagden, *Peoples and Empires: A Short History of European Migration, Exploration, and Conquest, from Greece to the Present* (New York: Modern Library, 2003); "Stoicism, Cosmopolitanism, and the Legacy of European Imperialism," *Constellations* 7, no. 1 (2000): 3–22; *Facing Each Other: The World's Perception of Europe and Europe's Perception of the World*, 2 vols. (Aldershot: Ashgate/Variorum, 2000); *Lords of All the World: Ideologies of Empire in Spain, Britain and France, c. 1500–c. 1800* (New Haven, Conn.: Yale University Press, 1995); *The Uncertainties of Empire: Essays in Iberian and Ibero-American Intellectual History* (Aldershot: Ashgate/Variorum, 1994); *European Encounters with the New World: From Renaissance to Romanticism* (New Haven, Conn.: Yale University Press, 1993); *Spanish Imperialism and the Political Imagination: Studies in European and Spanish-American Social and Political Theory, 1513–1830* (New Haven, Conn.: Yale University Press, 1990); *The Fall of Natural Man: The American Indian and the Origins of Comparative Ethnology* (Cambridge: Cambridge University Press, 1986). For useful overviews, see Harry Liebersohn, "Anthropology before Anthropology," in *A New History of Anthropology*, ed. Henrika Kucklick (Malden, Mass.: Blackwell Publishing, 2008); Thomas Hyllans Eriksen and Finn Sivert Nielsen, *A History of Anthropology* (London: Pluto Press, 2001); Mary Douglas, *Implicit Meanings: Essays in Anthropology* (London: Routledge, 1993); Murray Leaf, *Man, Mind, and Science: A History of Anthropology* (New York: Columbia University Press, 1979).

61. For a good discussion of the issues involved, see Carl B. Boyer and Uta C. Merzbach, *A History of Mathematics*, 2nd ed. (New York: John Wiley & Sons, 1991), 483–503.

62. As the work of Jeremy Gray makes clear, we moderns are two steps beyond Euclid. See Gray, *Ideas of Space*.

63. Max Jammer, *Concepts of Space: The History of Theories of Space in Physics* (Cambridge, Mass.: Harvard University Press, 1954), 144–53.

64. See Chapter 2.

65. Fernand Braudel, *The Mediterranean and the Mediterranean World in the Age of Philip II*, 2 vols. (New York: Harper & Row, 1972).

66. Ibid., 2: 678–81.

67. Amir Alexander has noted that early mathematicians borrowed from the voyages of discovery what he calls a "rhetoric of discovery." I would add that the borrowing went in both directions. Amir R. Alexander, *Geometrical Landscapes: The Voyages of Discovery and the Transformation of Mathematical Practice* (Stanford, Calif.: Stanford University Press, 2002).

68. Columbus knew latitude and longitude well and used this knowledge to sail both south and west, in an effort to maximize the territory that he hoped to acquire. See the important work by Nicolás Wey Gómez, *The Tropics of Empire: Why Columbus Sailed South to the Indies* (Cambridge, Mass.: MIT Press, 2008).

69. Bernhardus Varenius and Isaac Newton, *Bernhardi Vareni geographia generalis in qua affectiones generales Telluris explicantur, summâ curâ quam plurimis in locis emendata, & XXXIII schematibus novis, aere incisis, unà cum tabb. aliquot quae desiderabantur* (Cantabrigiae: Ex officina Joann Hayes . . . sumptibus Henrici Dickinson, 1672).

70. I am building on the work of Edward Said and Mary Louise Pratt, although obviously from a different perspective. Mary Louise Pratt, *Imperial Eyes: Travel Writing and Transculturation* (London: Routledge, 1992); Edward W. Said, *Orientalism* (New York: Vintage Books, 1979).

71. Quoted in Simon Schaffer, "Instruments, Surveys and Maritime Empire," in *Empire, the Sea and Global History: Britain's Maritime World, c. 1760–c. 1840*, ed. David Cannadine (Houndmills: Palgrave Macmillan, 2007), 85.

72. Ancient Greek geometry did not reach China before the seventeenth century. See the seminal work by Peter M. Engelfriet, *Euclid in China: The Genesis of the First Chinese Translation of Euclid's Elements, Books I-VI (Jihe Yuanben, Beijing, 1607) and Its Reception up to 1723*, ed. W. L. Idema, (Leiden: Brill, 1998).

73. Friedrich Schiller, "Spruch Des Konfucius," in *Musen-Almanach für das Jahr 1800*, ed. Friedrich Schiller (Tübingen: J. G. Cotta'sche Buchhandlung, 1799), 209–10.

74. That geometry was an essential sign of civilization was an ancient conceit within the Western tradition. Alexander, *Geometrical Landscapes*, 198.

75. The first Chinese translation of Euclid's *Elements* met with a slow and fitful reception. Engelfriet, *Euclid in China*, 86–97.

76. Genesis 1:1–2, in *New American Standard Bible* (Carol Stream, Ill.: Creation House, 1971).

77. Both these texts remain overlooked outside of the specialist literature. See Olivier Depré, "The Ontological Foundations of Hegel's Dissertation of 1801," in *Hegel and the Philosophy of Nature*, ed. Stephen Houlgate (Albany: State University of New York Press, 1998); Alan L. Paterson, "GWF Hegel: Geometrical Studies Introduction," *Hegel Bulletin* 29, no. 1–2 (2008): 118–31.

78. G. W. F. Hegel, "Disseratio philosophica de orbitis planetarum" (Habilitation, University of Jena, 1801); G. W. F. Hegel, *System der Wissenschaft: Erster Theil, Die Phänomenologie des Geistes* (Bamberg: Joseph Anton Goebhardt, 1807). The former has not been translated. The latter is available in English as *Phenomenology of Spirit*, trans. Arnold V. Miller and J. N. Findlay (Oxford: Oxford University Press, 1977).

79. On Hegel's geometric education, see Terry P. Pinkard, *Hegel: A Biography* (Cambridge: Cambridge University Press, 2000), 4. A contemporary translation of Hegel's foray into geometry is available in G. W. F. Hegel, "Geometrical Studies," *Hegel Bulletin* 29, no. 1–2 (2008): 132–53.

80. Georg Wilhelm Friedrich Hegel, *Lectures on the Philosophy of World History, Introduction: Reason in World History*, trans. H. B. Nisbet (Cambridge: Cambridge University Press, 1975), 54–55.

81. I have been influenced, on this score, by Jeremy Gray's views on the history of mathematics. See Gray, *Ideas of Space*; as well as his *Plato's Ghost: The Modernist Transformation of Mathematics* (Princeton, N.J.: Princeton University Press, 2008). Gray puts the full arrival of mathematical "modernism" in the 1890s, which is consistent both with my overall argument and with the arguments of Stephen Kern, whose work I discuss in the conclusion.

82. Nicole Oresme, "Traité de la Sphère; Aristote, Du Ciel et du Monde [De Caelo Et De Mundo], Traduction Française par Nicole Oresme," in *Département des Manuscrits* (Bibliothéque Nationale de France, 1410–1420).

83. See especially Gray, *Ideas of Space*.

Chapter 1

1. *Biblia Latina* (Mainz: Johann Gutenberg and Johannes Fust, 1454); Euclid, *Elementa geometriae* (Venice: Erhard Ratdolt, 1482).

2. On this point, see also Alexandre Koyré, *From the Closed World to the Infinite Universe* (New York: Harper & Brothers, 1957).

3. Nicholas of Cusa, *Opera omnia: Iussu et auctoritate Academiae Litterarum Heidelbergensis ad codicum fidem edita*, 28 vols., vol. 3 (Hamburg: Meiner, 1959), 144.

4. See, for example, Marcia L. Colish, *Medieval Foundations of the Western Intellectual Tradition, 400–1400* (New Haven, Conn.: Yale University Press, 1997); William J. Bouwsma, *The Waning of the Renaissance, 1550–1640* (New Haven, Conn.: Yale University Press, 2000); Anthony Levi, *Renaissance and Reformation: The Intellectual Genesis* (New Haven, Conn.: Yale University Press, 2002); Steven E. Ozment, *The Age of Reform, 1250–1550: An Intellectual and Religious History of Late Medieval and Reformation Europe* (New Haven, Conn.: Yale University Press, 1980). Moreover, even the great Hans Blumenberg's intensely philosophical works of intellectual history do not confront Euclid's career. Blumenberg's *The Legitimacy of the Modern Age* (Cambridge, Mass.: MIT Press, 1985), which is the profoundest discussion of modernity's rise that is available in print today, mentions Euclid a total of zero times.

5. Among others, see Roy Porter, *The Creation of the Modern World: The Untold Story of the British Enlightenment* (New York: W. W. Norton, 2000); Jonathan I. Israel, *Radical Enlightenment: Philosophy and the Making of Modernity 1650–1750* (Oxford: Oxford University Press, 2001); *Enlightenment Contested: Philosophy, Modernity, and the Emancipation of Man, 1670–1752* (Oxford: Oxford University Press, 2006); Ernst Cassirer, *The Philosophy of the Enlightenment* (Princeton, N.J.: Princeton University Press, 1968); Peter Gay, *The Enlightenment: An Interpretation*, 2 vols. (New York: Vintage Books, 1968); Norman Hampson, *The Enlightenment* (Harmondsworth: Penguin, 1968); Paul Hazard, *La Pensée européenne au XVIIIe siècle: de Montesquieu à Lessing* (Paris: Boivin, 1946); *La Crise de la conscience européene (1680–1715)* (Paris: Boivin, 1935); David A. Bell, *Lawyers and Citizens: The Making of a Political Elite in Old Regime France* (New York: Oxford University Press, 1994); Dena Goodman, *The Republic of Letters: A Cultural History of the French Enlightenment* (Ithaca, N.Y.: Cornell University Press, 1994).

6. See, for example, Marvin Perry, *An Intellectual History of Modern Europe* (Boston: Houghton Mifflin, 1993); Roland N. Stromberg, *European Intellectual History since 1789*, 6th ed. (Englewood Cliffs, N.J.: Prentice Hall, 1994).

7. Laurence W. Dickey, *Hegel: Religion, Economics, and the Politics of Spirit, 1770–1807* (Cambridge: Cambridge University Press, 1987); Terry P. Pinkard, *Hegel: A Biography* (Cambridge: Cambridge University Press, 2000); John E. Toews, *Hegelianism: The Path Toward Dialectical Humanism, 1805–1841* (Cambridge: Cambridge University Press, 1980).

8. Pinkard, *Hegel: A Biography*, 4.

9. For two specialized studies, see Antonio Moretto, "Hegel on Greek Mathematics and the Modern Calculus," in *Hegel and Newtonianism*, ed. M. J. Petry (Dordrecht: Springer, 1993); Alan L. Paterson, "GWF Hegel: Geometrical Studies Introduction," *Hegel Bulletin* 29, no. 1–2 (2008): 118–31.

10. The standard work on late nineteenth-century space remains Stephen Kern, *The Culture of Time and Space, 1880–1918* (Cambridge, Mass.: Harvard University Press, 1983). Kern's idea that space and time became less rigid and more fluid concepts within modern culture has heavily influenced how I see space's early modern history. See this book's conclusion for more on this issue.

11. The pioneering work on this matter is Sachiko Kusukawa, *The Transformation of Natural Philosophy: The Case of Philip Melanchthon* (Cambridge: Cambridge University Press, 1995).

12. Euclid, *Elementa geometriae ex Evclide singulari prudentia collecta a Ioanne Vogelin . . . cum praefatione Philippi Melanthonis* (Vitebergae, 1536) (unpaginated preface).

13. Paul L. Rose, *The Italian Renaissance of Mathematics: Studies on Humanists and Mathematicians from Petrarch to Galileo* (Genève: Droz, 1975), 39–48; "Humanist Culture and Renaissance Mathematics: The Italian Libraries of the Quattrocento," *Studies in the Renaissance* 20 (1973): 46–105.

14. On the significance of print, see the fundamental monograph by Elizabeth Eisenstein, *The Printing Press as an Agent of Change: Communications and Cultural Transformations in Early Modern Europe*, 2 vols. (Cambridge: Cambridge University Press, 1979).

15. See Vincent Jullien, *Philosophie naturelle et géométrie au XVIIe siècle* (Paris: Honoré Champion Éditeur, 2006), 257; Carl B. Boyer and Uta C. Merzbach, *A History of Mathematics*, 2nd ed. (New York: John Wiley & Sons, 1991), 119. As the evidence presented in the rest of this chapter shows, there can be little doubt that Euclid's great work diffused widely over many centuries. Still, one must keep in mind that no dedicated census of the *Elements*' early modern print career has been done. Such a study would be an excellent idea.

16. Owen Chadwick, *The Secularization of the European Mind in the Nineteenth Century* (Cambridge: Cambridge University Press, 1990); Nicholas J. Demerath III, "Secularization and Sacralization: Deconstructed and Reconstructed," in *The Sage Handbook of the Sociology of Religion*, ed. James A. Beckford and Nicholas J. Demerath III (Los Angeles: SAGE Publications, 2007); Philip S. Gorski, "Historicizing the Secularization Debate: Church, State, and Society in Late Medieval and Early Modern Europe, Ca. 1300 to 1700," *American Sociological Review* (2000): 138–67; Jonathan Sheehan, "Enlightenment, Religion, and the Enigma of Secularization: A Review Essay," *American Historical Review* 108, no. 4 (2003); Keith Thomas, *Religion and the Decline of Magic: Studies in Popular Beliefs in Sixteenth and Seventeenth-Century England*, 2nd ed. (London: Weidenfeld & Nicolson, 1971); Max Weber, *The Protestant Ethic and the Spirit of Capitalism*, trans. Talcott Parsons (London: Routledge, 1992); Steve Bruce, *God Is Dead: Secularization in the West* (Oxford: Blackwell, 2002).

17. Hans Blumenberg's seminal historical-philosophical work has influenced my approach greatly, even though he reads secularization as coming from religion itself. For his critical examination of traditional views of secularization, especially those expressed by Karl Löwith, see Blumenberg, *The Legitimacy of the Modern Age*, 37–52.

18. Ray C. Jurgensen, Richard G. Brown, and John W. Jurgensen, *Geometry* (Evanston, Ill.: McDougal Littell, 2000), 1.

19. Ibid., 5. This is merely one text among many. I have cited it for two reasons. First, it went through multiple editions, with the first appearing in 1963 as Ray C. Jurgensen, *Modern Geometry: Structure and Method*. Second, its 1982 edition enjoys the distinction of being the one that I used (all too incompetently) as a student in high school: Ray C. Jurgensen, Richard G. Brown, and Alice M. King, *Geometry*, teacher's ed.

20. Euclid, *Euclid's Elements: All Thirteen Books Complete in One Volume*, trans. Thomas L. Heath (Santa Fe: Green Lion Press, 2002), 1.

21. Alexandre Koyré's discussion of space, for instance, was mediated by his interest in the history of science. Koyré, *Closed World*. See also Max Jammer, *Concepts of Space: The History of Theories of Space in Physics*, 2nd ed. (Cambridge, Mass.: Harvard University Press, 1969). In the history of art, see Kirsti Andersen, *The Geometry of an Art: The History of the Mathematical Theory of Perspective from Alberti to Monge* (New York: Springer, 2007); Samuel Y. Edgerton, *The Heritage of Giotto's Geometry: Art and Science on the Eve of the Scientific Revolution* (Ithaca, N.Y.: Cornell University Press, 1991). In the history of mathematics, see Jeremy J. Gray, *Ideas of Space: Euclidean, Non-Euclidean, and Relativistic*, 2nd ed. (Oxford: Clarendon Press, 1989).

22. For a philosophical discussion of humanity's position within the cosmos, see the classic work by Max Scheler, *Die Stellung des Menschen im Kosmos* (Darmstadt: O. Reichl, 1928).

23. See Leon Battista Alberti, *The Mathematical Works of Leon Battista Alberti* (Basel: Springer Science & Business Media, 2010), 17–34; Branko Mitrović, "Leon Battista Alberti and the Homogeneity of Space," *Journal of the Society of Architectural Historians* (2004): 424–39; Samuel Y. Edgerton, "Alberti's Perspective: A New Discovery and a New Evaluation," *Art Bulletin* 48, no. 3–4 (1966): 367–78; Andersen, *The Geometry of an Art*.

24. I had access to a subsequent edition: *Biblia: Das ist: Die gantze heilige Schrifft: Deudsch, auffs new zugericht*, trans. Martin Luther, 2 vols. (Gedrückt zu Wittemberg: Durch Hans Lufft, 1545).

25. On Regiomontanus and mathematics, see the essays in Menso Folkerts, ed., *The Development of Mathematics in Medieval Europe: The Arabs, Euclid, Regiomontanus* (Aldershot: Ashgate, 2006).

26. In general, see the following works by James Hankins: *Plato in the Italian Renaissance*, 2 vols. (Leiden: Brill, 1990); "The Myth of the Platonic Academy of Florence," *Renaissance Quarterly* 44, no. 3 (1991): 429–75; *The Cambridge Companion to Renaissance Philosophy* (Cambridge: Cambridge University Press, 2007); James Hankins and Ada Palmer, *The Recovery of Ancient Philosophy in the Renaissance: A Brief Guide* (Florence: L. S. Olschki, 2008).

27. One excellent example of how this process, albeit from the late sixteenth century, is Henry Savile's progression from the study of Ptolemy's *Almagest* to Euclid's *Elements*. See Robert Goulding, "Henry Savile Reads His Euclid," in *For the Sake of Learning: Essays in Honor of Anthony Grafton*, ed. Ann Blair, Anja-Silvia Goeing, and Anthony Grafton (Leiden: Brill, 2016), 781–82.

28. On the *Timaeus*'s history, see James Hankins, "The Study of the *Timaeus* in Early Renaissance Italy," in *Natural Particulars: Nature and the Disciplines in Renaissance Europe*, ed. Anthony Grafton and Nancy G. Siraisi (Cambridge, Mass.: MIT Press, 1999); Thomas Leinkauf and Carlos G. Steel, *Platons Timaios als Grundtext der Kosmologie in Spätantike, Mittelalter und Renaissance = Plato's Timaeus and the Foundations of Cosmology in Late Antiquity, the Middle Ages and Renaissance* (Leuven: Leuven University Press, 2005); J. H. Waszink, *Studien zum Timaioskommentar des Calcidius* (Leiden: Brill, 1964). For a modern edition, see Plato, *The Collected Dialogues of Plato, Including the Letters* (Princeton, N.J.: Princeton University Press, 1961).

29. For an overview, see Wilbur R. Knorr, *The Ancient Tradition of Geometric Problems* (Boston: Birkhäuser, 1986), 49–88.

30. On the Plethon and the Byzantine migration more broadly, see the essays in John Monfasani, *Byzantine Scholars in Renaissance Italy: Cardinal Bessarion and Other Émigrés: Selected Essays* (Aldershot: Variorum, 1995).

31. Plato, *Dialogues*, 1151–1211.

32. Ibid., 40–98.

33. Ibid., 575–844.

34. In general, see C. A. Patrides, *The Cambridge Platonists* (London: Edward Arnold, 1969); G. A. J. Rogers, Jean-Michel Vienne, and Yves Charles Zarka, *The Cambridge Platonists in Philosophical Context: Politics, Metaphysics, and Religion* (Dordrecht: Kluwer Academic Publishers, 1997); A. Rupert Hall, *Henry More and the Scientific Revolution* (Cambridge: Cambridge University Press, 1996). See also Antoni Malet, "Isaac Barrow on the Mathematization of Nature: Theological Voluntarism and the Rise of Geometrical Optics," *Journal of the History of Ideas* 58, no. 2 (1997): 265–87; "Renaissance Notions of Number and Magnitude," *Historia Mathematica* 33 (2006): 63–81; Richard S. Westfall, *Never at Rest: A Biography of Isaac Newton* (Cambridge: Cambridge University Press, 1983).

35. For an overview of the other "returns," see Jill Kraye, "The Revival of Hellenistic Philosophies," in *The Cambridge Companion to Renaissance Philosophy*, ed. James Hankins (Cambridge: Cambridge University Press, 2007).

36. Richard H. Popkin, *The History of Scepticism: From Savonarola to Bayle*, rev. and expanded ed. (Oxford: Oxford University Press, 2003), 84.

37. On Boethius and Euclid, see the following: Menso Folkerts, *Boethius Geometrie II: Ein mathematisches Lehrbuch des Mittelalters* (Wiesbaden: Steiner, 1970); "The Importance of the Pseudo-Boethian Geometria During the Middle Ages," in *Boethius and the Liberal Arts: A Collection of Essays*, ed. Michael Masi (Bern: Peter Lang, 1981); "Euclid in Medieval Europe," in *The Development of Mathematics in Medieval Europe: The Arabs, Euclid, Regiomontanus* ed. Menso Folkerts (Aldershot: Ashgate, 2006).

38. John L. Flood, "Luther and Tyndale as Bible Translators: Achievement and Legacy," in *Landmarks in the History of the German Language*, ed. Geraldine Horan, Nils Langer, and Sheila Watts (Bern: Peter Lang, 2009); Willem Jan Kooiman, *Luther and the Bible* (Philadelphia: Muhlenberg Press, 1961); Richard Griffiths, *The Bible in the Renaissance: Essays on Biblical Commentary and Translation in the Fifteenth and Sixteenth Centuries* (Aldershot: Ashgate, 2001); Roland H. Worth, *Bible Translations: A History Through Source Documents* (Jefferson, N.C.: McFarland, 1992).

39. *Biblia Latina*, fol. 5r.

40. The Luther Bible originally appeared in 1534. I had access to an edition of 1545. *Biblia: Das ist: Die gantze heilige Schrifft,* 1, fol. Ir.

41. Gordon Campbell, *Bible: The Story of the King James Version, 1611–2011* (Oxford: Oxford University Press, 2011), 47–64.

42. *The Holy Bible Conteyning the Old Testament, and the New: Newly Translated . . . Appointed to Be Read in Churches* (London: By Robert Barker, printer to the Kings most excellent Maiestie, 1611), fol. 39.

43. Folkerts, *Mathematics in Medieval Europe,* 2; John Murdoch, "Euclid: Transmission of the Elements," in *Dictionary of Scientific Biography,* ed. Charles Coulston Gillispie (New York: Charles Scribner's Sons, 1971), 449.

44. Euclid and Proclus, *Eukleidou stoicheion bibl. 15. ekton theonos synousion . . . in qua de disciplinis mathematicis nonnihil,* trans. Simon Grynäus and Johann Herwagen (Basileae: Apud Ioan. Heruagium, 1533).

45. On the editions, see Euclid, *The Thirteen Books of Euclid's Elements,* trans. Thomas Little Heath, 2nd ed., 3 vols., vol. 1 (New York: Dover Publications, 1956), 100–103.

46. Peter Weidhaas, *A History of the Frankfurt Book Fair,* trans. Carolyn Gossage and W. A. Wright (Toronto: Dundurn Press, 2007), 14.

47. Euclid, *Euclide megarense philosopho solo introduttore delle scientie mathematice . . . sera capace à poterlo intendere.,* trans. Gabriele Tadino et al. (Stampato in Vinegia: Per Venturino Roffinelli, 1543); *Das sibend, acht und neunt Buch, des hoch berümbten Mathematici Euclidis Megarensis . . . auß dem Latein ins Teütsch gebracht . . .* trans. Johann Scheubel (Augspurg: Valentin Otmar, 1555); *Les Six Premiers Livres des Eléments d'Euclide,* trans. Pierre Forcadel de Béziers (Paris, 1564); *The Elements of Geometrie of the Most Auncient Philosopher Euclide of Megara . . . of the Best Mathematiciens, Both of Time Past, and in This Our Age,* trans. Henry Billingsley (London: John Daye, 1570); *Los seis libros primeros dela geometria de Evclides . . . canonigo dela Sancta Yglesia de Seuilla,* trans. Rodrigo Zamorano (Seuilla: Alonso de la Barrera, 1576).

48. Euclid, *Elementa geometriae (1536); Analyseis geo-||metricæ sex librorum|| Euclidis.|| . . . pro schola argentinensi.||,* ed. Christian Herlin (Straßburg: Josias Rihel, 1566).

49. Euclid, *Euclid's Elements of Geometry, from the Latin Translation of Commandine . . . Now Done into English* (London: Thomas Woodward, 1723).

50. Euclid, *Die sechs erste Bücher Euclidis, vom Anfang oder Grund der Geometrj . . . Durch Wilhelm Holtzman, genant Xylander/ von Augspurg,* trans. Guillelmus Xylander (Basel: Ioanns Oporini, 1562); *Les Six Premiers Livres des Élémens d'Euclide, traduicts et commentez par J. Errard,* trans. J. Errard (Paris, 1598).

51. Most of the information in this paragraph comes from the very useful overview of translations in *Elements,* 1, 91–113.

52. Euclid, *Elementa geometriae (1536).*

53. Marin Mersenne, *Euclidis elementorum libri, Apollonii Pergae conica, sereni de sectione* (Paris, 1626). On Mersenne's relationship to Skepticism, see Popkin, *History of Scepticism,* 112–25.

54. Euclid, *Euclide megarense philosopho solo introduttore delle scientie mathematice,* trans. Niccolò Tartaglia (Vinegia: Venturino Rossinelli, 1543); Peter Ramus, *Euclides* (Paris: Thomas Richard, 1549).

55. Euclid, *Euclidis Elementorum, libri XIV* (Rome: Vincentius Accoltus, 1574); *Elementorvm libri XC. breviter demonstrati, operâ Is. Barrow, Cantabrigensis,* trans. Isaac Barrow (Cantabrigiae: Guilielmo Nealand, 1655). On Clavius's reputation, see Sabine Rommevaux, *Clavius, un clé pour Euclide au XVI siècle* (Paris: Vrin, 2005), 15–17; Frederick A. Homann, "Christopher Clavius and the Renaissance of Euclidean Geometry," *Archivum Historicum Societatis Iesu Roma* 52, no. 4 (1983): 233–46. On Clavius himself, see Romano Gatto, "Christoph Clavius' 'Ordo servandus in addiscendis disciplinis mathematicis' and the Teaching of Mathematics in Jesuit Colleges at the Beginning of the Modern Era," *Science & Education* 15 (2006): 235–58; James M. Lattis, *Between Copernicus and Galileo: Christoph Clavius and the Collapse of Ptolemaic Cosmology* (Chicago: University of Chicago Press, 1995). On Barrow and Clavius, see Mordechai Feingold, ed., *Before Newton: The Life and Times of Isaac Barrow* (Cambridge: Cambridge University Press, 1990), 351. On Barrow, see Malet, "Isaac Barrow on the

Mathematization of Nature"; G. A. J. Rogers, "Locke, Newton, and the Cambridge Platonists on Innate Ideas," *Journal of the History of Ideas* (1979): 191–205.

56. Euclid, *Elementa geometriae* (unpaginated).

57. Euclid, *Evclidis Elementorvm libri XV: Acceßit XVI. de solidorvm regvlarivm . . . multarum rerum accessione locupletati*, ed. Christoph Clavius, Nunc tertio ed., summaq', diligentia recogniti, atque emendati ed. (Colonia: Ioh. Baptistae Ciottus, 1591).

58. Euclid, *Euclide megarense*, Fol IIIIr.

59. Euclid, *Die sechs erste Bücher Euclidis*, 1.

60. Euclid, *Six Premiers Livres*, 1.

61. Euclid, *Los seis libros primeros*, Fol. 9.

62. Euclid, *Teutsch-redender Euclides, oder acht Bücher von denen Anfängen der Mesz-Kunst . . . in Teutscher Sprach eingerichtet und bewiesen durch A. E. B. V. P.*, trans. Anton Ernst Burckhard von Birckenstein (Wienn: Philipp Fievers, Buch- und Kunst-Händler, 1694).

63. Ibid., 3.

64. Rose, *Italian Renaissance*, 29.

65. Ibid.

66. Leonardo da Vinci, *The Notebooks of Leonardo Da Vinci / Arranged and Rendered into English with Introductions by Edward Maccurdy* (New York: Reynal & Hitchcock, 1938), 39.

67. Nicolaus Copernicus, *De lateribvs et angvlis triangulorum, tum planorum rectilineorum, tum sphæricorum . . . additus est canon semissium subtensarum rectarum linearum in circulo* (Vittembergae: Johannes Lufft, 1542).

68. Rose, *Italian Renaissance* 130.

69. Albrecht Dürer, *Underweysung der Messung/ mit dem Zirkel und Richtscheyt/ in Linien Ebnen unnd gantzen Corporen* (Norimbergae, 1525), unpaginated.

70. Johannes Kepler, *Prodromus dissertationvm cosmographicarvm, continens mysterivm cosmographicvm . . . astronomiæ restauratoris D. Nicolai Copernici* (Tvbingae: Georg Gruppenbach, 1596). Also quoted in Boyer and Merzbach, *History of Mathematics*, 55.

71. John Aubrey, *Brief Lives: A Modern English Version* (Woodbridge: Boydell Press, 1982), 151–52.

72. See John Locke, "Instructions for the Education of Edward Clarke's Children, C. 8 February 1686," in *Electronic Enlightenment*, ed. Robert McNamee (Oxford: Oxford University Press, 2011), http://www.e-enlightenment.com; Jeremy Bentham, "To Sir Samuel Bentham, 20–26 August 1773," ibid.; Adam Smith, "To John Petty, 1st Earl of Shelburne, 4 April 1759," ibid.

73. Thomas Reid, "To James Gregory, 1786," in *Electronic Enlightenment*, ed. Robert McNamee (Oxford: Oxford University Press, 2011), http://www.e-enlightenment.com.

74. I am borrowing from the following: Walter Mignolo, *The Darker Side of the Renaissance: Literacy, Territoriality, and Colonization* (Ann Arbor: University of Michigan Press, 1995); *The Darker Side of Western Modernity: Global Futures, Decolonial Options* (Durham, N.C.: Duke University Press, 2011).

75. Kevin J. Hayes, "How Thomas Jefferson Read the Qur'ān," *Early American Literature* 39, no. 2 (2004).

76. James Gilreath and Douglas L. Wilson, *Thomas Jefferson's Library: A Catalog with the Entries in His Own Order* (Washington, D.C.: Library of Congress, 1989), 94–95; Roger Kennedy, "Jefferson and the Indians," *Wintherthur Portfolio* 27, no. 2/3 (1992): 106.

77. Isaac Barrow, *The Usefulness of Mathematical Learning Explained and Demonstrated . . . Translated by the Revd. Mr. John Kirkby*, trans. John Kirkby (London: printed for Stephen Austen, 1734), xxvi.

78. Ibid., xxvii.

79. On this theme, see Gerhard Oestreich, *Antiker Geist und moderner Staat bei Justus Lipsius (1547–1606): Der Neustoizismus als politische Bewegung*, Schriftenreihe der Historischen Kommission bei der Bayerischen Akademie der Wissenschaften (Göttingen: Vandenhoeck & Ruprecht, 1989).

80. Euclid, *The Elements of Geometrie* (unpaginated translator's preface).

81. On Clavius and geometry, see Homann, "Christopher Clavius"; Peter Dear, *Discipline & Experience: The Mathematical Way in the Scientific Revolution* (Chicago: University of Chicago Press, 1995), 65–66; Rommevaux, *Clavius*; Peter Dear, *Revolutionizing the Sciences: European Knowledge and Its Ambitions, 1500–1700* (Princeton, N.J.: Princeton University Press, 2001), 66–67.

82. Euclid, *Euclidis Elementorum (1574)*.

83. *Evclidis Elementorvm libri XV . . . Christophori Clauijè Societ. Iesv, & aliorum collati, emendati & aucti*, ed. Christoph Clavius (Coloniae: Gosuinus Cholinus, 1607).

84. On the Jesuits and mathematics, see John L. Heilbron, *Elements of Early Modern Physics* (Berkeley: University of California Press, 1981), 95–96.

85. Robert Recorde, *The Pathway to Knowledge, Containing the First Principles of Geometrie . . . Much Necessary for All Sortes of Men* (London: Reynold Wolfe, 1551) (unpaginated preface).

86. Euclid, *Elementos geometricos de Evclides . . . Jacobo Kresa de la Compañia de Jesus, cathedrático de mathematicas en los estudios reales del Colegio Imperial de Madrid . . .*, trans. Jacobo Kresa (Brvsselas: Francisco Foppens, 1689).

87. On Euclidism, see Michele Sbacchi, "Euclidism and Theory of Architecture," *Nexus Network Journal* 3, no. 2 (2001). On homogenous space's implications, see Koyré, *Closed World*.

88. See the introduction to Abel Bürja, *Lehrbuch der Astronomie*, 5 vols., vol. 1 (Berlin: Schöne, 1794), xxxiii–xxxvii.

89. For a perfect example, see the primer in the (unpaginated) section one of Mauro Fiorentino, *Sphera volgare novamente tradotta con molte notande additioni di geometria, cosmographia, arte navicatoria, et stereometria, proportioni, et qvantità delli ellementi, distanze, grandeze, et movimienti di tutti li corpi celesti, cose certamente rade et maravigliose* (Venetia: Bartholomeo Zanetti, 1537.)

90. Bernard Lamy, *Les Élémens de géométrie ou de la mesure du corps. Qui comprennent les Élémens d'Euclide, les plus belles propositions d'Archimède & l'analise* (Paris: André Pralard, 1685).

91. Ibid. (unpaginated preface).

92. Renzo Baldasso, "Portrait of Luca Pacioli and Disciple: A New, Mathematical Look," *Art Bulletin* 92, no. 1/2 (2010).

93. Margaret Jacob was the first to identify this phenomenon. Margaret C. Jacob, *The Radical Enlightenment: Pantheists, Freemasons, and Republicans* (London: Allen & Unwin, 1981). For a work that builds on her insights, see Israel, *Radical Enlightenment*.

94. On Euclid's position in the history of mathematics, see Benno Artmann, *Euclid: The Creation of Mathematics* (New York: Springer, 1999). On the shift away from Euclidean space within mathematics, see Gray, *Ideas of Space*.

95. Boyer and Merzbach, *History of Mathematics*, 282–84; Morris Kline, *Mathematics in Western Culture* (New York: Oxford University Press, 1953), 99–109; Glen Van Brummelen, *The Mathematics of the Heavens and the Earth: The Early History of Trigonometry* (Princeton, N.J.: Princeton University Press, 2009), 247–72.

96. For overviews of this history, see David L. Wagner, ed., *The Seven Liberal Arts in the Middle Ages* (Bloomington: Indiana University Press, 1983); "The Seven Liberal Arts and Classical Scholarship," in *The Seven Liberal Arts in the Middle Ages*, ed. David L. Wagner (Bloomington: Indiana University Press, 1983).

97. For Kepler's attempts to apply geometry to celestial mechanics, see Kepler, *Mysterivm cosmographicvm*.

98. The literature on Galileo's approach to space is enormous. I cite one key work: Alexandre Koyré, *Galileo Studies* (Atlantic Highlands, N.J.: Humanities, 1978).

99. For a lively discussion of what these developments meant for European culture more broadly, see Amir R. Alexander, *Infinitesimal: How a Dangerous Mathematical Theory Shaped the Modern World* (New York: Scientific American, 2014).

100. For a general discussion, see Carl B. Boyer, *The History of the Calculus and Its Conceptual Development* (New York: Dover, 1959).

101. For a discussion of Euclid's approach to numbers, see Artmann, *Euclid*, 193–202.

102. Katherine Neal, *From Discrete to Continuous: The Broadening of Number Concepts in Early Modern England* (Dordrecht: Kluwer, 2002), 22.

103. Gray, *Ideas of Space*, 6–7.

104. On the early stages of this transition, see Neal, *From Discrete to Continuous*, 28–45.

105. For a brief overview of Euler's career, see Boyer and Merzbach, *History of Mathematics*, 406–22.

106. In general, see Gray, *Ideas of Space*; *Worlds out of Nothing: A Course in the History of Geometry in the 19th Century* (London: Spinger-Verlag, 2007).

107. Euclid, *Euclid's Elements*, 2.

108. On this process, see Goulding, "Henry Savile Reads His Euclid." See also Gray, *Ideas of Space*; Kern, *Culture*, 133.

109. On these developments, see Boyer and Merzbach, *History of Mathematics*, 572–97; Morris Kline, *Mathematical Thought from Ancient to Modern Times*, 3 vols., vol. 3 (New York: Oxford University Press, 1990), 861–81; Howard W. Eves, *An Introduction to the History of Mathematics*, 5th ed. (Philadelphia: Saunders College Publishing, 1983); Dirk J. Struik, *A Concise History of Mathematics*, 4th rev. ed. (New York: Dover Publications, 1987), 169–81.

110. For a general assessment of Riemann, see Detlef Laugwitz, *Bernhard Riemann, 1826–1866: Turning Points in the Conception of Mathematics* (Boston: Birkhäuser, 1999). Within the context of the history of mathematics, see Boyer and Merzbach, *History of Mathematics*, 496–99.

111. Bernhard Riemann, *Ueber die Hypothesen, welche der Geometrie zu Grunde liegen*, vol. 13, Abhandlungen der Königlichen Gesellschaft der Wissenschaften zu Göttingen (Göttingen, 1867).

112. Boyer and Merzbach, *History of Mathematics*, 572–97.

113. Laugwitz, *Riemann*, 269–75.

Chapter 2

1. Jósef Babicz, "The Celestial and Terrestrial Globes of the Vatican Library, Dating from 1477, and Their Maker Donnus Nicolaus Germanus (Ca 1420–Ca 1490)," *Der Globusfreund: Wissenschaftliche Zeitschrift für Globenkunde* 35/37 (1987): 155–56; "Donnus Nicolaus Germanus—Probleme seiner Biographie und sein Platz in der Rezeption der ptolemäischen Geographie," *Wolfenbütteler Forschungen* 7 (1980): 9–42.

2. For informative discussions of pre-Behaim globes, see Patrick Gautier Dalché, *La Géographie de Ptolémée en Occident (IVe–XVIe Siècle)*, Terrarum Orbis (Turnhout: Brepols, 2009), 246; "Avant Behaim: Les globes terrestres au XVe *siècle*," *Médiévales* 58, no. 1 (2010): 43–61; Jacques Paviot, "Ung Mapmonde Rond, En Guise De Pom(M)E: Ein Erdglobus von 1440–44, Hergestellet für Philipp den Guten, Herzog von Burgund," *Der Globusfreund* 43/44 (1995): 19–29; José Ruysschaert, "Du Globe terrestre attribué à Giulio Romano aux globes et au planisphère oubliés de Nicolaus Germanus," *Bolletino dei monumenti, musei e gallerie pontifice* 6 (1985): 93–104.

3. Elly Dekker and P. C. J. van der Krogt, *Globes from the Western World* (London: Zwemmer), 158.

4. Petrus Apian, *Cosmographiae introductio: Cum quibusdam gaeometriae ac astronomiae principiis ad eam rem necessariis* (Ingolstadt, 1529).

5. Peter Apian, *Petri Apiani cosmographia, per Gemmam Phrysium, apud louanienses medicum ac mathematicum insignem, denuo restitua. Additis de eadem re ipsius Gemmae Phry. libellis, quos sequens pagina docet* (Antuerpiae: Arnoldo Berckmanno, 1540).

6. Ute Obhof, "Der Erdglobus, der Amerika benannte: Die Überlieferung der Globensegmente von Martin Waldseemüller," *Der Globusfreund: Wissenschaftliche Zeitschrift für Globenkunde* 55/56 (2009): 13–22; Thomas Horst, "Der Niederschlag von Entdeckungsreisen auf Globen des frühen 16. Jahrhunderts," *Der Globusfreund: Wissenschaftliche Zeitschrift für Globenkunde* 55/56 (2009): 23–38; Edward H. Dahl and Jean-François Gauvin, *Sphaerae Mundi: Early Globes at the Stewart Museum* (Sillery, Que.: McGill-Queen's University Press, 2000); Tony Campbell, "A Descriptive Census of Willem Blaeu's Sixty-Eight Centimeter Globes," *Imago Mundi* 28 (1976): 21–50; H. M. Wallis, "The

Molyneux Globes," *British Museum Quarterly* 16, no. 4 (1952): 89–90; Alfred Kohler, "Die Entwicklung der Darstellung Afrikas auf deutschen Globen des 15. und 16. Jahrhunderts," *Der Globusfreund: Wissenschaftliche Zeitschrift für Globenkunde* 18/20 (1970): 85–96.

7. Edward Luther Stevenson, *Terrestrial and Celestial Globes: Their History and Construction, Including a Consideration of Their Value as Aids in the Study of Geography and Astronomy* (New Haven, Conn.: Pub. for the Hispanic Society of America by the Yale University Press, 1921). See also Alois Fauser, *Die Welt in Händen: Kurze Kulturgeschichte des Globus* (Stuttgart: Schuler, 1967); *Kulturgeschichte des Globus*, reprint ed. (Vienna: Vollmer, 1967); Oswald Muris and Gert Saarmann, *Der Globus im Wandel der Zeiten; Eine Geschichte der Globen* (Berlin: Columbus Verlag, 1961); Catherine Hofmann, *Le Globe & son image* (Paris: Bibliothèque Nationale de France, 1995); Dennis Cosgrove, *Apollo's Eye: A Cartographic Genealogy of the Earth in the Western Imagination* (Baltimore: Johns Hopkins University Press, 2001).

8. Dekker and Krogt, *Globes*, 158; David Woodward, "The Image of the Spherical Earth," *Perspecta* 25 (1989): 3–15.

9. See the critical comments in Patrick Gautier Dalché, "The Reception of Ptolemy's *Geography*," in *The History of Cartography*, ed. J. B. Harley and David Woodward (Chicago: University of Chicago Press, 1987), 297; Elly Dekker, "Globes in Renaissance Europe," in ibid., 140.

10. Bott and Willers, *Focus Behaim Globus: Germanisches Nationalmuseum, Nürnberg, 2. Dezember 1992 bis 28. Februar 1993*; Oswald Muris, "Der Globus des Martin Behaim," *Mitteilungen der Geographischen Gessellschaft Wien* 97 (1955): 169–82; Friedrich Wilhelm Ghillany, *Geschichte des Seefahrers Ritter Martin Behaim* (Nürnberg: Bauer and Raspe, J. Merz, 1853); Siegmund Günther, *Martin Behaim. Zeichnungen von Otto E. Lau* (Bamberg: Buchnersche Verlagsbuchhandlung, 1890); Ernest George Ravenstein, *Martin Behaim: His Life and His Globe* (London: George Philip & Son, 1908); Andreas Reichenbach, *Martin Behaim. Ein deutscher Seefahrer aus dem fünfzehnten Jahrhundert* (Wurzen: C. Kiesler, 1889).

11. Babicz, "The Celestial and Terrestrial Globes of the Vatican Library"; "Donnus Nicolaus Germanus."

12. Gautier Dalché, *Géographie de Ptolémée*; "Avant Behaim"; Paviot, "Ung Mapmonde Rond"; Ruysschaert, "Du Globe terrestre attribué à Giulio Romano."

13. Horst, "Niederschlag"; Alfred Kohler, "Die Entwicklung der Darstellung Afrikas." For an example of how historians understand Columbus's significance for the history of space, see Simek, *Heaven and Earth*, 1–4, 23–24.

14. A notable exception is Dekker, "Globes in Renaissance Europe," 140–44.

15. Luitpold Dussler, *Raphael: Kritisches Verzeichnis der Gemälde, Wandbilder und Bildteppiche* (Munich: Bruckmann, 1966), 68–82; Christiane L. Joost-Gaugier, *Raphael's Stanza della Segnatura: Meaning and Invention* (Cambridge: Cambridge University Press, 2002), 1–17; *Italian Renaissance Art: Understanding Its Meaning* (Chichester: Wiley-Blackwell, 2013), 152–83.

16. There is some debate about the identity of the figures depicted, since Raphael included no names. In identifying the celestial globe holder as Hipparchus I am following Noel M. Swerdlow, "Essay Review: Ptolemy's Geography, an Annotated Translation of the Theoretical Chapters by J. Lennart Berggren and Alexander Jones," *Annals of Science* 60, no. 3 (2010): 313. There are, however, alternative views. Since the seventeenth century, it has been common to identify the globe bearer as Zoroaster. For an example, see Roger Jones and Nicholas Penny, *Raphael* (New Haven, Conn.: Yale University Press, 1983), 77. This reading makes little sense, given that the image's other figures are all Greek speakers, while Zoroaster spoke Persian. Christiane L. Joost-Gaugier, in contrast, has argued that the celestial globe bearer is the geographer Strabo (64 BC–24 AD), who did speak Greek: Christiane L. Joost-Gaugier, "Ptolemy and Strabo and Their Conversation with Appelles and Protogenes: Cosmography and Painting in Raphael's School of Athens," *Renaissance Quarterly* 51, no. 3 (1998): 761–87. As Swerdlow argues, however, this view cannot be reconciled with the history of astronomy, since Strabo's work included no stellar coordinates. Moreover, Strabo explicitly rejected the geometric approach to geography that Ptolemy later cultivated. On Strabo's relationship to Ptolemaic methods,

see Katherine Clarke, *Between Geography and History: Hellenistic Constructions of the Roman World* (Oxford: Clarendon Press, 1999), 215–16.

17. Here, I am following Joost-Gaugier, "Ptolemy and Strabo."

18. Woodward, "Image," 7–8. See also Simek, *Heaven and Earth*, 10–13.

19. Alois Schlachter and Friedrich Gisinger, *Der Globus, seine Entstehung und Verwendung in der Antike nach den Literarischen Quellen und den Darstellungen in der Kunst* (Leipzig: B. G. Teubner, 1927).

20. Dekker and Krogt, *Globes*, 12–13.

21. Ibid., 9–10; Elly Dekker, *Illustrating the Phaenomena Celestial Cartography in Antiquity and the Middle Ages* (Oxford: Oxford University Press, 2013), 69–79.

22. Pascal Arnaud, "L'Image du globe dans le monde romain: Science, iconographie, symbolique," *Mélanges de l'Ecole Française de Rome. Antiquité* 96, no. 1 (1984): 53–116.

23. Woodward, "Image," 8; Matteo Fiorini, *Sfere terrestri e celesti di autore italiano, oppure fatte o conservate in Italia* (Roma: La Società Geografica Italiana, 1899), 1–2. See also Jean-Marc Besse, *Les Grandeurs de la terre: Aspects du savoir géographique à la renaissance*, Collection Sociétés, Espaces, Temps (Lyon: ENS, 2003), 58, 80.

24. Germaine Aujac, "Greek Cartography in the Early Roman World," in *The History of Cartography: Cartography in Prehistoric, Ancient and Medieval Europe and the Mediterranean*, ed. J. B. Harley and David Woodward (Chicago: University of Chicago Press, 1987), 171.

25. Giorgio Tabarroni, "Globi celesti e terrestri sulle monete romane," *Physis: Rivista di storia della scienza* 8, no. 3 (1965): 318–53. In general, see Dekker, *Illustrating the Phaenomena*. On Islam and celestial globes, see Emilie Savage-Smith, "Celestial Mapping," in *The History of Cartography: Cartography in Prehistoric, Ancient and Medieval Europe and the Mediterranean*, ed. J. B. Harley and David Woodward (Chicago: University of Chicago Press, 1987), 3–60; Emilie Savage-Smith and Andrea P. A. Belloli, *Islamicate Celestial Globes, Their History, Construction, and Use* (Washington, D.C.: Smithsonian Institution Press, 1985); Fiorini, *Sfere terrestri*, 27–30.

26. Quite similar sentiments are expressed in Brown, *The Story of Maps*, 83–85.

27. Willy Hartner, "The Astronomical Instruments of Cha-Ma-Lu-Ting, Their Identification, and Their Relations to the Instruments of the Observatory of Marāgha," *Isis* 41, no. 2 (1950). It is long past time for a complete history of this globe and its intellectual milieu to be written.

28. For overviews, see Olaf Pedersen, "Astronomy," in *Science in the Middle Ages*, ed. David C. Lindberg (Chicago: University of Chicago Press, 1978); *Early Physics and Astronomy: A Historical Introduction*, rev. ed. (Cambridge: Cambridge University Press, 1993); W. G. L. Randles, *The Unmaking of the Medieval Christian Cosmos, 1500–1700: From Solid Heavens to Boundless Aether* (Aldershot: Ashgate, 1999).

29. Pedersen, "Astronomy," 305–8.

30. For a more specific discussion, see Bruce Eastwood, *Ordering the Heavens: Roman Astronomy and Cosmology in the Carolingian Renaissance* (Leiden: Brill, 2007), 187–217.

31. On Gerbert d'Aurillac, see Marco Zuccato, "Gerbert of Aurillac and a Tenth-Century Jewish Channel for the Transmission of Arabic Science to the West," *Speculum* 80, no. 3 (2005). On Notker Labeo, see Anna A. Grotans, *Reading in Medieval St. Gall* (Cambridge: Cambridge University Press, 2006), 80.

32. For an overview of how this issue related to the medieval production of celestial globes, see Dekker, *Illustrating the Phaenomena*, 192–207.

33. Stephen C. McCluskey, *Astronomies and Cultures in Early Medieval Europe* (Cambridge: Cambridge University Press, 1998), 29–50; Edward Grant, "Science and Theology in the Middle Ages," in *God and Nature: Historical Essays on the Encounter Between Christianity and Science*, ed. David C. Lindberg and Ronald L. Numbers (Berkeley: University of California Press, 1986), 49–50.

34. McCluskey, *Astronomies and Cultures*, 123–27.

35. For an overview, see Stephen Brown, "The Intellectual Context of Later Medieval Philosophy: Universities, Aristotle, Arts, Theology," in *Medieval Philosophy*, ed. John Marenbon (London: Routledge, 1998), 188–203. On Aristotle and medieval cosmology, see Grant, *Science and Religion*, 41–46; "The Medieval Cosmos," 148–50; "Science and Theology in the Middle Ages," 52–53.

36. The three works mentioned are available in Aristotle, *The Complete Works of Aristotle: The Revised Oxford Translation*, 2 vols., vol. 1 (Princeton, N.J.: Princeton University Press, 1984), 315–446, 447–511, 555–625.

37. For a good overview, see the translator's introduction to Claudius Ptolemy, *Ptolemy's Almagest*, trans. G. J. Toomer (Princeton, N.J.: Princeton University Press, 1998). See also G. J. Toomer, "Ptolemy," in *Dictionary of Scientific Biography*, ed. Charles C. Gillispie and Frederic L. Holmes (New York: Scribner, 1981); Alain Bernard, "The Significance of Ptolemy's *Almagest* for Its Early Readers," *Revue de Synthèse* 6, no. 4 (2010): 495–521.

38. Gautier Dalché, *Géographie de Ptolémée*, 20.

39. Ptolemy, *Ptolemy's Almagest*, 45–47.

40. Alexander Jones, "Ptolemy's Mathematical Models and Their Meaning," in *Mathematics and the Historian's Craft: The Kenneth O. May Lectures*, ed. Glen van Brummelen and Michael Kinyon (New York: Springer, 2005), 28; "The Stoics and the Astronomical Sciences," in *The Cambridge Companion to the Stoics*, ed. Brad Inwood (Cambridge: Cambridge University Press, 2003), 342–44.

41. David E. Hahm, *The Origins of Stoic Cosmology* (Columbus: Ohio State University Press, 1977), 136–84; Shmuel Sambursky, *Physics of the Stoics* (London: Routledge & Kegan Paul, 1959), 30.

42. On medieval thought's separation of Heaven and Earth, see Eastwood, *Ordering the Heavens*, 187–217.

43. The fragments are available in Hans Friedrich August von Arnim, *Stoicorum veterum fragmenta*, 4 vols. (Lipsiae: in aedibus B. G. Teubneri, 1903).

44. Ptolemy, *Ptolemy's Almagest*, 36.

45. Gerd Grasshoff, *The History of Ptolemy's Star Catalogue* (New York: Springer-Verlag, 1990), 9–16; Olaf Pedersen and Alexander Jones, *A Survey of the Almagest: With Annotation and New Commentary by Alexander Jones* (New York: Springer, 2011), 257.

46. These spheres were generally "inside out," by which is meant that they put the apparently concave surface onto the convex surface of a sphere. Over time, celestial spheres would appear that presented an actual view from "above." Both the process behind and the meaning of this shift require further study.

47. Dekker, "Globes in Renaissance Europe," 138.

48. For the original, see Aristotle, *Complete Works* 1, 315–447. For useful discussions, see Ian Mueller, "Aristotle on Geometrical Objects," *Archiv für Geschichte der Philosophie* 52 (1970): 156–71; Engelfriet, *Euclid in China*, 23–45.

49. On alternative traditions of classical spatial thought, see David Sedley, "Philoponus' Conception of Space," in *Philoponus and the Rejection of Aristotelian Science*, ed. Richard Sorabji (London: Duckworth, 1987), 181–82. On the pivotal role of Carolingian thought and its emphasis on bringing order to the Heavens, see Eastwood, *Ordering the Heavens*, 372–75.

50. Pedersen, "Astronomy," 320–21.

51. Lynn Thorndike, *The Sphere of Sacrobosco and Its Commentators* (Chicago: University of Chicago Press, 1949), 76–77.

52. David Woodward, "Roger Bacon's Terrestrial Coordinate System," *Annals of the Association of American Geographers* 80, no. 1 (1990): 109–22.

53. Anne D. Hedeman, "Gothic Manuscript Illustration: The Case of France," in *A Companion to Medieval Art: Romanesque and Gothic in Northern Europe*, ed. Conrad Rudolph (Malden, Mass.: Blackwell, 2006), 425–27; John Lowden, *The Making of the Bibles Moralisées*, 2 vols., vol. 1 (University Park: Pennsylvania State University Press, 2000), 1–11.

54. On these maps, see Evelyn Edson, *The World Map, 1300–1492: The Persistence of Tradition and Transformation* (Baltimore: Johns Hopkins University Press, 2007); David Woodward, "Medieval *Mappaemundi*," in *The History of Cartography: Cartography in Prehistoric, Ancient and Medieval Europe and the Mediterranean*, ed. J. B. Harley and David Woodward (Chicago: University of Chicago Press, 1987); "Reality, Symbolism, Time, and Space in Medieval World Maps," *Annals of the Association of American Geographers* 75, no. 4 (1985): 510–21.

55. Jim Bennett has called for including instruments more fully in the history of science. I fully endorse his point. See Jim Bennett, "Practical Geometry and Operative Knowledge," *Configurations* 6, no. 2 (1998): 195–222.

56. W. R. Laird, "Archimedes Among the Humanists," *Isis* 82, no. 4 (1991): 631.

57. Johannes Regiomontanus, *Epytoma Joannis de Monte Regio in Almagestum Ptolomei* (Venice, 1496). On Regiomontanus in general, see Rudolf Mett, *Regiomontanus: Wegbereiter des neuen Weltbildes* (Stuttgart: B. G. Teubner, 1996); Ernst Zinner, *Leben und Wirken des Joh. Müller von Königsberg Genannt Regiomontanus*, ed. Helmut Rosenfeld and Otto Zeller, 2nd ed., Milliaria, 10,1 (Osnabrück: Otto Zeller, 1968); John D. North, *Cosmos: An Illustrated History of Astronomy and Cosmology* (Chicago: University of Chicago Press, 2008), 207–8.

58. Michael Hoskin, ed., *The Cambridge Concise History of Astronomy* (Cambridge: Cambridge University Press, 1999), 39.

59. Johannes Schöner, *Opusculum . . . summa cura & diligentia collectum, accomodatum ad recenter elaboratum ab eodem globum descriptionis terrenae* (Norimbergae: Johann Petreius, 1533).

60. This process is evident in the images provided in Dennis Cosgrove, "Images of Renaissance Cosmography, 1450–1650," in *The History of Cartography*, ed. J. B. Harley and David Woodward (Chicago: University of Chicago Press, 1987), 62–63.

61. Armillary spheres underwent repeated changes in design philosophy. For a general overview, see the informative discussion in Elly Dekker, Silke Ackermann, and Kristen Lippincott, *Globes at Greenwich: A Catalogue of the Globes and Armillary Spheres in the National Maritime Museum, Greenwich* (Oxford: Oxford University Press and the National Maritime Museum, 1999), 5–6.

62. Christoph Clavius, *Christophori Clavii bambergensis in sphaeram Ioannis de Sacro Bosco commentarius* (Romae: Apud Victorium Helianum, 1570).

63. Mauro Fiorentino, *Sphera volgare novamente tradotta con molte notande additioni di geometria, cosmographia, arte navicatoria, et stereometria, proportioni, et qvantita delli ellementi, distanze, grandeze, et movimienti di tutti li corpi celesti, cose certamente rade et maravigliose* (Venetia: Bartholomeo Zanetti, 1537).

64. For an analysis of not only this fresco, but also the entire room in which it is located, see Joost-Gaugier, *Raphael's Stanza della Segnatura*, 50–52. The fresco is also called the "Urania," although Joost-Gaugier notes that there is no evidence of this fresco's actual title.

65. On Renaissance receptions of classical works in general, see Anthony Grafton, *Defenders of the Text: The Traditions of Scholarship in an Age of Science, 1450–1800* (Cambridge, Mass.: Harvard University Press, 1991), 23–46; *Commerce with the Classics: Ancient Books and Renaissance Readers* (Ann Arbor: University of Michigan Press, 1997).

66. J. A. May, "The Geographical Interpretation of Ptolemy in the Renaissance," *Tijdschrift voor Economische en Sociale Geografie* 73, no. 6 (1982): 350–61; Erich Polaschek, "Ptolemy's 'Geography' in a New Light," *Imago Mundi* 14 (1959): 17–37; Woodward, "Image," 11–12.

67. On the persistence of Ptolemy's geographic thought, see Gautier Dalché, *Géographie de Ptolémée*.

68. O. A. W. Dilke, "Cartography in the Byzantine Empire," in *The History of Cartography: Cartography in Prehistoric, Ancient and Medieval Europe and the Mediterranean*, ed. J. B. Harley and David Woodward (Chicago: University of Chicago Press, 1987), 256–75; in the same work, Gerald R. Tibbetts, "The Beginnings of a Cartographic Tradition," 106.

69. Gautier Dalché, "The Reception of Ptolemy's *Geography*," 291–94; Germaine Aujac, *Claude Ptolémée, astronome, astrologue, géographe: connaissance et représentation du monde habité* (Paris: Editions du CTHS, 1993), 173.

70. On this text, see Anthony Grafton, April Shelford, and Nancy G. Siraisi, *New Worlds, Ancient Texts: The Power of Tradition and the Shock of Discovery* (Cambridge, Mass.: Harvard University Press, 1992), 48–50. The original, Claudius Ptolemy, "Geographia," is in the New York Public Library, Manuscripts and Archives Division, 1460–70.

71. For all the editions listed in this paragraph, see Henry N. Stevens, *Ptolemy's Geography. A Brief Account of All the Printed Editions Down to 1730* (London: Henry Stevens, Son and Stiles, 1908).

72. Claudius Ptolemy, *Klaudiou Ptolemaiou alexandreōs philosophou en tis malisa te paideumenou. Peri tēs geōgraphikēs biblia oktō, meta pasēs akribeias entiptothenta* (Basel: Frobenius, 1533).

73. *Claudii Ptolemaei geographicae enarrationis libri octo* (Strasbourg: Koberger, 1525). On the significance of this translation, see Gautier Dalché, "The Reception of Ptolemy's *Geography*," 356–58.

74. Claudius Ptolemy, *Claudii Ptolemaei alexandrini geographicae enarrationis libri octo . . . exemplaria à Michaele Villanovano iam primum recogniti . . .* ed. Michael Servetus, trans. Willibald Pirckheimer (Lugduni: Trechsel, 1535).

75. *Geographia universalis, vetvs et nova, complectens Clavdii Ptolemæi alexandrini enarrationis libros VIII . . . in quo varij gentium & regionum ritus & mores explicantur. Pr[a]efixus est* (Basileae: Heinrich Petri, 1540); *Tabvlae geographicae Cl: Ptolomei: Ad mentem autoris restitutae & emendate* (Cologne: Godefridus Kempen, 1578). Mercator's version was abbreviated.

76. Babicz, "The Celestial and Terrestrial Globes of the Vatican Library," 219–24; "Donnus Nicolaus Germanus"; Gautier Dalché, *Géographie De Ptolémée*; Dekker, "Globes in Renaissance Europe," 146; Gautier Dalché, "The Reception of Ptolemy's *Geography*," 320–21.

77. Stevens, *Ptolemy's Geography*, 38–42.

78. Paviot, "Ung Mapmonde Rond," 25. See also Gautier Dalché, "Avant Behaim."

79. Numa Broc, *La Géographie de la Renaissance (1420–1620)* (Paris: Bibliothèque Nationale, 1980), 9–11; Muris, "Der Globus des Martin Behaim"; Muris and Saarmann, *Der Globus*, 49.

80. Chet A. Van Duzer and Johann Schöner, *Johann Schoner's Globe of 1515: Transcription and Study* (Philadelphia: American Philosophical Society, 2010), 202; Margriet Hoogvliet, "The Medieval Texts of the 1486 Ptolemy Edition by Johann Reger of Ulm," *Imago Mundi* 54 (2002): 7–18.

81. James Sykes, "Der Erdglobus in Raphaels 'Die Schule von Athen,'" *Der Globusfreund: Wissenschaftliche Zeitschrift für Globenkunde* 55/56 (2009): 58–60.

82. On mathematics' role in the reception, see Gautier Dalché, "The Reception of Ptolemy's *Geography*," 336–42.

83. On Euclid's initial return, see H. L. L. Busard, ed., *The First Latin Translation of Euclid's Elements Commonly Ascribed to Adelard of Bath: Books I–VIII and Books X.36–XV.2* (Toronto: Pontifical Institute of Mediaeval Studies, 1983); Marshall Clagett, "The Medieval Latin Translations from the Arabic of the Elements of Euclid, with Special Emphasis on the Versions of Adelard of Bath the Medieval Latin Translations from the Arabic of the Elements of Euclid, with Special Emphasis on the Versions of Adelard of Bath," *Isis* 44, no. 1/2 (1953): 16–42; Folkerts, *Mathematics in Medieval Europe*; John Murdoch, "The Medieval Euclid: Salient Aspects of the Translations of the *Elements* by Adelard of Bath and Campanus of Novara," *Revue de Synthèse* 89, no. 2 (1968): 67–94.

84. The *Elements'* way of making arguments had an important effect on medieval philosophical argumentation. Charles Burnett, "Scientific Speculations," in *A History of Twelfth-Century Western Philosophy*, ed. Peter Dronke (Cambridge: Cambridge University Press, 1988), 151–76.

85. Stephen Wagner, "Seven Liberal Arts and Classical Scholarship," in *The Seven Liberal Arts in the Middle Ages*, ed. Stephen Wagner (Bloomington: Indiana University Press, 1983), 12; Stephen K. Victor, *Practical Geometry in the High Middle Ages: Artis Cuiuslibet Consummatio and the Pratike De Geometrie* (Philadelphia: American Philosophical Society, 1979), 2–5; O. A. W. Dilke, *The Roman Land Surveyors: An Introduction to the Agrimensores* (Newton Abbot: David and Charles, 1971); Edward Grant, *The Nature of Natural Philosophy in the Late Middle Ages* (Washington, D.C.: Catholic University of America Press, 2010), 21. Medieval geometry was a practical discipline, as opposed to a theoretical one, and long emphasized measurement of things over the theory of empty space.

86. The great humanist Lorenzo Valla seems to have been particularly enamored of Euclid. See Rose, "Humanist Culture and Renaissance Mathematics," 97–98. See also Laird, "Archimedes Among the Humanists," 634–36.

87. On the Renaissance Papacy and mathematics, see Rose, *Italian Renaissance*, 36–44. On Christopher Columbus, see Fiorini, *Sfere Terrestri*, 66–67. On Dürer, see Jane Campbell Hutchison, *Albrecht Dürer: A Biography* (Princeton, N.J.: Princeton University Press, 1990), 237.

88. Carl B. Boyer and Uta C. Merzbach, *A History of Mathematics*, 2nd ed. (New York: John Wiley & Sons, 1991), 119.

89. See the excellent English translation of the text in J. L. Berggren and Alexander Jones, eds., *Ptolemy's Geography: An Annotated Translation of the Theoretical Chapters* (Princeton, N.J.: Princeton University Press, 2000).

90. Ibid., 57–59.

91. Ibid., 58.

92. On the medieval cartographic imagination in general, see P. D. A. Harvey, *The Hereford World Map: Medieval World Maps and Their Context* (London: British Library, 2006); Naomi Reed Kline, *Maps of Medieval Thought: The Hereford Paradigm* (Woodbridge: Boydell Press, 2001); Woodward, "Medieval *Mappaemundi*"; Evelyn Edson, *Mapping Time and Space: How Medieval Mapmakers Viewed Their World* (London: British Library, 1999); *The World Map, 1300–1492.*

93. On religion and maps, see Pauline Moffitt Watts, "The European Religious Worldview and Its Influence on Mapping," in *The History of Cartography*, ed. J. B. Harley and David Woodward (Chicago: University of Chicago Press, 1987), 382–400. On the broader significance of the shift away from these maps, see Besse, *Les Grandeurs de la Terre*, 54–59.

94. Gautier Dalché, "The Reception of Ptolemy's *Geography*" (this issue is discussed on p. 291; the quote comes from n. 41 on the same page).

95. Berggren and Jones, *Ptolemy's Geography*, 60.

96. Ibid., 81–82.

97. Ibid., 110.

98. Ibid., 86–93.

99. Ibid.

100. Aujac, *Claude Ptolémée*, 165–66.

101. On the problematic nature of the term "Donis," see Gautier Dalché, "The Reception of Ptolemy's *Geography*," 320n.242.

102. The oldest globe manual is (probably) Gemma Frisius, *De principiis astronomiae et cosmographiae, deque usu globi ab eodem editi. Item de orbis divisione, & insulis, rebusque nuper inventis.* (Antwerp: Gregorius Bontius, 1530). It appeared, however, as one volume and, as the title would indicate, is not limited to globes. For a discussion of this work, see Dekker, "Globes in Renaissance Europe," 143–45. These were not Schöner's first works on global space; see Johannes Schöner, *Luculentissima quaedā Terrae totius descriptio: Cū multis utilissimis cosmographiæ iniciis. nouaq, & q ante fuit verior Europæ nostræ formatio* (Norimbergae: Impressum ī excusoria officina I. Stuchssen, 1515); Johannes Schöner, *Solidi ac sphaerici corporis sive globi astronomici canones, usum et expeditam praxim ejus dem expromentes* (Norimbergae: Ioannes Stuchs, 1516).

103. *Opusculum geographicum; globi stelliferi, sive sphaerae stellarum fixarum usus, & explicationes . . . ac studio in lucem edita fuere anno Christi M. D. XXXIII* (Norimbergae: Petreius, 1533).

104. Neither volume is paginated. I therefore rely on chapter headings. *Globi stelliferi* (Prolegomena).

105. Ibid. (Caput Primum).

106. On Sacrobosco, see Thorndike, *The Sphere of Sacrobosco*; Olaf Pedersen, "In Quest of Sacrobosco," *Journal for the History of Astronomy* 16, no. 3 (1985): 175–220; Woodward, "Image," 9.

107. Schöner, *Opusculum geographicum*. For simplicity's sake, I have included chapter citations in the text.

108. On the Renaissance's manner of seeing climate in terms of terrestrial zones, see Broc, *Géographie*; Besse, *Les Grandeurs de la Terre*, 46–53.

109. If the exact time when an eclipse is observed at one point is known, this can be compared to the time when an eclipse is observed at another point, with the difference being a longitudinal value.

110. Schöner was a respected mathematician, and his mathematical works continued to be published even after his death. Johannes Schöner, *Opera mathematica Ioannis Schoneri in vnvm volvmen congesta* (Norimbergae: Impressa in officina I. Montani & V. Neuberi, 1551). It is also worth noting, given what I argued in the previous chapter, that Philipp Melanchthon wrote the introduction to this text.

111. In general, A. Rupert Hall, *The Scientific Revolution, 1500–1800: The Formation of the Modern Scientific Attitude* (London: Longman, 1962); Steven Shapin, *The Scientific Revolution* (Chicago: University of Chicago Press, 1996); A. Rupert Hall, *The Revolution in Science, 1500–1750,* 3rd ed. (London: Longman, 1983); Robert S. Westman, *The Copernican Achievement,* vol. 7 (Berkeley: University of California Press, 1975); Betty Jo Teeter Dobbs and Margaret C. Jacob, *Newton and the Culture of Newtonianism* (Atlantic Highlands, N.J.: Humanities Press, 1995); Jacob, *The Cultural Meaning of the Scientific Revolution.* For critical perspectives, see David C. Lindberg and Robert S. Westman, *Reappraisals of the Scientific Revolution* (Cambridge: Cambridge University Press, 1990); Margaret J. Osler, ed., *Rethinking the Scientific Revolution* (Cambridge: Cambridge University Press, 2000); Toby E. Huff, *The Rise of Early Modern Science: Islam, China, and the West* (Cambridge: Cambridge University Press, 1993).

112. Pierre Duval, *La Connoissance et l'usage des globes; et des cartes de géographie* (Paris: Chez l'Autheur, 1654), 5–8; Robert Hues, *Learned Treatise of Globes: Both Celestiall and Terestriall: With Their Several Uses. Written First in Latine, by Mr. Robert Hues: And by Him So Published.,* trans. John Chilmead (London: Andrew Kemb, 1659), unpaginated preface; John Newton, *Mathematical Elements, in III Parts, the First, Being a Discourse of Practical Geometry . . . According to the Stereographick, or Circular Projection* (London: R. and W. Leybourn, 1660), 1–3.

113. Johann Wolfgang Müller, *Anweisung zur Kenntnis und dem Gebrauch der künstlichen Himmels- und Erdkugeln . . . der Schulen und Liebhaber der Sphaerologie,* 2 vols., vol. 1 (Nürnberg: Johann Georg Klinger, 1791); *Anweisung zur Kenntnis und dem Gebrauch der künstlichen Himmels- und Erdkugeln . . . der Schulen und Liebhaber der Sphaerologie,* 2 vols., vol. 2 (Nürnberg: Johann Georg Klinger, 1792).

114. On Klinger and his globes, see P. C. J. van der Krogt, *Old Globes in the Netherlands: A Catalogue of Terrestrial and Celestial Globes Made Prior to 1850 and Preserved in Dutch Collections* (Utrecht: HES, 1984), 171–78; Dekker, Ackermann, and Lippincott, *Globes at Greenwich,* 388–90.

115. Johann Ludwig Hocker, *Einleitung zur Kenntnis und Gebrauch der Erd- und Himmels-Kugel auf das deutlichste und leichteste in Frag und Antwort* (Nürnberg: Peter Conrad Monath, 1734).

116. Müller, *Anweisung,* 1, 8.

117. *Anweisung,* 2, 2–26.

118. On this topic, see the fundamental study by Mary Terrall, *The Man Who Flattened the Earth: Maupertuis and the Sciences in the Enlightenment* (Chicago: University of Chicago Press, 2002).

119. Müller, *Anweisung,* 2, 27–80.

120. On these developments, see Boyer and Merzbach, *History of Mathematics,* 572–97; Morris Kline, *Mathematical Thought from Ancient to Modern Times,* 3 vols., vol. 3 (New York: Oxford University Press, 1990), 861–81; Howard Whitley Eves, *An Introduction to the History of Mathematics,* 5th ed. (Philadelphia: Saunders College Publishing, 1983); Dirk J. Struik, *A Concise History of Mathematics,* 4th rev. ed. (New York: Dover Publications, 1987), 169–81.

121. Augustus de Morgan, *The Globes, Celestial and Terrestrial* (London: Malby and Co., 1845).

122. Ibid., iii–iv.

123. Ibid., 2–3.

124. Robert Hues, *Tractatus de globis coelesti et terrestri eorumque usu* (Amstelodami: Henricus Hondius, 1627). This work also appeared in an English version, *Learned Treatise of Globes.* That we are dealing with multiple shifts in European thought's deepest recesses is confirmed by contemporaneous anticipations of fundamental change within geometry. As I explained in the previous chapter, in 1621 the Oxford mathematician Henry Savile became the first person to note that the *Elements'* parallel postulate was inconsistent with spherical space.

125. De Morgan, *Globes,* 3.

126. Ibid., 2.

Chapter 3

1. For a useful overview of the literature, see Jane Campbell Hutchison, *Albrecht Dürer: A Guide to Research* (New York: Garland, 2000). (For simplicity's sake, I drop the "I" for the rest of this chapter.)

2. Key works of analysis include Hartmut Böhme, *Albrecht Dürer: Melencolia I: Im Labyrinth der Deutung*, Orig.-Ausg. ed., Fischer-Taschenbücher (Frankfurt am Main: Fischer Taschenbuch Verl., 1989); Raymond Klibansky, Erwin Panofsky, and Fritz Saxl, *Saturn und Melancholie: Studien zur Geschichte der Naturphilosophie und Medizin, der Religion und der Kunst*, trans. Christa Buschendorf (Frankfurt am Main: Suhrkamp, 1990); Erwin Panofsky, *Albrecht Dürer*, 3rd ed. (London: Oxford University Press, 1948); "Artist, Scientist, Genius: Notes on the 'Renaissance-Dämmerung,'" in *The Renaissance: Six Essays*, ed. Wallace K. Ferguson et al. (New York: Harper & Row, 1962); Erwin Panofsky and Fritz Saxl, *Dürers 'Melencolia I': Eine quellen- und typengeschichtliche Untersuchung* (Leipzig: B. G. Teubner, 1923); Ernst Rebel, *Albrecht Dürer: Maler und Humanist* (München: Bertelsmann, 1996); Peter-Klaus Schuster, *Melencolia I: Dürers Denkbild*, 2 vols. (Berlin: Gebr. Mann Verlag, 1991); Paul Weber, *Beiträge zu Dürers Weltanschauung: Eine Studie über die drei Stiche 'Ritter, Tod und Teufel', 'Melancholie', und 'Hieronymus Im Gehäus'* (Nendeln: Kraus, 1979); Heinrich Wölfflin, *Die Kunst Albrecht Dürers*, 9., durchges. Aufl. ed., Pantheon-Colleg (München: Bruckmann, 1984); Friedrich Winkler, *Albrecht Dürer: Leben und Werk* (Berlin: Mann, 1957); Jane Campbell Hutchison, *Albrecht Dürer: A Biography* (Princeton, N.J.: Princeton University Press, 1990).

3. This work originally appeared in Latin. An excellent translation is available in William S. Heckscher, "Melancholia (1541) an Essay in the Rhetoric of Description by Joachim Camerarius," in *Joachim Camerarius: (1500–1574); Beiträge zur Geschichte des Humanismus im Zeitalter der Reformation*, ed. Frank Baron (München: Wilhelm Fink Verlag, 1978).

4. For the last of these interpretations, see Schuster, *Melencolia I*.

5. Rebel, *Albrecht Dürer*; Hutchison, *Albrecht Dürer*; Panofsky, *Albrecht Dürer*; Wölfflin, *Die Kunst*; Winfried Schleiner, *Melancholy, Genius, and Utopia in the Renaissance* (Wiesbaden: Harrassowitz, 1991); Jean Starobinski, *Histoire du traitement de la mélancolie des origines à 1900* (Basel: Geigy, 1960); Patrick Doorly, "Dürer's 'Melencolia I': Plato's Abandoned Search for the Beautiful," *Art Bulletin* 86, no. 2 (2004): 255–76.

6. See, for example, Aby Moritz Warburg, *Die Erneuerung der heidnischen Antike: Kulturwissenschaftliche Beiträge zur Geschichte der europäischen Renaissance; mit einem Anhang unveröffentlichter Zusätze* (Leipzig: Teubner, 1932). Of particular significance are the essays "The Emergence of the Antique as a Stylistic Ideal in Early Renaissance Painting" (1914) and "Dürer and Italian Antiquity" (1905). On Warburg's broader influence, see Carl Landauer, "Erwin Panofsky and the Renascence of the Renaissance," *Renaissance Quarterly* 47, no. 2 (1994): 255–81; Margaret Iversen, "Retrieving Warburg's Tradition," *Art History* 16, no. 4 (1993): 541–53.

7. Raymond Klibansky, Erwin Panofsky, and Fritz Saxl, *Saturn and Melancholy: Studies in the History of Natural Philosophy, Religion, and Art* (London: Nelson, 1964).

8. Johannes Röll, "'Das Problem ist das vom Nachleben der Antike': Fritz Saxl 1890–1948," *Pegasus: Berliner Beiträge zum Nachleben der Antike* 1 (1999): 27–32.

9. For the classic work in this tradition, see Panofsky and Saxl, *Dürers 'Melencolia I.'*

10. See the collection of useful essays in Dagmar Eichberger and Charles Zika, eds., *Dürer and His Culture* (Cambridge: Cambridge University Press, 1998).

11. For an example of Dürer's reliance on Christianity, see Joseph L. Koerner, *The Moment of Self-Portraiture in German Renaissance Art* (Chicago: University of Chicago Press, 1993), 63–79.

12. The broader scholarship often mentions geometry, but without examining the significance of Dürer's encounter with Euclid. For example, scholars have made much of Dürer's use of geometry in his *Münchener Selbstildnis* of 1500. Yet Dürer's understanding of geometry in this image is largely two-dimensional—and one reason for this is that he had not yet read the *Elements*, which occurred only after 1507. For an example of this kind of analysis, see Klaus H. Jürgens, "Neue Forschungen zu dem Münchener Selbstbildnis des Jahres 1500 von Albrecht Dürer," *Kunsthistorisches Jahrbuch Graz* 19–20, no. 19–20; 21 (1983–1984; 1985): 167–90; 143–64.

13. The few scholars who have pursued space concentrate more on how Dürer used geometric space rather than what its use meant. See Eberhard Schröder, *Dürer, Kunst und Geometrie: Dürers künstlerisches Schaffen aus der Sicht seiner "Underweysung"* (Basel: Birkhäuser Verlag, 1980); Wilhelm Klingenberg, *Mathematik und Melancholie: Von Albrecht Dürer bis Robert Musil* (Stuttgart: Steiner,

1997); Terence Lynch, "The Geometric Body in Dürer's Engraving Melencolia I," *Journal of the Warburg and Courtauld Institutes* 45 (1982): 226–32.

14. On this theme in medieval thought, see Marcia L. Colish, "Early Scholastic Angelology," *Recherches de théologie ancienne et médiévale* 62 (1980): 80–109.

15. The text is available in Jonathan Barnes, ed., *The Complete Works of Aristotle*, 2 vols., vol. 2 (Princeton, N.J.: Princeton University Press, 1984), 1319–1527.

16. See, for example, Hans Weitzel, "Zum Polyeder auf A. Dürers Stich Melencolia I—Ein Nürnberger Skizzenblatt mit Darstellungen Archimedischer Körper," *Sudhoffs Archiv* 91, no. 2 (2007): 129–73; Joseph E. Hofmann, "Dürers Verhältnis zur Mathematik," in *Albrecht Dürers Umwelt: Festschrift Zum 500. Geburtstag Albrecht Dürers Am 21. Mai 1971*, ed. Otto Herding (Nürnberg: Selbstverl. des Vereins für Geschichte der Stadt Nürnberg, 1971); Lynch, "Geometric Body."

17. On the bat and the dog, see Colin T. Eisler, *Dürer's Animals* (Washington, D.C.: Smithsonian Institution Press, 1991), 78, 166–75.

18. For examples, see the pseudo-Aristotelian *Problemata* in Barnes, *The Complete Works*, 2: 1319–1527; and Marsilio Ficino, *Three Books on Life: A Critical Edition and Translation*, trans. Carol V. Kaske and John R. Clark (Binghamton, N.Y.: Center for Medieval and Early Renaissance Studies, 1989).

19. On Ghent, see Steven P. Marrone, *Truth and Scientific Knowledge in the Thought of Henry of Ghent* (Cambridge: Medieval Academy of America, 1985); Gordon Wilson, "Henry of Ghent's 'Quodlibet I': Initial Departures from Thomas Aquinas," *History of Philosophy Quarterly* 16, no. 2 (1999): 167–80; Steven P. Marrone, "Henry of Ghent and Duns Scotus on the Knowledge of Being," *Speculum* 63, no. 1 (1988): 22–57; Jean Paulus, *Henri de Gand: Essai sur les tendances de sa métaphysique* (Paris: J. Vrin, 1938); J. V. Brown, "Abstraction and the Object of the Human Intellect According to Henry of Ghent," *Vivarium* 11, no. 1 (1973): 80–104.

20. Henry of Ghent, *Henrici de Gandavo opera omnia*, 38 vols., vol. 6 (Leuven: Leuven University Press, 1978), 64.

21. Starobinski, *Histoire*, 23–25.

22. Charles Edward Trinkaus, *In Our Image and Likeness: Humanity and Divinity in Italian Humanist Thought*, 2 vols., vol. 1 (Notre Dame: University of Notre Dame Press, 1995), 3–5.

23. The key medieval text in this tradition is John Scotus Eriugena's ninth-century work, *Commentary on the Dionysian Celestial Hierarchy*. For a modern translation, see Johannes Scotus Eriugena, *Eriugena's Commentary on the Dionysian Celestial Hierarchy*, trans. Paul Rorem (Toronto: Pontifical Institute of Mediaeval Studies, 2005).

24. Colish makes clear that angels are a philosophical necessity. Colish, "Angelology."

25. St. Thomas Aquinas, *The Summa Theologica of St. Thomas Aquinas*, trans. Fathers of the English Dominican Province, 25 vols., vol. 1 (London: Burns, Oates & Washbourne, 1911), 126–36.

26. For useful overviews, see Ali Benmakhlouf, "Nature et cosmos: Incursions en philosophie ancienne et médiévale," *Revue de Métaphysique et de Morale*, no. 3 (2004): 343–52; Pierre Duhem, *Le Système du monde; histoire des doctrines cosmologiques de Platon à Copernic* (Paris: A. Hermann, 1913); Bruce Eastwood, *Ordering the Heavens: Roman Astronomy and Cosmology in the Carolingian Renaissance* (Leiden: Brill, 2007); Edward Grant, "Cosmology," in *Science in the Middle Ages*, ed. David C. Lindberg (Chicago: University of Chicago Press, 1978); Henry Guerlac, "Copernicus and Aristotle's Cosmos," *Journal of the History of Ideas* 29, no. 1 (1968): 109–13; Charles W. Misner, "Cosmology and Theology," in *Cosmology, History, and Theology*, ed. Wolfgang Yourgrau, Allen duPont Breck, and Hannes Alfvén (New York: Plenum Press, 1977); John D. North, *Cosmos: An Illustrated History of Astronomy and Cosmology* (Chicago: University of Chicago Press, 2008).

27. On measurement, see Alfred W. Crosby, *The Measure of Reality: Quantification and Western Society, 1250–1600* (Cambridge: Cambridge University Press, 1998), 100–108.

28. Eisler, *Dürer's Animals*, 98.

29. Ibid., 163–66. See also Böhme, *Albrecht Dürer*, 15.

30. Erika Rummel, "Et cum theologo bella poeta gerit: The Conflict Between Humanists and Scholastics Revisited," *Sixteenth Century Journal* 23, no. 4 (1992): 713–26; Charles G. Nauert, "The

Clash of Humanists and Scholastics: An Approach to Pre-Reformation Controversies," *Sixteenth Century Journal* 4, no. 1 (1973): 1–18.

31. For Koyré's analysis, see Alexandre Koyré, *Galileo Studies* (Atlantic Highlands, N.J.: Humanities Press, 1978). In general, see Stillman Drake, *Galileo at Work: His Scientific Biography* (Chicago: University of Chicago Press, 1978); A. Rupert Hall, *From Galileo to Newton* (New York: Dover Publications, 1981); Stillman Drake, *Galileo: A Very Short Introduction* (Oxford: Oxford University Press, 2001); Gerald Rottman, *The Geometry of Light: Galileo's Telescope, Kepler's Optics* (Baltimore: Gerald Rottman, 2008); Mark A. Peterson, *Galileo's Muse: Renaissance Mathematics and the Arts* (Cambridge, Mass.: Harvard University Press, 2011).

32. The text is available in Ottavio Gigli, ed., *Studi sulla Divina Commedia, di Galileo Galilei, Vincenzo Borghini ed altri* (Florence: Felice Le Monnier, 1855), 3–34. For scholarly discussions, see Alice K. Turner, *The History of Hell* (New York: Harcourt Brace, 1993), 133–35. As Mark Peterson has explained, Dante understood Euclid in an "unspatial way." This was manifestly not the case for Galileo, which meant that Hell had to go. Peterson, *Galileo's Muse*, 75.

33. Copernicus completed a sketch of his heliocentric system, called *Commentariolus*, already around 1510. For this text, see Nicolaus Copernicus, Georg Joachim Rhäticus, and Edward Rosen, *Three Copernican Treatises: The Commentariolus of Copernicus, the Letter Against Werner, the Narratio Prima of Rheticus* (New York: Columbia University Press, 1939).

34. Martin Bailey, "Dürer's Comet," *Apollo-London-Apollo Magazine Limited* 141 (1995): 19–32. See also Schuster, *Melencolia I*, 149.

35. Aristotle, *The Complete Works of Aristotle: The Revised Oxford Translation*, 2 vols., vol. 1 (Princeton, N.J.: Princeton University Press, 1984), 555–625.

36. For the best overview of the slow break with Aristotle's vision of celestial space, see Craig Martin, *Renaissance Meteorology: Pomponazzi to Descartes* (Baltimore: Johns Hopkins University Press, 2011).

37. Jaques Lefèvre d'Étaples, *Quatuor libri meteororum Aristotelis paraphrasi Iacobi Stapulensis explanati* (Valentius Schumann, 1512).

38. Schuster, *Melencolia I*, 17.

39. Carl B. Boyer, *The Rainbow: From Myth to Mathematics* (Princeton, N.J.: Princeton University Press, 1987).

40. Gregorius Reisch, *Margarita philosophica* (Argentorati [Strasbourg]: Schottus, 1504).

41. Jean-Marc Besse, *Les Grandeurs de la Terre: aspects du savoir géographique à la Renaissance* (Lyon: ENS, 2003), 91.

42. Masha'allah Ibn-Athari, *De scientia motus orbis* (Norimbergae: Johannes Stabius, 1505).

43. Nor was this the only time that these two worked together. See Günther Hamann, "Der Behaim-Globus als Vorbild der Stabius Dürer Karte von 1515," *Der Globusfreund* 25–27 (1977–1979): 135–47.

44. On Islamic appropriations of ancient science, see George Saliba, *Islamic Science and the Making of the European Renaissance* (Cambridge, Mass.: MIT Press, 2007).

45. Hamann, "Der Behaim-Globus."

46. On the role that numbers played in classical philosophical debates, see the introduction to Proclus, *A Commentary on the First Book of Euclid's Elements*, trans. Glenn R. Morrow (Princeton, N.J.: Princeton University Press, 1970). Although my take differs from that of the following scholars, see also Brian P. Copenhaver, "Hermes Trismegistus, Proclus, and the Question of a Philosophy of Magic in the Renaissance," in *Hermeticism and the Renaissance: Intellectual History and the Occult in Early Modern Europe* ed. Ingrid Merkel and Allen George Debus (Washington, D.C.: Folger Books, 1988), 86; Lucas Siorvanes, *Proclus: Neo-Platonic Philosophy and Science* (Edinburgh: Edinburgh University Press, 1996), 36–37.

47. Frances A. Yates, *The Occult Philosophy in the Elizabethan Age* (London: Routledge & Kegan Paul, 1979), 44.

48. For an overview, see Paola Zambelli, "Scholastic and Humanist Views of Hermeticism and Witchcraft," in *Hermeticism and the Renaissance: Intellectual History and the Occult in Early Modern Europe* ed. Ingrid Merkel and Allen George Debus (Washington, D.C.: Folger Books, 1988).

49. On the history of the *Hermetic Corpus*, see the introduction to Brian P. Copenhaver, ed., *Hermetica: The Greek Corpus Hermeticum and the Latin Asclepius in a New English Translation* (Cambridge: Cambridge University Press, 1995), xxxiii–xxxvii.

50. Yates, *Occult Philosophy*, 49–52; Karl Giehlow, "Dürers Stich 'Melencolia' und der Maximilianische Humanistenkreis," *Mitteilungen der Gesellschaft für verfielfaltigende Kunst*, no. 2, 3, 4 (1903–4): 29–41, 6–21, 57–78; Wölfflin, *Die Kunst*, 207–8.

51. See also Böhme, *Albrecht Dürer*, 55–56.

52. Heinrich Cornelius Agrippa, *De occulta philosophia. libri tres* (Louvain, 1531), 249.

53. Luca Pacioli, "De viribus quantitatis" (Biblioteca Universitaria di Bologna, 1508). See also Carl B. Boyer and Uta C. Merzbach, *A History of Mathematics*, 2nd ed. (New York: John Wiley & Sons, 1991), 266–68.

54. Giulia Bartrum and Günter Grass, *Albrecht Dürer and His Legacy: The Graphic Work of a Renaissance Artist* (London: British Museum Press, 2002), 136; Hutchison, *Albrecht Dürer: A Guide to Research*, 237.

55. Luca Pacioli, *De viribus quantitatis. Facsimile ad uso professionale* (Sansepolcro: Aboca Edizioni, 2009).

56. Bret Rothstein, "Making Trouble: Strange Wooden Objects and the Early Modern Pursuit of Difficulty," *Journal for Early Modern Cultural Studies* 13, no. 1 (2013): 103.

57. See the famous cosmological drawing in Johannes Kepler, *Prodromus dissertationvm cosmographicarvm, continens mysterivm cosmographicvm . . . totiusq[ue] astronomiæ Restauratoris D. Nicolai Copernici* (Tvbingae: Georg Gruppenbach, 1596). I discuss this image in the conclusion; see Figure 29.

58. The mathematician Benno Artmann has explained the mathematical significance of this series of books. Benno Artmann, *Euclid: The Creation of Mathematics* (New York: Springer, 1999), 161–202.

59. Euclid borrowed his approach to number from Aristotle and separated geometry from arithmetic within the *Elements* itself. He thus never made the shift toward understanding number as a continuous quantity. Significantly, for my purposes here, it was not until the sixteenth century that European mathematics made that shift. See Katherine Neal, *From Discrete to Continuous: The Broadening of Number Concepts in Early Moder England* (Dordrecht: Kluwer, 2002), 5.

60. Yates, *Occult Philosophy*, 50–52.

61. Giehlow, "Dürers Stich."

62. Rebel, *Albrecht Dürer*, 275–304. William R. Newman and Anthony Grafton, *Secrets of Nature: Astrology and Alchemy in Early Modern Europe* (Cambridge, Mass.: MIT Press, 2001), 3.

63. Agrippa, *De occulta philosophia*, 249–53.

64. Vincent Jullien, *Philosophie naturelle et géométrie au XVIIe siècle* (Paris: Honoré Champion Éditeur, 2006), 257.

65. On Dürer and mathematics, see Boyer and Merzbach, *History of Mathematics*, 268.

66. Hutchison, *Albrecht Dürer*, 95–96.

67. Hans Rupprich, *Schriftlicher Nachlass*, vol. 2 (Berlin: Deutscher Verein für Kunstwissenschaft, 1956), 19, 159.

68. Albrecht Dürer, *Underweysung der Messung/ Mit dem Zirkel und Richtscheyt/ in Linien Ebnen unnd gantzen Corporen* (Norimbergae, 1525) (unpaginated).

69. Ibid. (my emphasis).

70. Luca Pacioli, *Summa de arithmetica geometria proportioni & proportionalita: Continentia de tutta l'opera.* (Venetijs: Paganino de Paganini, 1494). I had access to a later edition: *Summa de arithmetica geometria: proportioni: Et proportionalità: Nouamente impressa* (Toscolano, 1523).

71. Euclid, *Euclidis megarensis philosophi acutissimi mathematicorumque . . . scientia rarissimus iudicio castigatissimo detersit emendauit* (Venice: A. Paganius, 1509).

72. Niccolò Tartaglia, *Nova scientia inventa da Nicolò Tartalea* (Vinegia: Stephano da Sabio, 1537). I had access to a later edition: *La Nova scientia, con una gionta al terzo libro* (Venice: Camillo Castelli, 1583).

73. Paul L. Rose, *The Italian Renaissance of Mathematics: Studies on Humanists and Mathematicians from Petrarch to Galileo* (Genève: Droz, 1975), 74.

74. Euclid, *Euclide megarense philosopho solo introduttore delle scientie mathematice... ouer suffragio di alcun'altra scientia con facilita, sera capace à poterlo intendere.*, trans. Gabriele Tadino et al. (Stampato in Vinegia: Per Venturino Roffinelli, 1543).

75. For excellent overviews, see James Hankins, *Plato in the Italian Renaissance*, 2 vols. (Leiden: Brill, 1990); "The Study of the *Timaeus* in Early Renaissance Italy," in *Natural Particulars: Nature and the Disciplines in Renaissance Europe*, ed. Anthony Grafton and Nancy G. Siraisi (Cambridge, Mass.: MIT Press, 1999); Miguel A. Granada, "New Visions of the Cosmos," in *The Cambridge Companion to Renaissance Philosophy*, ed. James Hankins (Cambridge: Cambridge University Press, 2007).

76. For a definitive statement on this issue, see Charles B. Schmitt, *Aristotle and the Renaissance* (Cambridge: Published for Oberlin College by Harvard University Press, 1983).

77. I put this idea forward in response to A. C. Crombie's suggestive essay "Mathematics and Platonism in the Sixteenth-Century Italian Universities and in Jesuit Educational Policy," in *Prismata: Naturwissenschaftsgeschichtliche Studien: Festschrift für Willy Hartner*, ed. Willy Hartner, Yasukatsu Maeyama, and Walter G. Saltzer (Wiesbaden: Franz Steiner Verlag, 1977). Moreover, as Jill Kraye has noted, multiple classical currents influenced the development of Renaissance thought. Jill Kraye, "The Revival of Hellenistic Philosophies," in *The Cambridge Companion to Renaissance Philosophy*, ed. James Hankins (Cambridge: Cambridge University Press, 2007). See also Michele Sbacchi, "Euclidism and Theory of Architecture," *Nexus Network Journal* 3, no. 2 (2001): 25–38.

78. Nick Mackinnon, "The Portrait of Fra Luca Pacioli," *Mathematical Gazette* 77, no. 479 (1993): 130–219; Sergio Guarino, "La Formazione veneziana di Jacopo de'Barbari," in *Giorgione e la cultura venetra tra '400 e '500: mito, allegoria, analisi iconologica*, ed. Maurizio Calvesi (Rome: De Luca, 1981).

79. Rebel, *Albrecht Dürer*, 181–86; Hutchison, *Albrecht Dürer*, 71–72.

80. Renzo Baldasso, "Portrait of Luca Pacioli and Disciple: A New, Mathematical Look," *Art Bulletin* 92, no. 1/2 (2010): 83–102.

81. On these universities, see Paul F. Grendler, *The Universities of the Italian Renaissance* (Baltimore: Johns Hopkins University Press, 2002), 3–40; James H. Overfield, *Humanism and Scholasticism in Late Medieval Germany* (Princeton, N.J.: Princeton University Press, 1984), 3–60.

82. Edmond Vansteenberghe, *Le Cardinal Nicolás de Cues (1401–1464): L'action—la pensée* (Paris: H. Champion, 1920), 469–72.

83. In general, F. Edward Cranz, Thomas M. Izbicki, and Gerald Christianson, *Nicholas of Cusa and the Renaissance* (Aldershot: Ashgate, 2000). On Italy, see Giuseppe Saitta, *Nicolò Cusano e l'umanesimo italiano; con altri saggi sul Rinascimento italiano* (Bologna: Tamari, 1957). On Germany, see Lewis William Spitz, *The Religious Renaissance of the German Humanists* (Cambridge, Mass.: Harvard University Press, 1963); Overfield, *Humanism and Scholasticism*.

84. The best discussion is available in Vansteenberghe, *Le Cardinal Nicolás de Cues*, 469–72.

85. Christopher M. Bellitto, Thomas M. Izbicki, and Gerald Christianson, *Introducing Nicholas of Cusa: A Guide to a Renaissance Man* (New York: Paulist Press, 2004); Vansteenberghe, *Le Cardinal Nicolás de Cues*, 465.

86. Nicholas of Cusa, *Opera* (Paris: in Aidibus Ascensianis, 1514).

87. Ernst Cassirer, *The Individual and the Cosmos in Renaissance Philosophy*, trans. Mario Domandi (Mineola: Dover Publications, 2000); Johann Kreuzer, *Gestalten Mittelalterlicher Philosophie: Augustinus, Eriugena, Eckhart, Tauler, Nikolaus v. Kues* (München: Fink, 2000); Karl Hermann Kandler, *Nikolaus von Kues: Denker zwischen Mittelalter und Neuzeit* (Göttingen: Vandenhoeck & Ruprecht, 1995).

88. Pauline M. Watts, *Nicolaus Cusanus, a Fifteenth-century Vision of Man* (Leiden: Brill, 1982); Kandler, *Nikolaus von Kues*; Cranz, Izbicki, and Christianson, *Nicholas of Cusa and the Renaissance*; Manfred Groten, "Nikolaus von Kues: Vom Studenten zum Kardinal—Lebensweg und Lebenswelt eines spätmittelalterlichen Intellektuellen," in *Nicholas of Cusa: A Medieval Thinker for the Modern Age*, ed. Kazuhiko Yamaki (Richmond, Surrey: Curzon Press, 2002); Kazuhiko Yamaki, *Nicholas of Cusa: A Medieval Thinker for the Modern Age* (Richmond, Surrey: Curzon Press, 2002); Bellitto, Izbicki, and Christianson, *Introducing Nicholas of Cusa*; Peter J. Casarella, *Cusanus: The Legacy of Learned Ignorance* (Washington, D.C.: Catholic University of America Press, 2006); Dermot Moran, "Nicholas of Cusa

and Modern Philosophy," in *The Cambridge Companion to Renaissance Philosophy*, ed. James Hankins (Cambridge: Cambridge University Press, 2007); Ronald Levao, *Renaissance Minds and Their Fictions: Cusanus, Sidney, Shakespeare* (Berkeley: University of California Press, 1985), 22–23; Jasper Hopkins, *A Concise Introduction to the Philosophy of Nicholas of Cusa* (Minneapolis: University of Minnesota Press, 1978), 15; Kreuzer, *Gestalten mittelalterlicher Philosophie*; Saitta, *Nicolò Cusano e l'umanesimo italiano*.

89. Heiko Augustinus Oberman, *The Harvest of Medieval Theology: Gabriel Biel and Late Medieval Nominalism* (Cambridge, Mass.: Harvard University Press, 1963); *Spätscholastik und Reformation* (Zürich: EVZ-Verlag, 1965); Gerhard Ritter, *Studien zur Spätscholastik*, 3 vols. (Heidelberg: C. Winter, 1921).

90. Rose, *Italian Renaissance*, 36–44.

91. This is an old tradition that runs up through Cusa. See David Albertson, *Mathematical Theologies: Nicholas of Cusa and the Legacy of Thierry of Chartres* (New York: Oxford University Press, 2014), 222–52. See also Kandler, *Nikolaus von Kues*, 47.

92. Nicholas of Cusa, *Nicholas of Cusa on Learned Ignorance: A Translation and an Appraisal of De docta ignorantia*, trans. Jasper Hopkins (Minneapolis: A. J. Banning Press, 1990).

93. Ibid., 90.

94. Ibid., 61–62, 75–79. See also Kandler, *Nikolaus von Kues*, 62.

95. See the very fine essay: Elizabeth Brient, "Transitions to a Modern Cosmology: Meister Eckhart and Nicholas of Cusa on the Intensive Infinite," *Journal of the History of Philosophy* 37, no. 4 (1999): 575–600.

96. Hermes Trismegistus, *Liber viginti quattuor philosophorum* (Turnhout: Brepols, 1997); Françoise Hudry, *Le Livre des XXIV philosophes* (Grenoble: J. Millon, 1989); Françoise Hudry and Marius Victorinus, *Le Livre des XXIV philosophes: Résurgence d'un texte du IVe siècle* (Paris: J. Vrin, 2009).

97. Cusa, *On Learned Ignorance*, 64–66.

98. Ibid., 64, 66.

99. Trismegistus, *Liber viginti*.

100. On this big question, I am following Cassirer. See Cassirer, *The Individual and the Cosmos*. See also Lewis White Beck, *Early German Philosophy: Kant and His Predecessors* (South Bend, Ind.: St. Augustine's Press, 1999).

101. Cusa, *On Learned Ignorance*, see esp. 96–102. See also Alexandre Koyré, *From the Closed World to the Infinite Universe* (New York: Harper & Brothers, 1957), 16–18. On the complexity of this theological issues involved, see Elizabeth Brient, *The Immanence of the Infinite: Hans Blumenberg and the Threshold to Modernity* (Washington, D.C.: Catholic University of America Press, 2002), 204–22, and Brient, "Transitions to a Modern Cosmology," 576–77. See also Kandler, *Nikolaus von Kues*, 62–65.

102. Cusa, *On Learned Ignorance*, 116.

103. Ibid., 117.

104. *Opera omnia: iussu et auctoritate Academiae Litterarum Heidelbergensis ad codicum fidem edita*, 28 vols., vol. 3 (Hamburg: Meiner, 1959), 144.

105. Ibid., 143–44.

106. Nicholas of Cusa, *Der Laie über den Geist: Lateinisch-Deutsch*, ed. Ernst Hoffmann, trans. Renate Steiger (Hamburg: Felix Meiner Verlag, 1995), 74.

107. Ibid., 26.

108. Ibid., 50.

109. See the critical discussion of this tendency in the literature in Angus Gowland, "The Problem of Early Modern Melancholy," *Past & Present* 191 (2006): 117.

110. On time, see Gerhard Dohrn-van Rossum, *History of the Hour: Clocks and Modern Temporal Orders* (Chicago: University of Chicago Press, 1996), 289–323.

111. Yates, *Occult Philosophy*, 56.

112. Schuster, *Melencolia I*, 331.

Chapter 4

1. Sebastian Münster, *Cosmographia: Beschreibung aller Lender . . . Alles mit Figuren und schönen Landt Tafeln erklert/ und füre Augen gestelt* (Basel: Heinrich Petri, 1544).

2. In this sense, German cosmographers' contributions to the emergence of a truly European culture were even more profound than has been assumed. In general, see Christine R. Johnson, *The German Discovery of the World: Renaissance Encounters with the Strange and Marvelous* (Charlottesville: University of Virginia Press, 2008).

3. Frank Büttner, *Sammeln, Ordnen, Veranschaulichen: Zur Wissenskompilatorik in der Frühen Neuzeit* (Münster: Lit, 2003); Jean-Marc Besse, *Les Grandeurs de la Terre: Aspects du savoir géographique à la renaissance* (Lyon: ENS, 2003); Frédéric Tinguely, "Le Vertige cosmographique à la Renaissance," *Archives Internationales d'Histoire des Sciences* 59, no. 163 (2009); James H. Overfield, *Humanism and Scholasticism in Late Medieval Germany* (Princeton, N.J.: Princeton University Press, 1984); Lewis W. Spitz, *The Religious Renaissance of the German Humanists* (Cambridge, Mass.: Harvard University Press, 1963); Karl Heinz Burmeister, "Sebastian Münster: Versuch eines biographischen Gesamtbildes" (Basel: Helbing und Lichtenhahn, 1963); *Neue Forschungen zu Sebastian Münster: Mit einem Anhang von Ernst Emmerling: Graphische Bildnisse Sebastian Münsters* (Ingelheim: Historischer Verein, 1971); Günther Wessel, *Von Einem, der Daheim blieb, die Welt zu entdecken: Die Cosmographia des Sebastian Münster, oder, wie man sich vor 500 Jahren die Welt vorstellte* (Frankfurt am Main: Campus, 2004); Matthew McLean, *The "Cosmographia" of Sebastian Münster: Describing the World in the Reformation* (Aldershot: Ashgate, 2007).

4. Bernhardus Varenius, *Geographia generalis in qua affectiones generales Telluris explicantur* (Amstelodami: Apud Ludovicum Elzevirium, 1650).

5. J. N. L. Baker, "The Geography of Bernhard Varenius," *Transactions and Papers (Institute of British Geographers)* 21 (1955): 51–60; Manfred Büttner, *Die Geographia Generalis vor Varenius: Geograph. Weltbild u. Providentiallehre* (Wiesbaden: F. Steiner, 1973); Manfred Büttner and Józef Babicz, *Zur Entwicklung der Geographie vom Mittelalter bis zu Carl Ritter* (Paderborn: F. Schöningh, 1982); Siegmund Günther, *Varenius* (Amsterdam: Meridan, 1970); Klaus Lehmann, "Der Bildungsweg des Jungen Bernhard Varenius," in *Bernhard Varenius (1622–1650)*, ed. Margret Schuchard (Leiden: Brill, 2007); Margret Schuchard, *Bernhard Varenius (1622–1650)* (Leiden: Brill, 2007).

6. Denis Cosgrove has described, in a definitive work, the modern cartographic gaze as "Apollo's Eye." See Denis E. Cosgrove, *Apollo's Eye: A Cartographic Genealogy of the Earth in the Western Imagination* (Baltimore: Johns Hopkins University Press, 2001). Based on what I have argued above, I would add that the core issues are larger than the earth's discovery, as the development of "Apollo's Eye" comprised many and varied discussions about not only the ontological status of unseen spaces and places, but also the nature of the epistemological claims that the "human viewer" was justified in making about both the cosmos and God.

7. Münster's earlier works were *Cosmographei. Mappa Evropae, eygentlich fürgebildet/ . . . Mappen vnd Landtaffeln zumachen/ durch Sebastianum Munsterum|| an Tag geben.* (Frankfurt am Main: Egenolff, 1537); *Canones super novum instrumentum luminarium . . . Caput Draconis* (Baslieae: Cratander, 1534); and *Erklerung des newen Instruments der Sunnen . . . An alle Liebhaber der Künstenn im Hilff zu thun zu warer unnd rechter Beschreybung Teutscher Nation* (Oppenheym: Kobel, 1528).

8. On this point I have drawn inspiration especially from McLean, *The "Cosmographia" of Sebastian Münster*, 5–44.

9. On Germany and cosmography, see Johnson, *German Discovery.* In general, see Dennis Cosgrove, "Images of Renaissance Cosmography, 1450–1650," in *The History of Cartography*, ed. J. B. Harley and David Woodward (Chicago: University of Chicago Press, 1987); Frank Lestringant, *L'Atelier du cosmographe: Ou L'Image du monde à la Renaissance* (Paris: A. Michel, 1991); Wessel, *Von Einem, der Daheim blieb*; McLean, *The "Cosmographia" of Sebastian Münster*; Lesley B. Cormack, "The World at Your Fingertips: English Renaissance Globes as Cosmographical, Mathematical and Pedagogical Instruments," *Archives Internationales d'Histoire des Sciences* 59, no. 163 (2009): 485–97; van der Krogt, "Gerard Mercator and His Cosmography"; Mosley, "The Cosmographer's Role in the Sixteenth Century"; Tinguely, "Le Vertige cosmographique à la Renaissance."

10. More recent work has noted Münster's exposure to Euclid, although not quite with the emphasis that I place upon it here. McLean, *The "Cosmographia" of Sebastian Münster*, 29–38.

11. See, for example, Lewis W. Spitz, *The Northern Renaissance* (Englewood Cliffs, N.J.: Prentice Hall, 1972); *Religious Renaissance; The Renaissance and Reformation Movements* (Chicago: Rand McNally, 1971); Arie Johan Vanderjagt et al., *Christian Humanism: Essays in Honour of Arjo Vanderjagt* (Leiden: Brill, 2009).

12. Terrence Heath, "Logical Grammar, Grammatical Logic, and Humanism in Three German Universities," *Studies in the Renaissance* 18 (1971): 9–64; Paul F. Grendler, "The Universities of the Renaissance and Reformation," *Renaissance Quarterly* 57, no. 1 (2004): 1–12.

13. In general, see Wessel, *Von Einem, der Daheim blieb*; Burmeister, *Neue Forschungen zu Sebastian Münster; Johannes Campensis und Sebastian Münster: Ihre Stellung in der Geschichte der Hebräischen Sprachstudien* (Louvain: Bibliothèque de l'Université, 1970); "Sebastian Münster: Versuch eines biographischen Gesamtbildes." For background on the educational network that formed Münster, see Paul L. Nyhus, "The Franciscans in South Germany, 1400–1530: Reform and Revolution," *Transactions of the American Philosophical Society* (1975): 1–47.

14. In Tübingen, Münster may have encountered the 1550 edition of Euclid's *Elements* by Johann Scheubel. See McLean, *The "Cosmographia" of Sebastian Münster*, 38. Given what I argued in Chapter 2, it is also significant that Stöffler also made a celestial globe. See Günther Oestmann, "Johannes Stoefflers Himmelsglobus," *Der Globusfreund: Wissenschaftliche Zeitschrift für Globenkunde* 43/44 (1995): 71–74.

15. Münster emerged from what Mack Walker has described as "German home towns," whose influence on German culture and politics reached well into the nineteenth century. Mack Walker, *German Home Towns: Community, State, and General Estate, 1648–1871* (Ithaca, N.Y.: Cornell University Press, 1971).

16. Münster's maps were, thus, a way of expressing his own identity. Along complementary lines, see David Woodward, "Reality, Symbolism, Time, and Space in Medieval World Maps," *Annals of the Association of American Geographers* 75, no. 4 (1985): 510–21; J. B. Harley, *The New Nature of Maps: Essays in the History of Cartography* (Baltimore: Johns Hopkins University Press, 2001).

17. On humanism and the German universities, see Overfield, *Humanism and Scholasticism*, 102–19, and on changes in university curricula, see 298–326. In general, see Spitz, *Religious Renaissance*, 1–19.

18. Münster, *Cosmographia* (unpaginated).

19. Peter Weidhaas, *A History of the Frankfurt Book Fair*, trans. Carolyn Gossage and W. A. Wright (Toronto: Dundurn Press, 2007), 37–53.

20. Münster, *Cosmographia* (unpaginated).

21. Ibid.

22. Ibid.

23. Ibid.

24. Gerhard Oestreich identified this trend long ago, although along different lines. Gerhard Oestreich, *Neostoicism and the Early Modern State*, ed. Brigitta Oestreich and H. G. Koenigsberger (Cambridge: Cambridge University Press, 1982); *Antiker Geist und moderner Staat bei Justus Lipsius (1547–1606): Der Neustoizismus als politische Bewegung* (Göttingen: Vandenhoeck & Ruprecht, 1989); Marc Raeff, "The Well-Ordered Police State and the Development of Modernity in Seventeenth- and Eighteenth-Century Europe: An Attempt at a Comparative Approach," *American Historical Review* 80, no. 5 (1975): 1221–43; *The Well-Ordered Police State: Social and Institutional Change Through Law in the Germanies and Russia, 1600–1800* (New Haven, Conn.: Yale University Press, 1983).

25. This is not the sin of knowledge against the divine that merited Adam and Eve's expulsion from Eden, but a simpler *human* sinfulness. On Adam and Eve, see Elaine H. Pagels, *Adam, Eve, and the Serpent* (New York: Random House, 1988).

26. André Thevet, *La Cosmographie vniverselle d'André Thevet Cosmographe dv Roy: Illvstrée de diverses figvres des choses plvs remarqvables vevës par l'auteur, & incogneuës de noz anciens & modernes*, 2 vols. (Paris: Pierre l'Huillier, 1575).

27. Ibid., unpaginated preface.

28. In general, see Büttner and Babicz, *Zur Entwicklung der Geographie*; Büttner, *Die Geographia Generalis vor Varenius*; Günther, *Varenius*; Baker, "The Geography of Bernhard Varenius"; Margarita Bowen, *Empiricism and Geographical Thought: From Francis Bacon to Alexander von Humboldt* (Cambridge: Cambridge University Press, 1981).

29. Bernhardus Varenius, *Descriptio regni Iaponiae: Cum quibusdam affinis materiæ, ex variis auctoribus collecta et in ordinem redacta*, 2 vols. (Amstelodami: apud Ludovicum Elzevirium, 1649); *Geographia generalis.*

30. Jonathan I. Israel, *The Dutch Republic: Its Rise, Greatness, and Fall, 1477–1806* (Oxford: Oxford University Press, 1995), 575–81.

31. On this history, see Johnson, *German Discovery*, 47–87.

32. Werner Kundert, "Hermann Conring als Professor der Universität Helmstedt," in *Hermann Conring (1606–1681): Beiträge zu Leben und Werk*, ed. Michael Stolleis (Berlin: Duncker & Humblot, 1983).

33. Philippus Cluverius and Josephus Vorstius, *Philippi Cluuerii introductionis in universam geographiam, tam veterem quàm novam, libri VI., cui adjuncta est D. Heinsii oratio in obitum eiusdem P. Cluuerii [ed. J. Vorstius]. ed. ultima* (Amsterdam, 1629), 1–2.

34. On Jungius's central position in reforming the categories that drove German traditions in natural philosophy, see Christoph Meinel, *In physicis futurum saeculum respicio: Joachim Jungius und die naturwissenschaftliche Revolution des 17. Jahrhunderts* (Göttingen: Vandenhoeck & Ruprecht, 1984). For an overview, see Daniel Garber, "Philosophia, Historia, Mathematica: Shifting Sands in the Disciplinary Geography of the Seventeenth Century," in *Scientia in Early Modern Philosophy*, ed. Tom Sorell, G. A. J. Rogers, and Jill Kraye (Dordrecht: Springer, 2010).

35. Quoted in Lehmann, "Der Bildungsweg des Jungen Bernhard Varenius," 75–76.

36. René Descartes, "To Frans Burman, 16 April 1648," in *Electronic Enlightenment*, ed. Robert McNamee (Oxford: Oxford University Press, 2011), http://www.e-enlightenment.com.

37. Varenius's published works had a lasting effect on classroom teaching in Königsberg. Werner Stark, "Historical and Philological References on the Question of a Possible Hierarchy of Human 'Races,' 'Peoples,' or 'Populations' in Immanuel Kant—a Supplement," in *Reading Kant's Geography*, ed. Stuart Elden and Eduardo Mendieta (Albany: State University of New York Press, 2011), 71.

38. Samuel Chappuzeau, *Idée du monde, ou introduction facile & méthodique à la cosmographie et à l'histoire* (Celle: André Holven, 1690). Along the same lines, see also Nicholas Sanson, *Description de tout l'universe, en plusieurs cartes, & en divers traitez de geographie et d'histoire; où sont décrits succinctement & avec une methode belle & facile ses empires, ses peuples, ses colonies, leur moeurs, langues, religions, richesses, &c, et ce qu'il y a de plus beau & plus rares dans toutes ses parties & dans ses isles* (Amsterdam: Chez François Halma, 1700).

39. Chappuzeau, *Idée du monde*, 1–2.

40. Varenius, *Geographia Generalis*, preface (unpaginated).

41. Ibid.

42. Ibid.

43. Ibid., 94.

44. Ibid., 1–8, 9–18.

45. Ibid., 1.

46. Ibid., 14–15 (my emphasis).

47. Ibid., 49.

48. In 1685, Allain Manesson-Mallet took exactly the same tack in five-volume work he published in Paris in 1685. See Allain Manesson-Mallet, *Description de l'univers: Contenant les differents systemes du monde, les cartes generales & particulieres de la geographie ancienne & moderne: les plans & les profils des principales villes & des autres lieux plus considerables de la Terre; avec les portraits des souverains qui y commandent, leurs blasons, titres & livrées: et les moeurs, religions, gouvernemens & divers habillemens de chaque nation . . . ,* 5 vols. (Paris: Denys Thierry, 1685), 72–76. This fascinating work is particularly worthy of further study.

49. Dirk van Miert, *Humanism in an Age of Science: The Amsterdam Athenaeum in the Golden Age, 1632–1704* (Leiden: Brill, 2009), 56.

50. Varenius, *Geographia Generalis*, 48–62.

51. Ibid., 335–36. A key component of Varenius's critique is his realization that the world's waters are not identical but have different characteristics, depending on what minerals are dissolved in them. Water as a "philosophical" category was thus subject to the passage of historical time—and this sort of thing could only occur within a consistent space. See, for example, ibid., 292–300.

52. Ibid., 66–67.

53. Ibid., 313–38.

54. See especially ibid., 333–35.

55. Ibid., 315.

56. Anthony Grafton has traced the rise of a uniform historical chronology in the work of the sixteenth-century French thinker J. J. Scaliger. See Grafton, "Joseph Scaliger and Historical Chronology: The Rise and Fall of a Discipline," *History and Theory* 14, no. 2 (1975): 156–85. Along the lines that I am arguing, it is significant that Scaliger not only wound up teaching at Leiden, but also understood the history of mathematics and the history of astronomy very well. See Anthony Grafton, *Joseph Scaliger: A Study in the History of Classical Scholarship* (Oxford: Clarendon Press, 1983), 193–211.

57. Edmond Halley, "To John Wallis, 25 February 1687," in *Electronic Enlightenment*, ed. Robert McNamee (Oxford: Oxford University Press, 2011), http://www.e-enlightenment.com.

58. Pierre Duval, *Traité de geographie qui donne la connoissance et l'usage du globe et de la carte avecque le figures necessaires pour sujet* (Paris: Chez l'Auteur, en l'Isle du Palais sur le Quay de l'Orloge, au coin de la Rue de Harlay, 1672), unpaginated preface.

59. Bernhardus Varenius, *Bernhardi Varenii geographia generalis in qua affectiones generales Telluris explicantur. Adjecta est appendix, . . . A Jacobo Jurin* (Cantabrigiæ: Impensis Cornelii Crownfield, 1712); Bernhardus Varenius, *Bernhardi Varenii geographia generalis in qua affectiones generales Telluris explicantur . . . Aliquot quae desiderabantur.* ed. Isaac Newton (Cantabrigiae: Ex officina Joann Hayes, 1672); Bernhardus Varenius, *Bernhardi Varenj Med. D. geographia generalis in qua affectiones generales Telluris explicantur . . . Aliquot quæ desiderabantur aucta & illustrata*, ed. Isaac Newton, editio secunda auctior & emendatior ed. (Cantabrigiæ: ex officina Joann Hayes, 1681).

60. Schuchard, *Bernhard Varenius (1622–1650).*

61. William Warntz, "Newton, the Newtonians, and the Geographia Generalis Varenii," *Annals of the Association of American Geographers* 79, no. 2 (1989): 165–91; Anne Marie Godlewska, *Geography Unbound: French Geographic Science from Cassini to Humboldt* (Chicago: University of Chicago Press, 1999).

62. A. Rupert Hall, *Isaac Newton, Adventurer in Thought* (Oxford: Blackwell, 1992), 21–22; Mordechai Feingold, ed., *Before Newton: The Life and Times of Isaac Barrow* (Cambridge: Cambridge University Press, 1990), 240–41.

63. Isaac Newton, *Philosophiae naturalis principia mathematica* (Londini: J. Societatis Regiae ac Typis J. Streater, 1687).

64. William Robertson, *The History of America*, 3 vols. (Dublin: Price, 1777).

65. Ibid. (table of contents, unpaginated).

66. George H. Guttridge, *The Correspondence of Edmund Burke III* (Cambridge: Cambridge University Press, 1961), 351.

67. Peter M. Engelfriet, *Euclid in China: The Genesis of the First Chinese Translation of Euclid's Elements, Books I–VI (Jihe Yuanben, Beijing, 1607) and Its Reception up to 1723*, ed. W. L. Idema (Leiden: Brill, 1998), 289–90.

68. Thomas Hobbes, *Elementorum philosophiae sectio prima de corpore* (London: Andrea Cook, 1655), 4–5.

69. Manesson-Mallet, *Description de l'univers*, 2 (see also n. 48, above).

Chapter 5

1. The classic view of Hobbes as the first modern political thinker is expressed in Leo Strauss, *The Political Philosophy of Hobbes, Its Basis and Its Genesis* (Chicago: University of Chicago Press, 1961),

viii. See also Patricia Springborg, "The Enlightenment of Thomas Hobbes," *British Journal for the History of Philosophy* 12, no. 3 (2004): 513–34; Fiammetta Palladini, "Pufendorf Disciple of Hobbes: The Nature of Man and the State of Nature: The Doctrine of *Socialitas*," *History of European Ideas* 34 (2008): 26–60; Deborah Baumgold, *Contract Theory in Historical Context: Essays on Grotius, Hobbes, and Locke* (Leiden: Brill, 2010).

2. Thomas Hobbes, *Leviathan, or the Matter, Forme, & Power of a Common-Wealth Ecclesiasticall and Civill* (London: Printed for Andrew Crooke by William Wilson, at the Green Dragon in St. Pauls Church-Yard, 1651).

3. On Münster's relationship to Euclid, see Matthew McLean, *The "Cosmographia" of Sebastian Münster: Describing the World in the Reformation* (Aldershot: Ashgate, 2007), 33, 38.

4. The few works that fully confront the significance of Hobbes's geometry are Hardy Grant, "Geometry and Politics: Mathematics in the Thought of Thomas Hobbes," *Mathematics Magazine* 63, no. 3 (1990): 147–54; William Sacksteder, "Hobbes: The Art of the Geometricians," *Journal of the History of Philosophy* 18, no. 2 (1980): 131–46; "Hobbes: Geometrical Objects," *Philosophy of Science* 48, no. 4 (1981): 573–90. See also the important chapter "More Geometrico" in Gordon Hull, *Hobbes and the Making of Modern Political Thought* (London: Continuum, 2009), 51–69.

5. For an example of the literature, see Isabel F. Knight, *The Geometric Spirit: The Abbé de Condillac and the French Enlightenment* (New Haven, Conn.: Yale University Press, 1968).

6. John Aubrey, *Brief Lives: A Modern English Version* (Woodbridge: Boydell Press, 1982), 151–52.

7. An excellent example is Garrett Ward Sheldon, *The History of Political Theory: Ancient Greece to Modern America* (New York: Peter Lang, 1988). The chapters on Hobbes, Locke, and Rousseau are in direct sequence. Further examples: Michael Oakeshott, *Lectures in the History of Political Thought* (Exeter: Imprint Academic, 2006); Dick Howard, *The Primacy of the Political: A History of Political Thought from the Greeks to the French & American Revolutions* (New York: Columbia University Press, 2010). More broadly, see Matthew D. Mendham, "Gentle Savages and Fierce Citizens Against Civilization: Unraveling Rousseau's Paradoxes," *American Journal of Political Science* 55, no. 1 (2011): 170–87; Barry Hindess, "Locke's State of Nature," *History of the Human Sciences* 20, no. 3 (2007): 1–20; Jonathan Marks, "Who Lost Nature? Rousseau and Rousseauism," *Polity* 34, no. 4 (2002): 479–502; Tom Sorell, "Hobbes and Aristotle," in *Philosophy in the Sixteenth and Seventeenth Centuries: Conversations with Aristotle*, ed. Constance Blackwell and Sachiko Kusukawa (Aldershot: Ashgate, 1999); Julia Simon, "Natural Man and the Lessons of History: Rousseau's Chronotypes," *Clio* 26 (1997): 473–84; P. N. Miller, "Citizenship and Culture in Early Modern Europe," *Journal of the History of Ideas* 57, no. 4 (1996): 725–42; James Tully, *An Approach to Political Philosophy: Locke in Contexts* (Cambridge: Cambridge University Press, 1993); John T. Scott, "The Theodicy of the Second Discourse: The 'Pure State of Nature' and Rousseau's Political Thought," *American Political Science Review* 86, no. 3 (1992): 696–711; Jeremy Waldron, "John Locke: Social Contract Versus Political Anthropology," *Review of Politics* 51, no. 1 (1989): 3–28; Victor Gourevitch, "Rousseau's Pure State of Nature," *Interpretation* 16, no. 1 (1988): 23–59; Richard Ashcraft, *Revolutionary Politics & Locke's Two Treatises of Government* (Princeton, N.J.: Princeton University Press, 1986); Mark Glat, "John Locke's Historical Sense," *Review of Politics* 43, no. 1 (1981): 3–21; Perez Zagorin, ed., *Culture and Politics from Puritanism to the Enlightenment* (Berkeley: University of California Press, 1980); Malcolm Jack, "One State of Nature: Mandeville and Rousseau," *Journal of the History of Ideas* 39, no. 1 (1978): 119–24; Robert A. Goldwin, "Locke's State of Nature in Political Society," *Western Political Quarterly* 29, no. 1 (1976); William Pickles, "The Notion of Time in Rousseau's Political Thought," in *Hobbes and Rousseau: A Collection of Critical Essays*, ed. Maurice William Cranston and R. S. Peters (Garden City, N.Y.: Anchor Books, 1972); Maurice William Cranston and R. S. Peters, *Hobbes and Rousseau: A Collection of Critical Essays* (Garden City, N.Y.: Anchor Books, 1972); Christopher W. Morris, *The Social Contract Theorists: Critical Essays on Hobbes, Locke, and Rousseau* (Lanham, Md.: Rowman and Littlefield, 1999); James Devine, "The Positive Political Economy of Individualism and Collectivism: Hobbes, Locke, and Rousseau," *Politics and Society* 28, no. 2 (2000): 265–304; Piotr Hoffman, *Freedom, Equality, Power: The Ontological Consequences of the Political Philosophies of Hobbes, Locke, and Rousseau* (New York:

Peter Lang, 1999); Patrick Riley, *Will and Political Legitimacy: A Critical Exposition of Social Contract Theory in Hobbes, Locke, Rousseau, Kant, and Hegel* (Cambridge, Mass.: Harvard University Press, 1982); John Morrow, *A History of Political Thought: A Thematic Introduction* (New York: New York University Press, 1998); Leo Strauss and Joseph Cropsey, *History of Political Philosophy*, 3rd ed. (Chicago: University of Chicago Press, 1987); Carl Schmitt, *Der Leviathan in der Staatslehre des Thomas Hobbes: Sinn und Fehlschlag eines politischen Symbols* (Hamburg: Hanseatische Verlagsanstalt, 1938).

8. Richard Ashcraft's fine analysis of Locke already gives away the game in the title, *Revolutionary Politics & Locke's Two Treatises of Government*. Not coincidentally, Ashcraft mentions Euclid a total of zero times and geometry all of once (on page 3)—and then without any reference to Locke.

9. See Michael Levin, "What Makes a Classic in Political Theory?," *Political Science Quarterly* (1973): 462–76; Steven M. Dworetz, *The Unvarnished Doctrine: Locke, Liberalism, and the American Revolution* (Durham, N.C.: Duke University Press, 1989); Carol Blum, *Rousseau and the Republic of Virtue: The Language of Politics in the French Revolution* (Ithaca, N.Y.: Cornell University Press, 1989); Richard Ashcraft and Maurice M. Goldsmith, "Locke, Revolution Principles, and the Formation of Whig Ideology," *Historical Journal* 26, no. 4 (1983): 773–800. Probably one of the finest works on early modern political thought does not mention Euclid once. See Richard Tuck, *Philosophy and Government, 1572–1651* (Cambridge: Cambridge University Press, 1993).

10. Sorell, "Hobbes and Aristotle," 364.

11. Ibid.

12. James J. Hamilton, "Hobbes's Study and the Hardwick Library," *Journal of the History of Philosophy* 16, no. 4 (1978): 445–53.

13. Thucydides, *Eight Bookes of the Peloponnesian Warre*, trans. Thomas Hobbes (London: Imprinted for Richard Mynne, 1634).

14. Paolo Rossi, *Francis Bacon: From Magic to Science* (London: Routledge & Kegan Paul, 1968); Michel Malherbe and Jean-Marie Pousseur, *Francis Bacon, science et méthode: Actes du colloque du Nantes* (Paris: Vrin, 1985); Stephen Gaukroger, *Francis Bacon and the Transformation of Early-Modern Philosophy* (Cambridge: Cambridge University Press, 2001); Joseph Agassi, *The Very Idea of Modern Science: Francis Bacon and Robert Boyle* (Dordrecht: Springer, 2013).

15. Aubrey, *Brief Lives*, 151; Robin Bunce, "Thomas Hobbes' Relationship with Francis Bacon—an Introduction," *Hobbes Studies* 16, no. 1 (2003): 41–83.

16. Linda Levy Peck, "Hobbes on the Grand Tour: Paris, Venice, or London?," *Journal of the History of Ideas* 57, no. 1 (1996): 177–83; Tom Sorell, "Seventeenth-Century Materialism: Gassendi and Hobbes," in *The Renaissance and Seventeenth-Century Rationalism*, ed. G. H. R. Parkinson (London: Routledge, 2003), 221–22.

17. Lisa T. Sarasohn, "Thomas Hobbes and the Duke of Newcastle: A Study in the Mutuality of Patronage Before the Establishment of the Royal Society," *Isis* (1999): 715–37; Stephen Clucas, "The Atomism of the Cavendish Circle: A Reappraisal," *Seventeenth Century* 9, no. 2 (1994): 247–73; Robert Hugh Kargon, *Atomism in England from Hariot to Newton* (Oxford: Clarendon Press, 1966), 54–62.

18. Shailesh A. Shirali, "Marin Mersenne, 1588–1648," *Resonance* 18, no. 3 (2013): 226–40; Peter Dear, *Mersenne and the Learning of the Schools* (Ithaca, N.Y.: Cornell University Press, 1988); "Mersenne's Suggestion: Cartesian Meditation and the Mathematical Model of Knowledge in the Seventeenth Century," in *Descartes and His Contemporaries: Meditations, Objections, and Replies*, ed. Roger Ariew and Marjorie Grene (Chicago: University of Chicago Press, 1995).

19. Hobbes had used the term "elements" in an earlier work, too: Thomas Hobbes, *Humane Nature: Or, the Fundamental Elements of Policie. Being a Discoverie of the Faculties, Acts, and Passions of the Soul of Man, from Their Original Causes, According to Such Philosophical Principles as Are Not Commonly Known or Asserted. By Tho. Hobbs of Malmsbury* (London: printed by T. Newcomb, for Fra: Bowman of Oxon, 1650). This work had been written in 1640 and was published without permission in 1650. Unlike Hobbes's *Elements of Philosophy*, this text is purely Atomist in its foundation and does not engage geometry in a sustained manner, although Hobbes's skepticism toward geometric points is already apparent in the work. (See Hobbes, *Humane Nature*, 307.)

20. Along these lines, see also Gianni Paganini, "How Did Hobbes Think of the Existence and Nature of God?: *De motu, loco et tempore* as a Turning Point in Hobbes's Philosophical Career," in *The Bloomsbury Companion to Hobbes.*, ed. S. A. Lloyd (London: Bloomsbury, 2013).

21. Thomas Hobbes, *Religio elementorvm philosophiæ sectio tertia de cive* (Parisis: s.n., 1642).

22. Thomas Hobbes, *Elementorum philosophiae sectio prima de corpore* (London: Andrea Cook, 1655).

23. Thomas Hobbes, *Elementorum philosophiae sectio secunda de homine* (Londini: Typis T. Childe. sumptibus Andr. Crooke, & væneunt sub insigni viridis Draconis in Cæmetirio Paulino, 1658).

24. See the preface to Hobbes, *De corpore* (unpaginated).

25. As always, the classic example is the work of Leo Strauss. See Leo Strauss, *The Political Philosophy of Hobbes: Its Basis and Its Genesis* (Chicago: University of Chicago Press, 1963). See also David Runciman, *Political Hypocrisy: The Mask of Power, from Hobbes to Orwell and Beyond* (Princeton, N.J.: Princeton University Press, 2008); Tomasz Mastnak, *Hobbes's Behemoth: Religion and Democracy* (Exeter: Imprint Academic, 2009); Mark E. Button, *Contract, Culture, and Citizenship: Transformative Liberalism from Hobbes to Rawls* (University Park: Pennsylvania State University Press, 2008); Mary G. Dietz, ed., *Thomas Hobbes and Political Theory* (Lawrence: University Press of Kansas, 1990).

26. Thomas Hobbes, *Behemoth; or an Epitome of the Civil Wars of England, from 1640 to 1660* (London: s.n., 1679).

27. Hobbes, *De corpore*, 1.

28. Tullio Gregory, "The Platonic Inheritance," in *A History of Twelfth-century Western Philosophy*, ed. Peter Dronke (Cambridge: Cambridge University Press, 1988).

29. Plato, *The Collected Dialogues of Plato, Including the Letters* (Princeton, N.J.: Princeton University Press, 1961); James Hankins, "The Study of the *Timaeus* in Early Renaissance Italy," in *Natural Particulars: Nature and the Disciplines in Renaissance Europe*, ed. Anthony Grafton and Nancy G. Siraisi (Cambridge, Mass.: MIT Press, 1999).

30. J. A. Bennett, *The Divided Circle: A History of Instruments for Astronomy, Navigation, and Surveying* (Oxford: Phaidon, 1987), 27–37; Charles H. Cotter, *The Astronomical & Mathematical Foundations of Geography* (London: Hollis & Carter, 1966), 51–76; Katherine Neal, "Mathematics and Empire, Navigation and Exploration: Henry Briggs and the Northwest Passage Voyages of 1631," *Isis* 93, no. 3 (2002): 435–53.

31. Owen Gingerich, "From Copernicus to Kepler: Heliocentrism as Model and as Reality," *Proceedings of the American Philosophical Society*, 117, no. 6 (1973): 513–22; Hans Blumenberg, *Die Genesis der kopernikanischen Welt* (Frankfurt am Main: Suhrkamp, 1975); Robert S. Westman, *The Copernican Achievement*, vol. 7 (Berkeley: University of California Press, 1975); Jürgen Hamel, *Nicolaus Copernicus: Leben, Werk und Wirkung* (Heidelberg: Spektrum, 1994); Gudrun Wolfschmidt, *Nicolaus Copernicus (1473–1543): Revolutionär wider Willen* (Stuttgart: Verlag für Geschichte der Naturwissenschaften und der Technik, 1994); Owen Gingerich and James MacLachlan, *Nicolaus Copernicus: Making the Earth a Planet* (New York: Oxford University Press, 2005).

32. Nicolaus Copernicus, *De lateribvs et angvlis triangulorum, tum planorum rectilineorum, tum sphæricorum . . . additus est canon semissium subtensarum rectarum linearum in circulo* (Vittembergae: Johannes Lufft, 1542); *De revolutionibus orbium coelestium, libri VI* (Norimbergae: Ioh. Petreium, 1543).

33. I cannot explain the nature of the debate on this point. See the following two classic contributions: George Saliba, "The First Non-Ptolemaic Astronomy at the Maraghah School," *Isis* 70, no. 4 (1979): 571–76; I. N. Veselovsky, "Copernicus and Nasīr Al-Dīn Al-Tusī," *Journal for the History of Astronomy* 4, no. 2 (1973): 128–30. I am more in sympathy with the latter than the former. Also useful are these more recent interventions: F. Jamil Ragep, "Copernicus and His Islamic Predecessors: Some Historical Remarks," *History of Science* 45, no. 1 (2007): 65–81; Viktor Blasjö, "A Critique of the Arguments for Maragha Influence on Copernicus," *Journal for the History of Astronomy* 45, no. 2 (2014): 183–95. Again, my views align more with those of the latter than of the former.

34. Here, I am following Hans Blumenberg's seminal work on the rise of Copernicanism: Hans Blumenberg, *The Genesis of the Copernican World*, trans. Robert M. Wallace (Cambridge, Mass.: MIT

Press, 1987). (I have already cited the original German version in n. 33.) Blumenberg's arguments loom especially large in the next chapter.

35. Hobbes, *Leviathan*, 62.

36. On this score, see the useful discussion in Douglas Michael Jesseph, *Squaring the Circle: The War between Hobbes and Wallis* (Chicago: University of Chicago Press, 1999), 8–9. This work is simply fundamental for understanding the debate between the titans.

37. Marianne Gœury, "L'Atomisme épicurien du temps à la lumière de la physique d'Aristote," *Les Études philosophiques*, no. 4 (2013): 535–52; Thomas Bénatouïl, "Les Critiques épicuriennes de la géométrie," in *Construction: Festschrift for Gerhard Heinzmann*, ed. Gerhard Heinzmann (London: College Publications, 2010); Jill Kraye, "The Revival of Hellenistic Philosophies," in *The Cambridge Companion to Renaissance Philosophy*, ed. James Hankins (Cambridge: Cambridge University Press, 2007); Jeffrey S. Purinton, "Magnifying Epicurean Minima," *Ancient Philosophy* 14, no. 1 (1994): 115–46; Diskin Clay, *Lucretius and Epicurus* (Ithaca, N.Y.: Cornell University Press, 1983); Ivars Avotins, "On Some Epicurean and Lucretian Arguments for the Infinity of the Universe," *Classical Quarterly* 33, no. 2 (1983): 421–27; David Konstan, "Problems in Epicurean Physics," *Isis* (1979): 394–418; John M. Rist, *Epicurus: An Introduction* (Cambridge: Cambridge University Press, 1972); Gregory Vlastos, "Plato's Supposed Theory of Irregular Atomic Figures," *Isis* (1967): 204–9; "Minimal Parts in Epicurean Atomism," *Isis* (1965): 121–47; A. A. Long, *Hellenistic Philosophy: Stoics, Epicureans, Sceptics* (Berkeley: University of California Press, 1986).

38. Aristidis Arageorgis, "Aristotle and the Atomists vis-à-vis the Mathematicians," *Philosophical Inquiry* 39, no. 1 (2015): 164–80.

39. Kargon, *Atomism in England*; Catherine Wilson, *Epicureanism at the Origins of Modernity* (Oxford: Oxford University Press, 2008); Christoph Meinel, "Early Seventeenth-Century Atomism: Theory, Epistemology, and the Insufficiency of Experiment," *Isis* (1988): 66–103; Stephen Greenblatt, *The Swerve: How the World Became Modern* (New York: W.W. Norton & Company, 2011); Hilary Gatti, *Giordano Bruno and Renaissance Science* (Ithaca, N.Y.: Cornell University Press, 1999).

40. Jeremy J. Gray, *Ideas of Space: Euclidean, Non-Euclidean, and Relativistic*, 2nd ed. (Oxford: Clarendon Press, 1989), 177.

41. Antoni Malet and Cozzoli Daniele, "Mersenne and Mixed Mathematics," *Perspectives on Science* 18, no. 1 (2010): 1–8; Daniel Garber, "On the Frontlines of the Scientific Revolution: How Mersenne Learned to Love Galileo," *Perspectives on Science* 12, no. 2 (2004): 135–63.; Dear, *Mersenne and the Learning of the Schools*.

42. Marin Mersenne, *Euclidis Elementorum libri, Apollonii Pergae Conica, sereni de sectione* (Paris, 1626); *L'Impieté des deistes, athées, et libertins de ce temps, combatuë, & renuersee de point en point par raisons tirées de la philosophie, & de la theoglogie* (Paris: Pierre Bilaine, 1624); *Universae geometriae mixtaeque mathematicae synopsis, et bini refractionum demonstratarum tractatus* (Parisiis: Antonium Bertier, 1644).

43. Hobbes, *Leviathan*, 62.

44. Hobbes, *De corpore*, 57.

45. Ibid.

46. For an excellent discussion of this conflict, see Jesseph, *Squaring the Circle*, 16–22.

47. Hobbes, *De corpore*, 122.

48. Space limitations prevent me from entering into Hobbes's views on optics. I note here, however, that Hobbes's approach to seeing emphasized its uncertainty. See, for example, *Elements of Policie*, 9–10.

49. John Wallis, *Elenchus geometriæ Hobbianæ, sive, geometricorum, quæ in ipsius Elementis Philosophiæ . . . proferuntur, refutatio* (Oxoniensis,1655).

50. Ibid., 4.

51. The best one-volume work on this issue is Michael Heyd, *Be Sober and Reasonable: The Critique of Enthusiasm in the Seventeenth and Early Eighteenth Centuries* (Leiden: Brill, 1995). See also P. Casini, "Newton's 'Principia' and the Philosophers of the Enlightenment," *Notes and Records of the*

Royal Society of London 42, no. 1 (1988): 35–52; J. G. A. Pocock, "Enthusiasm: The Antiself of Enlightenment," *Huntington Library Quarterly* 60, no. 1–2 (1999): 7–28.

52. Carl B. Boyer and Uta C. Merzbach, *A History of Mathematics*, 2nd ed. (New York: John Wiley & Sons, 1991), 348–52.

53. Peter Pesic, "Secrets, Symbols, and Systems: Parallels Between Cryptanalysis and Algebra, 1580–1700," *Isis* (1997): 674–92; David Eugene Smith, "John Wallis as a Cryptographer," *Bulletin of the American Mathematical Society* 24, no. 2 (1917): 82–96.

54. John Wallis, *De sectionibus conicis, nova methodo expositis, tractatus* (Oxonii: Thomas Robinson, 1655); John Wallis, Leonard Lichfield, and Thomas Robinson, *Johannis Wallisii, Ss. Th. D. geometriæ professoris Saviliani in celeberrimâ Academia Oxoniensi, operum mathematicorum pars altera . . . ecclipseos solaris observatio* (Oxonii: Typis Leon: Lichfield Academiae typographi, impensis Tho. Robinson, 1656).

55. Thomas Hobbes, *Six Lessons to the Professors of the Mathematiques One of Geometry the Other of Astronomy, in the Chaires Set up by the Noble and Learned Sir Henry Savile in the University of Oxford* (London: Printed by J. M. for Andrew Crook, 1656).

56. Ibid.

57. Ibid.

58. John Wallis, *Due Correction for Mr Hobbes: Or Schoole Discipline, for Not Saying His Lessons Right: In Answer to His Six Lessons, Directed to the Professors of Mathematicks* (Oxford: Printed by Leonard Lichfield printer to the University for Tho: Robinson, 1656).

59. There is a roundup in Jesseph, *Squaring the Circle*, 8–9. See also Quentin Skinner, *Visions of Politics*, 3 vols., vol. 3 (Cambridge: Cambridge University Press, 2002), 25–26.

60. John Pocock has written: "It would be delightful to conclude by finding an ideological aspect to the difference between Hobbes' mathematics and those the Oxford circle, but I do not know if this can be done." J. G. A. Pocock, *Virtue, Commerce, and History: Essays on Political Thought and History, Chiefly in the Eighteenth Century* (Cambridge: Cambridge University Press, 1985), 62. For one possible response, see Jesseph, *Squaring the Circle*, 48–72.

61. England had not been completely without university-level mathematics teachers before Oxford and Cambridge founded their respective chairs, but such positions tended to be short-term lecture stints rather than long-term and well-paid professorships. Mordechai Feingold, "Isaac Barrow and the Foundation of the Lucasian Professorship," in *From Newton to Hawking: A History of Cambridge University's Lucasian Professors of Mathematics*, ed. Kevin C. Knox and Richard Noakes (Cambridge: Cambridge University Press, 2003), 46–48.

62. Hobbes, *Leviathan*, 205–6.

63. Ibid., 5.

64. This work initially circulated in manuscript and was published without authorization in 1650. I had access to a posthumous edition. *Hobbs's Tripos, in Three Discourses: The First, Humane Nature . . . The Second, De Corpore Politico . . . The Third, of Liberty and Necessity*, 3rd ed. (London, 1684).

65. Hobbes, *Six Lessons*, 1, 17.

66. Ibid. (unpaginated).

67. Ibid., 1.

68. Ibid., 2.

69. Ibid.

70. Ibid., 4.

71. Wallis, *Elenchus*, 4.

72. Ibid., 5.

73. Hobbes, *Six Lessons*, 5.

74. Jean-Jacques Rousseau, *Oeuvres complètes*, 3 vols., vol. 1 (Paris: Éditions du Seuil, 1967), 212.

75. John W. Yolton, *Locke and the Compass of Human Understanding: A Selective Commentary on the Essay* (Cambridge: Cambridge University Press, 1970), 101–2.

76. John Locke, *Two Treatises of Government: In the Former, the False Principles, and Foundation of Sir Robert Filmer, and His Followers, Are Detected and Overthrown. The Latter Is an Essay Concerning the*

True Original, Extent, and the End of Civil Government (London: A. Churchill, 1690); Jean-Jacques Rousseau, *Du Contrat social, ou, principes du droit politique*, Ed. sans cartons, à laquelle on a ajoûté une lettre de l'auteur au seul ami qui-lui reste dans le monde ed. (Amsterdam: Chez Marc-Michel Rey, 1762).

77. For Locke's State of Nature, see Locke, *Two Treatises*, 167–77. For Locke's views on human rationality in the State of Nature, see *Two Treatises*, 56. On Rousseau's State of Nature, see Rousseau, *Du Contrat social*, 5–6

78. James L Axtell, "Locke, Newton, and 'The Elements of Natural Philosophy,'" *Paedagogica Europaea* (1965); "Locke's Review of the 'Principia,'" *Notes and Records of the Royal Society of London* (1965).

79. John Locke, *An Essay Concerning Humane Understanding* (London: T. Basset, 1690).

80. Ibid., 266.

81. On Locke's political machinations while in exile, see Ashcraft, *Revolutionary Politics*, 406–66.

82. John Locke, *Some Thoughts Concerning Education* (London: Printed for A. and J. Churchill, 1693).

83. John Locke, "Instructions for the Education of Edward Clarke's Children, C. 8 February 1686," in *Electronic Enlightenment*, ed. Robert McNamee (Oxford: Oxford University Press, 2011), http://www.e-enlightenment.com.

84. Locke, *Some Thoughts*, 96–97, 186.

85. Ibid., 159.

86. Ibid., 186.

87. Ibid., 228–29.

88. I am extending the arguments in the classic study by Margaret C. Jacob, *The Newtonians and the English Revolution, 1689–1720* (Ithaca, N.Y.: Cornell University Press, 1976).

89. József Teleki, "To Jean-Jacques Rousseau, 26 February 1778," in *Electronic Enlightenment*, ed. Robert McNamee (Oxford: Oxford University Press, 2011), http://www.e-enlightenment.com.

90. For an interesting anthropological reading of the meaning behind Cook's demise, see Marshall D. Sahlins, "The Apotheosis of Captain Cook," *Kroeber Anthropological Society Papers* 53/54 (1978): 1–31; "Captain Cook at Hawaii," *Journal of the Polynesian Society* 98 (1989): 371–423.

91. Rousseau, *Oeuvres complètes*, 1: 212.

Chapter 6

1. My approach here is essentially a combination of Alexandre Koyré's perspective on the open universe with Hans Blumenberg's perspective on heliocentrism's anthropological significance. See Alexandre Koyré, *From the Closed World to the Infinite Universe* (New York: Harper and Brothers, 1957); Hans Blumenberg, *The Genesis of the Copernican World*, trans. Robert M. Wallace (Cambridge, Mass.: MIT Press, 1987).

2. For useful overviews of anthropology's history, see Hans Erich Bödeker, "Menschheit, Humanität, Humanismus," in *Geschichtliche Grundbegriffe*, ed. Otto Brunner, Werner Conze, and Reinhart Koselleck (Stuttgart: Klett-Cotta, 1982); Thomas Hyllans Eriksen and Finn Sivert Nielsen, *A History of Anthropology* (London: Pluto Press, 2001); Marvin Harris, *The Rise of Anthropological Theory: A History of Theories of Culture* (London: Routledge and Kegan Paul, 1968). Also important are these classic works: Wilhelm Dilthey, "Die Funktion der Anthropologie in der Kultur des 16. und 17. Jahrhunderts," in *Wilhelm Diltheys Gesammelte Schriften*, vol. 2 (Leipzig: B. G. Teubner, 1921), 416–92, which associates anthropology with philosophy's discovery of the subject, and also "Entwicklung des Individuums," in Jacob Burckhardt, *Die Kultur der Renaissance in Italien* (Berlin: Deutsche Buch-Gemeinshaft, 1961), 67–85, which is oriented toward the invention of the individual.

3. On the seventeenth century, see Alan Barnard, *History and Theory in Anthropology* (Cambridge: Cambridge University Press, 2000), 15–26; Murray Leaf, *Man, Mind, and Science: A History of Anthropology* (New York: Columbia University Press, 1979), 10–30.

4. On the sixteenth century, see Anthony Pagden, *The Fall of Natural Man: The American Indian and the Origins of Comparative Ethnology* (Cambridge: Cambridge University Press, 1986); Pagden,

European Encounters with the New World: From Renaissance to Romanticism (New Haven, Conn.: Yale University Press, 1993); Margaret T. Hodgen, *Early Anthropology in the Sixteenth and Seventeenth Centuries* (Philadelphia: University of Pennsylvania Press, 1964); John Howland Rowe, "The Renaissance Foundations of Anthropology," *American Anthropologist* 67, no. 1 (1965): 1–20; Pol-Pierre Gossiaux, ed., *L'Homme et la nature: Genéses de l'anthropologie à l'âge classique, 1580–1750: Anthologie* (Brussels: DeBoeck, 1993); Mario Erdheim, "Anthropologische Modelle des 16. Jahrhunderts: Oviedo (1478–1557), Las Casas (1475–1566), Sahagún (1499–1540), Montaigne (1533–1592)," in *Klassiker der Kulturanthropologie: Von Montaigne bis Margaret Mead*, ed. Wolfgang Marschall (München: C. H. Beck, 1990); Harry Liebersohn, "Anthropology Before Anthropology," in *A New History of Anthropology*, ed. Henrika Kucklick (Malden, Mass.: Blackwell Publishing, 2008); Donald M. Frame, *Montaigne's Discovery of Man: The Humanization of Humanist* (New York: Columbia University Press, 1955); Miguel León Portilla, *Bernardino de Sahagún, First Anthropologist* (Norman: University of Oklahoma Press, 2002). More broadly, see also John H. Elliott, *The Old World and the New, 1492–1650* (Cambridge: Cambridge University Press, 1970), 79–104; David A. Brading, *The First America: The Spanish Monarchy, Creole Patriots, and the Liberal State, 1492–1867* (Cambridge: Cambridge University Press, 1991), 166–83. On the eighteenth century, see John H. Zammito, *Kant, Herder, and the Birth of Anthropology* (Chicago: University of Chicago Press, 2002); Mareta Linden, *Untersuchungen zum Anthropologie Begriff des 18. Jahrhunderts* (Bern: Herbert Lang/Peter Lang, 1976); Michèle Duchet, *Anthropologie et histoire au siècle des lumières: Buffon, Voltaire, Rousseau, Helvétius, Diderot* (Paris: Albin Michel, 1995); Jörn Garber and Heinz Thoma, *Zwischen Empirisierung und Konstruktionsleistung: Anthropologie im 18. Jahrhundert* (Tübingen: Niemeyer, 2004); Larry Wolff and Marco Cipolloni, *The Anthropology of the Enlightenment* (Stanford, Calif.: Stanford University Press, 2007); E. E. Evans-Pritchard, *History of Anthropological Thought* (London: Faber and Faber, 1981); Han F. Vermeulen, "Enlightenment Anthropology," in *Encyclopedia of Social and Cultural Anthropology*, ed. Alan Barnard and Jonathan Spencer (London: Routledge, 1996); Katherine M. Faull, ed., *Anthropology and the German Enlightenment: Perspectives on Humanity* (Lewisburg, Pa.: Bucknell University Press, 1995); Hans-Jürgen Schings, *Der ganze Mensch: Anthropologie und Literatur im 18. Jahrhundert: DFG-Symposion 1992* (Stuttgart: Metzler, 1994); Henry Vyverberg, *Human Nature, Cultural Diversity, and the French Enlightenment* (New York: Oxford University Press, 1989); Sergio Moravia, *Beobachtende Vernunft: Philosophie und Anthropologie in der Aufklärung*, trans. Elisabeth Piras (Frankfurt am Main: Fischer Taschenbuch Verlag, 1989); Daniel Droixhe and Pol-Pierre Gossiaux, eds., *L'Homme des lumières et la découverte de l'autre*, Études sur le XXVIIIe *Siècle. Volume Hors Série; 3 (Brussels: Editions de l'Université de Bruxelles, 1985); Eberhard Berg, Zwischen den Welten: Über die Anthropologie der Aufklärung und ihr Verhältnis zu Entdeckungs-Reise und Welt-Erfahrung mit besonderem Blick auf das Werk Georg Forster*, Beiträge Zur Kulturanthropologie (Berlin: Reimer, 1982); Odo Marquard, "Zur Geschichte des philosophischen Begriffs 'Anthropologie' seit dem Ende des 18. Jahrhunderts," in *Collegium Philosophicum: Studien Joachim Ritter zum 60. Geburtstag*, ed. Ernst-Wolfgang Böckenförde (Basel: Schwabe and Co., 1965); E. A. Hoebel, "William Robertson: An 18th-Century Anthropologist-Historian," *American Anthropologist* 62 (1960): 648–55; Larry Wolff, *Inventing Eastern Europe: The Map of Civilization on the Mind of the Enlightenment* (Stanford, Calif.: Stanford University Press, 1994); Thomas Kleinknecht, "'Reise der Aufklärung': Selbstverortung, Empirie und epistemologischer Diskurs bei Herder, Lessing, Lichtenberg und Anderen," *Berichte zur Wissenschaftsgeschichte* 22 (1999): 95–111; Werner Krauss, *Zur Anthropologie des 18. Jahrhunderts: Die Frühgeschichte der Menschheit im Blickpunkt der Aufklärung* (Berlin: Akademie-Verlag, 1978).

5. José de Acosta, *Historia natural y moral de las Indias: En que se tratan de las cosas notables del cielo, elementos, metales, plantas y animales dellas, y los ritos, y ceremonías, leyes gobierno de los Indios*, ed. Edmundo O'Gorman, 2nd ed., Biblioteca Americana (Mexico City: Fondo de Cultura Económica, 1962), i–xcv; Hodgen, *Early Anthropology*, 8; Rowe, "The Renaissance Foundations of Anthropology," 1; Pagden, *Fall of Natural Man*, 146; Walter D. Mignolo, "José de Acosta's Historia Natural y Moral de Las Indias: Occidentalism, the Modern/Colonial World, and the Colonial Difference," in *Natural and Moral History of the Indies*, ed. Jane E. Mangan (Durham, N.C.: Duke University Press, 2002); Karl W.

Butzer, "From Columbus to Acosta: Science, Geography, and the New World," *Annals of the Association of American Geographers* 82, no. 3 (1992): 543–65.

6. On Pagden's view of Herder, see Pagden, *European Encounters*, 37–47, 173–80. Overall, see his *Fall of Natural Man*; *Spanish Imperialism and the Political Imagination: Studies in European and Spanish-American Social and Political Theory, 1513–1830* (New Haven, Conn.: Yale University Press, 1990); *The Uncertainties of Empire: Essays in Iberian and Ibero-American Intellectual History* (Aldershot: Ashgate/Variorum, 1994); *Lords of All the World: Ideologies of Empire in Spain, Britain and France, c. 1500–c. 1800* (New Haven, Conn.: Yale University Press, 1995).

7. Duchet, *Anthropologie et histoire*; Evans-Pritchard, *History of Anthropological Thought*; Claude Blanckaert, *Naissance de l'ethnologie?: Anthropologie et missions en Amérique XVI*[e]*–XVIII*[e] *siècle* (Paris: Ed. du Cerf, 1985); Wolff, *Inventing Eastern Europe*; Zammito, *Kant, Herder, and the Birth of Anthropology*.

8. On this issue, a useful collection of essays is Droixhe and Gossiaux, *L'Homme des lumières et la découverte de l'autre*. For Dilthey, see his "Auffassung und Analyse des Menschen im 15. und 16. Jahrhundert."; "Die Funktion der Anthropologie in der Kultur des 16. und 17. Jahrhunderts."

9. For works in this tradition, see Harris, *The Rise of Anthropological Theory*; Joel Kahn, "Culture: Demise or Resurrection?" *Critique of Anthropology* 9, no. 2 (1989): 5–25; Jerry D. Moore, *Visions of Culture: An Introduction to Anthropological Theories and Theorists* (Walnut Creek, Calif.: AltaMira Press, 1997); Marshall D. Sahlins, *Culture and Practical Reason* (Chicago: University of Chicago Press, 1976); Clarence J. Glacken, *Traces on the Rhodian Shore: Nature and Culture in Western Thought from Ancient Times to the End of the Eighteenth Century* (Berkeley: University of California Press, 1976).

10. Liebersohn, "Anthropology Before Anthropology."

11. See Harry Liebersohn, *Aristocratic Encounters: European Travelers and North American Indians* (Cambridge: Cambridge University Press, 1998); *The Travelers' World: Europe to the Pacific* (Cambridge, Mass.: Harvard University Press, 2006).

12. On the missionaries, see Pagden, *Fall of Natural Man*; Erdheim, "Klassiker der Kulturanthropologie." On the South Sea, see Erwin Ackerknecht, "George Forster, Alexander von Humboldt, and Ethnology," *Isis* 46, no. 2 (1955): 83–95; Berg, *Zwischen den Welten*; Tanja van Hoorn, *Dem Leibe Abgelesen: Georg Forster im Kontext der physischen Anthropologie des 18. Jahrhunderts* (Tübingen: M. Niemeyer, 2004); Hugh West, "The Limits of Enlightenment Anthropology: Georg Forster and the Tahitians," *History of European Ideas* 10, no. 2 (1989): 147–60; Liebersohn, *Travelers' World*, 186–224. On Eastern Europe, see Wolff, *Inventing Eastern Europe*.

13. Klaus E. Müller, *Geschichte der antiken Ethnologie*, Rowohlts Enzyklopädie (Reinbek bei Hamburg: Rowohlt Taschenbuch Verlag, 1997); John W. Bennett, "Comments on 'The Renaissance Foundations of Anthropology,'" *American Anthropologist* 68 (1966): 215–20; Clyde Kluckhohn, *Anthropology and the Classics* (Providence: Brown University Press, 1961); Krauss, *Anthropologie*, 20.

14. Joseph de Acosta, *Historia natvral y moral de las Indias, en qve se tratan las cosas notables del cielo, y elementos, metales, plantas y animales dellas: Y los ritos, y ceremonias, leyes, y gouierno, y guerras de los Indios* (Sevilla: de Leon, 1590); René Descartes, *L'Homme de René Descartes et un traitté de la formation du foetus du mesme auteur* (Paris: Charles Angot, 1664); Descartes, *Le Monde de Mr. Descartes, ou, le traité de la lumiere et des avtres principavx objets des sens* (Paris: Michel Bobin & Nicolas le Gras, 1664); Johann Gottfried Herder, *Ideen zur Philosophie der Geschichte der Menschheit*, 4 vols. (Riga: Johann Friedrich Hartknoch, 1784–91). A posthumous editor separated Descartes's original anthropological work into independent cosmological and anthropological texts, which is why two texts are cited. See Stephen Gaukroger, *Descartes: An Intellectual Biography* (Oxford: Clarendon Press, 1995), 221–22.

15. The historian Hans Blumenberg is an exception. See Blumenberg, *Genesis*, 37–47, 67–75.

16. There was an ancient tradition of heliocentrism; see John D. North, *Cosmos: An Illustrated History of Astronomy and Cosmology* (Chicago: University of Chicago Press, 2008), 84–85.

17. The sketch was entitled "Commentariolus." Jürgen Hamel, *Nicolaus Copernicus: Leben, Werk und Wirkung* (Heidelberg: Spektrum, 1994), 201–2. A translation is available in Noel M. Swerdlow,

"The Derivation and First Draft of Copernicus's Planetary Theory: A Translation of the Commentariolus with Commentary," *Proceedings of the American Philosophical Society* 117, no. 6 (1973): 423–512.

18. Johannes Kepler, *Prodromus dissertationvm cosmographicarvm, continens mysterivm cosmographicvm, . . . excellentissimi mathematici, totiusq[ue] astronomiæ restauratoris d. Nicolai Copernici* (Tvbingae: Georg Gruppenbach, 1596).

19. Acosta, *Historia natvral.*

20. For another candidate from the sixteenth century, see León Portilla, *Bernardino de Sahagún, First Anthropologist*. In the same tradition, but looking to the eighteenth century, see William N. Fenton and Elizabeth L. Moore, "J.-F. Lafitau (1681–1746), Precursor of Scientific Anthropology," *Southwestern Journal of Anthropology* 25, no. 2 (1969): 173–87; Michel de Certeau and James Hovde, "Writing vs. Time: History and Anthropology in the Works of Lafitau," *Yale French Studies* 59 (1980): 37–64. On this score, see also Krauss, *Anthropologie*, 44–54; Wilhelm E. Mühlmann, *Geschichte der Anthropologie*, 2nd ed. (Frankfurt am Main: Athenäum Verlag, 1968), 44.

21. José de Acosta, *Histoire naturelle et morale des Indes tant occidentales qu'orientales: Où il est traicté des choses remarquables du ciel, des élémens, métaux, plantes & animaux qui sont propres de ces païs . . .*, trans. Robert Regnault Cauxois (Paris: M. Orry, 1598); *The Naturall and Morall Historie of the East and West Indies: Intreating of the Remarkeable Things of Heaven, of the Elements, Mettalls, Plants and Beasts Which Are Proper to That Country: Together with the Manners, Ceremonies, Lawes, Governements, and Warres of the Indians*, trans. Edward Grimeston (Val: Sims, 1604); José de Acosta, *Historia naturale e morale delle Indie*, trans. Gio Paolo Galucci (Venetia: Pr. Bernardo Basa, 1596); Josephus de Acosta, *Historie naturael ende morael van de Westersche Indien: Waer inne ghehandelt Wordt van de merckelijcktse Dinghen des Hemels, Elementen, Metalen/Planten Ende ghedierten van Dien*, trans. Ian Huyghen van Linschoten (Enchuysen; Haerlem: Meyn, Rooman, 1598).

22. On Acosta, see Fermín del Pino, "Contribución del Padre Acosta a la Constitución de la Etnología su Evolucionismo," *Revista de Indias* (1978): 507–43; Acosta, *Historia natural y moral de las Indias: (Sevilla, Juan de Léon, 1590)*, Repr. facs. ed., Hispaniae Scientia (Valencia: Albatros Ed., 1977); Marcel Bataillon, "L'Unité du genre humaine du P. Acosta au P. Clavigero," in *Mélanges à la mémoire de Jean Sarrailh*, ed. Marcel Bataillon (Paris: Centre de Recherches de l'Institut d'Études Hispaniques, 1966); José R. Carracido, *El P. José de Acosta y su importancia en la literatura científica Española* (Madrid: Est. Tipográfico "Sucesores de Rivadeneyra," 1899); Fermín del Pino, "Los Reinos de Méjico y Cuzco en la obra Del P. Acosta," in *Economia y sociedad en Los Andes y Mesoamerica*, ed. José Alcina Franch (Madrid: Universidad Complutense de Madrid, 1979); "Culturas clásicas y americanas en la obra del Padre Acosta," in *America y la España del siglo XVI*, ed. Francisco de Solano and Fermin del Pino (Madrid: C.S.I.C., 1982); "La Historia Indiana del P. Acosta y su ponderación 'científica' del Perú," in *Dos Mundos, dos culturas, o de la historia (natural y moral) entre España y el Perú*, ed. Fermín del Pino Díaz (Madrid: Iberoamericana, 2004); Thayne R. Ford, "Stranger in a Foreign Land: Jose de Acosta's Scientific Realizations in Sixteenth-Century Peru," *Sixteenth Century Journal* 29, no. 1 (1998): 19–33; Mignolo, "José de Acosta's Historia Natural"; Pagden, *Fall of Natural Man*. On ancient anthropology, see Müller, *Geschichte der Antiken Ethnologie*; Kluckhohn, *Anthropology and the Classics*. On the Bible's effects on early astronomy, see Stephen C. McCluskey, *Astronomies and Cultures in Early Medieval Europe* (Cambridge: Cambridge University Press, 1998), 31–37.

23. On the philosophical significance of this change, see Max Scheler, *Die Stellung des Menschen im Kosmos* (Darmstadt: O. Reichl, 1928).

24. Laura Ammon, "Bernardino de Sahagún, Jose de Acosta and the Sixteenth-Century Theology of Sacrifice in New Spain," *Journal of Colonialism and Colonial History* 12, no. 2 (2011).

25. Joan-Pau Rubiés, "Theology, Ethnography, and the Historicization of Idolatry," *Journal of the History of Ideas* 67, no. 4 (2006): 571–96.

26. Fermín del Pino, "Edición de crónicas de indias e historia intellectual, o la distancia entre José de Acosta y José Alcina (1)," *Revista de Indias* 50, no. 190 (1990): 861–78; "Contribución del Padre Acosta"; "Los Reinos"; "Culturas clásicas"; "Historia Indiana"; Walter Mignolo, *The Darker Side of the Renaissance: Literacy, Territoriality, and Colonization* (Ann Arbor: University of Michigan Press, 1995);

"José de Acosta's Historia Natural"; *The Darker Side of Western Modernity: Global Futures, Decolonial Options* (Durham, N.C.: Duke University Press, 2011).

27. On Acosta's education as a Jesuit, see Claudio M. Burgaleta, *José de Acosta, S.J. (1540–1600): His Life and Thought* (Chicago: Jesuit Way, 1999), 12–31.

28. On Jesuit education, although with reference to France, see François de Dainville and Marie-Madeleine Compère, *L'Éducation des Jésuites: XVI*ᵉ*–XVIII*ᵉ *siècles* (Paris: Les Éditions de Minuit, 1978). On Aristotle and the Jesuits, see Ford, "Stranger in a Foreign Land," 22–23. On Aristotle and medieval thought, see Stephen Brown, "The Intellectual Context of Later Medieval Philosophy: Universities, Aristotle, Arts, Theology," in *Medieval Philosophy,* ed. John Marenbon (London: Routledge, 1998). On Aristotle in the Renaissance, see Charles B. Schmitt, *Aristotle and the Renaissance* (Cambridge, Mass.: Published for Oberlin College by Harvard University Press, 1983). On Aquinas and Acosta, see Burgaleta, *José de Acosta,* 19–21.

29. See also del Pino, "Culturas clásicas," 333.

30. Burgaleta, *José de Acosta,* 56–69.

31. Acosta, *Historia Natvral,* 13–115.

32. Ibid., 117–92, 193–299.

33. Ibid., 303–93, 395–449.

34. Ibid., 451–535.

35. Pagden, *Fall of Natural Man,* 6–8.

36. On Pliny and ethnography, see Trevor Murphy, *Pliny the Elder's Natural History: The Empire in the Encyclopedia* (Oxford: Oxford University Press, 2004), 77–128. On monsters, Rudolf Wittkower, "Marvels of the East: A Study in the History of Monsters," *Journal of the Warburg and Courtauld Institutes* 5 (1942): 159–97; Hodgen, *Early Anthropology,* 22–42.

37. Pagden, *Fall of Natural Man,* 146.

38. On the tension within Christian approaches to geography and cosmology, see W. G. L. Randles, *De la Terre plate au globe terrestre: Une mutation épistémologique rapide (1480–1520)* (Paris: Librairie Armand Colin, 1980).

39. Acosta, *Historia natvral,* 16.

40. Ibid., 16, 18, 19.

41. Ibid., 85–115.

42. Randles, *Terre plate,* 9–31; Numa Broc, *La Géographie de la Renaissance (1420–1620)* (Paris: Bibliothèque Nationale, 1980), 9–36.

43. On Aristotle's view of science, see R. J. Hankinson, "Philosophy of Science," in *The Cambridge Companion to Aristotle,* ed. Jonathan Barnes (Cambridge: Cambridge University Press, 1995). On Acosta and Aristotle, see Ford, "Stranger in a Foreign Land." For an overview of how Aristotelian categories were repeatedly reapplied in the Renaissance period, see Gianna Pomata and Nancy G. Siraisi, "Introduction," in *Historia: Empiricism and Erudition in Early Modern Europe,* ed. Gianna Pomata and Nancy G. Siraisi (Cambridge, Mass.: MIT Press, 2005), 1–8.

44. On Iberian astronomy, see W. G. L. Randles, *The Unmaking of the Medieval Christian Cosmos, 1500–1700: From Solid Heavens to Boundless Aether* (Aldershot: Ashgate, 1999), 6–7.

45. Olaf Pedersen, "Astronomy," in *Science in the Middle Ages,* ed. David C. Lindberg (Chicago: University of Chicago Press, 1978), 269–77; Menso Folkerts, ed., *The Development of Mathematics in Medieval Europe: The Arabs, Euclid, Regiomontanus* (Aldershot: Ashgate, 2006); Ernst Zinner, *Leben und Wirken des Joh. Müller von Königsberg genannt Regiomontanus,* ed. Helmut Rosenfeld and Otto Zeller, 2nd ed. (Osnabrück: Otto Zeller, 1968); Rudolf Mett, *Regiomontanus: Wegbereiter des neuen Weltbildes* (Stuttgart: B. G. Teubner, 1996); North, *Cosmos;* Michael A. Hoskin, ed., *The Cambridge Concise History of Astronomy* (Cambridge: Cambridge University Press, 1999), 84–86.

46. On Tycho, see Adam Mosley, *Bearing the Heavens: Tycho Brahe and the Astronomical Community of the Late Sixteenth Century* (Cambridge: Cambridge University Press, 2007).

47. Victor Navarro Brotons, "The Reception of Copernicus in Sixteenth-Century Spain: The Case of Diego de Zúñiga," *Isis* 86, no. 1 (1995): 52–78. Copernicus's work was quite widely read, even if it was not universally accepted.

48. The standard English translation is Claudius Ptolemy, *Ptolemy's Almagest*, trans. G. J. Toomer (Princeton, N.J.: Princeton University Press, 1998).

49. This difference in emphasis remained a source of tension within Christian Aristotelianism. Patrick Gautier Dalché, *La Géographie de Ptolémée en Occident (IV*[e]*–XVI*[e] *siècle)* (Turnhout: Brepols, 2009), 20. Ptolemy makes his differences with Aristotle very clear: Ptolemy, *Ptolemy's Almagest*, 35–37.

50. Acosta, *Historia natvral*, 103.

51. Mett, *Regiomontanus*; Blumenberg, *Genesis*, 35–51.

52. A contemporary translation of the *Geography* is available in J. L. Berggren and Alexander Jones, eds., *Ptolemy's Geography: An Annotated Translation of the Theoretical Chapters* (Princeton, N.J.: Princeton University Press, 2000). In general, see also Leo Bagrow, "The Origin of Ptolemy's Geographia," *Geografiska Annaler* 27 (1945): 9–31; J. A. May, "The Geographical Interpretation of Ptolemy in the Renaissance," *Tijdschrift voor Economische en Sociale Geografie* 73, no. 6 (1982): 350–61; Gautier Dalché, *Géographie de Ptolémée*; Zur Shalev and Charles Burnett, eds., *Ptolemy's Geography in the Renaissance* (London: Warburg Institute, 2011); Broc, *Géographie*, 9–19.

53. Acosta, *Historia natvral*, 118.

54. Ibid., 146.

55. Ibid., 171, 84.

56. Ibid., 300–301.

57. Ibid., 329–30.

58. Ibid., 345–56.

59. The devil's location remained an issue well into the early modern period. See Mark A. Peterson, *Galileo's Muse: Renaissance Mathematics and the Arts* (Cambridge, Mass.: Harvard University Press, 2011), and also Galileo's curious lectures on the matter, "La Figura, sito e grandezza dell'Inferno de Dante Alighieri," in *Studi sulla Divina Commedia, di Galileo Galilei, Vincenzo Borghini ed altri*, ed. Ottavio Gigli (Florence: Felice Le Monnier, 1855), 3–34.

60. This is especially the case in the seminal text: Pagden, *Fall of Natural Man*.

61. Del Pino, "Contribución del Padre Acosta," 508–13.

62. On Humboldt's approach to geography, see the fundamental work by Margarita Bowen, *Empiricism and Geographical Thought: From Francis Bacon to Alexander von Humboldt* (Cambridge: Cambridge University Press, 1981), 210–22.

63. Alexander von Humboldt, *Kosmos: Entwurf einer physischen Weltbeschreibung*, 5 vols., vol. 2 (Stuttgart: J. G. Cotta'scher Verlag, 1847), 298. See also the still useful historical overview in *Kritische Untersuchungen über die historische Entwickelung der geographischen Kenntnisse von der Neuen Welt und die Fortschritte der nautischen Astronomie in dem 15ten und 16ten Jahrhundert*, 2 vols., vol. 1 (Berlin: Nicolai'sche Buchhandlung, 1836), 34–83.

64. Humboldt was using categories that had already appeared in Immanuel Kant's work and ran, in turn, into the geography of the early nineteenth century. See Immanuel Kant, *Jmmanuel Kants physische Geographie* (Mainz, Hamburg: Vollmer, 1801), 1–89. The theoretical shift that stood behind this development is clearly explained in Bowen, *Empiricism and Geographical Thought*, 231–46.

65. On Humboldt and geography, see Anne Marie Godlewska, *Geography Unbound: French Geographic Science from Cassini to Humboldt* (Chicago: University of Chicago Press, 1999), 89–128; Bowen, *Empiricism and Geographical Thought*, 210–59.

66. In general, see Blumenberg, *Genesis*. On the cultural aspects of astronomical thought, see Rainer Baasner, *Das Lob der Sternkunst: Astronomie in der deutschen Aufklärung* (Göttingen: Vandenhoeck and Ruprecht, 1987); Sara J. Schechner, "The Material Culture of Astronomy in Daily Life: Sundials, Science, and Social Change," *Journal for the History of Astronomy* 32 (2001): 189–222; McCluskey, *Astronomies and Cultures*.

67. On observatories and instruments in the history of astronomy, see William B. Ashworth, Jr., "The Calculating Eye: Baily, Herschel, Babbage and the Business of Astronomy," *British Journal for the History of Science* 27, no. 4 (1994): 409–41; J. A. Bennett, "The English Quadrant in Europe: Instruments and the Growth of Consensus in Practical Astronomy," *Journal for the History of Astronomy* 23

(1992): 1–14; J. A. Bennett, Domenico Bertoloni Meli, *Sphaera Mundi: Astronomy Books in the Whipple Museum, 1478–1600* (Cambridge: Whipple Museum of the History of Science, 1994); Nicholas Jardine, "The Places of Astronomy in Early-Modern Culture," *Journal for the History of Astronomy* 29 (1998): 49–62; Pedersen, "Astronomy"; Derek Howse, "The Greenwich List of Observatories: A World List of Astronomical Observatories, Instruments and Clocks, 1670–1850," *Journal for the History of Astronomy* 14 (1986): 1–100. ; J. A. Bennett, *The Divided Circle: A History of Instruments for Astronomy, Navigation, and Surveying* (Oxford: Phaidon, 1987), 73–129.

68. On images in astronomical works, see Isabelle Pantin, "Kepler's *Epitome*: New Images for an Innovative Book," in *Transmitting Knowledge: Words, Images, and Instruments in Early Modern Europe*, ed. Sachiko Kusukawa and Ian Maclean (Oxford: Oxford University Press, 2006). On the Renaissance view of the earth and Copernicus, see Thomas Goldstein, "The Renaissance Concept of the Earth in Its Influence upon Copernicus," *Terrae Incognitae* 4, no. 1 (1972): 19–51.

69. Joseph Walker, *Astronomy's Advancement, or, News for the Curious; Being a Treatise of Telescopes: . . . &C. Out of the Best Astronomers, Ancient and Modern, Viz. Mr. Hooke, Mr. Boilleau, Mr. Hevelius, Father Kircher, &C.* (London: Philip Lea, 1684). The name "Boilleau" probably refers to the French astronomer Ismael Boulliau.

70. On material culture in astronomy, see Schechner, "Material Culture of Astronomy."

71. Fontenelle, *Entretiens sur la pluralité des mondes* (Paris: Chez la veuve C. Blageart, 1686).

72. On the gendered nature of science education in this period, see Massimo Mazzotti, "Newton for Ladies: Gentility, Gender and Radical Culture," *British Journal for the History of Science* 32, no. 2 (2004): 119–46.

73. Isaac Newton, *Philosophiae naturalis principia mathematica* (Londini: J. Societatis Regiae ac Typis J. Streater, 1687).

74. Johann Leonard Rost, *Atlas Portatilis Coelestis, oder compendiöse Vorstellung des gantzen Welt-Gebäudes . . . und durch mehr als anderthalb hundert Figuren erkläret* (Nürnberg: Johann Ernst Adelbulner, 1723).

75. Johann Heinrich Lambert, *Cosmologische Briefe über die Einrichtung des Weltbaues* (Augsburg: Eberhard Kletts Wittib., 1761), 2.

76. I had access to a combined edition from 1750: Young, *Night Thoughts.*

77. Ibid., 316.

78. Ibid.

79. Alexander Pope, *Essay on Man and Other Poems* (New York: Dover Publications, 1994), 4.

80. Ibid., 2, 15, 28.

81. Duchet, *Anthropologie et histoire*, 281–321; Wolff, *Inventing Eastern Europe.*

82. John N. Brown, "Voltaire and Astronomy," *Journal of the British Astronomical Association* 99, no. 1 (1989): 29–31; Duchet, *Anthropologie et histoire*; Sarah Hutton, "Émilie du Châtelet's Institutions de Physique as a Document in the History of French Newtonianism," *Studies in the History of the Philosophy of Science* 35, no. 3 (2004): 515–31; Mordechai Feingold, *The Newtonian Moment: Isaac Newton and the Making of Modern Culture* (New York: Oxford University Press, 2004), 94–117.

83. Voltaire, *Élémens de la philosophie de Neuton: mis à la portée de tout le monde* (Amsterdam: Desbordes, 1738); *Le Micromégas de Mr. de Voltaire avec une histoire des croisades & un nouveau plan de l'ésprit humain* (London: J. Robinson, 1752). I would like to thank Prof. Oscar Kenshur of Indiana University, Bloomington, for first recommending *Micromégas*. On Newton's influence on literature, see Karl Guthke, *The Last Frontier: Imagining Other Worlds, from the Copernican Revolution to Modern Science Fiction*, trans. Helen Atkins (Ithaca, N.Y.: Cornell University Press, 1990), 244–80.

84. Duchet, *Anthropologie et histoire*, 229–80.

85. Ibid., 230–33; Giovanni Solinas, "Newton and Buffon," *Vistas in Astronomy* 22, no. 4 (1979): 431–39; P. Casini, "Newton's 'Principia' and the Philosophers of the Enlightenment," *Notes and Records of the Royal Society of London* 42, no. 1 (1988): 35–52.

86. Georges-Louis Leclerc de Buffon, *Histoire naturelle, générale et particulière avec la description du cabinet du roy*, 29 vols., vol. 1 (Paris: L'Imprimerie Royale, 1749–1788), 1–64; 65–126. On the meaning of this second preface, see Jacques Roger, *Buffon: A Life in Natural History*, trans. Sara Lucille

Bonnefoi (Ithaca, N.Y.: Cornell University Press, 1997), 93–105. More broadly, see *The Life Sciences in Eighteenth-Century French Thought*, trans. Robert Ellrich (Stanford, Calif.: Stanford University Press, 1997), 426–74; Stanley L. Jaki, *Planets and Planetarians: A History of Theories of the Origin of Planetary Systems* (Edinburgh: Scottish Academic Press, 1978), 87–111.

87. On Goethe and anthropology, see Matthew Bell, *Goethe's Naturalistic Anthropology: Man and Other Plants* (Oxford: Clarendon Press, 1994). On Goethe and astronomy, see Aeka Ishihara, "Goethe und die Astronomie seiner Zeit. Eine astronomisch-literarische Landschaft um Goethe," *Goethe-Jahrbuch*, no. 117 (2000): 103–17; Stanley L. Jaki, "Goethe and the Physicists," *American Journal of Physics* 37, no. 2 (1969): 195–203.

88. Quoted in Ishihara, "Goethe und die Astronomie," 115.

89. Abraham G. Kästner, "Das Lob der Sternkunst," *Hamburgisches Magazin, oder gesammlete Schriften, zum Unterricht und Vergnügen* 1, no. 2 (1747): 215.

90. Immanuel Kant, *Immanuel Kant Werkausgabe*, ed. Wilhelm Weischedel, 12 vols., vol. 7 (Frankfurt am Main: Suhrkamp Verlag, 1974), 300.

91. *Allgemeine Naturgeschichte und Theorie des Himmels: Oder Versuch von der Verfassung und dem mechanischen Ursprunge des ganzen Weltgebäudes nach newtonischen Grundsätzen abgehandelt* (Königsberg: Petersen, 1755). A contemporary version is available in *Immanuel Kant Werkausgabe*, ed. Wilhelm Weischedel, 12 vols., vol. 1 (Frankfurt am Main: Suhrkamp Verlag, 1974), 219–396.

92. Johann Gottfried Herder, *Ideen zur Philosophie der Geschichte der Menschheit*, 4 vols. (Riga: Johann Friedrich Hartknoch, 1784–91). On Goethe, see *Herders Werke in Fünf Bänden*, ed. Regine Otto, 5th ed., 5 vols., vol. 4 (Berlin: Hermann Duncker, 1978), 470.

93. Max Scheler, *Die Stellung des Menschen im Kosmos* (Darmstadt: O. Reichl, 1928); Wolfhart Pannenberg, *Anthropologie in theologischer Perspektive* (Göttingen: Vandenhoeck and Ruprecht, 1983); Helmut Pfotenhauer, *Literarische Anthropologie: Selbstbiographien und ihre Geschichte, Am Leitfaden des Leibes* (Stuttgart: Metzler, 1987); Wolfgang Pross, "Herder und die Anthropologie der Aufklärung," in *Johann Gottfried Herder: Werke*, ed. Wolfgang Pross (Munich: Carl Hanser Verlag, 1987); Helmuth Plessner, "Der Mensch als Lebewesen," in *Philosophische Anthropologie*, ed. Werner Schüssler (München: Verlag Karl Alber Freiburg, 2000); Zammito, *Birth of Anthropology*; David Denby, "Herder: Culture, Anthropology and the Enlightenment," *History of the Human Sciences* 18, no. 1 (2005): 55–76.

94. Herder, *Ideen*, 1:3.

95. Ibid.

96. Ibid., 4. (On Bode and Hegel, see this book's introduction.)

97. Ibid.

98. Ibid., 2.

99. Ibid., 3.

100. Ibid., 4.

101. On Herder and science, see Hugh Barr Nisbet, *Herder and the Philosophy and History of Science* (Cambridge, Mass.: Modern Humanities Research Association, 1970), 140–47; *Herder and Scientific Thought* (Cambridge, Mass.: Modern Humanities Research Association, 1970); John Dillenberger, *Protestant Thought and Natural Science: A Historical Interpretation* (Notre Dame, Ind.: University of Notre Dame Press, 1988), 190–93.

102. Herder, *Ideen*, 1: 8–9.

103. Ibid., 17–18. On Herder and his relationship to the anthropology of the Enlightenment, see Pross, "Herder und die Anthropologie."

104. Herder, *Ideen*, 1: 22–23.

105. Ibid., 29–35, 35–59.

106. Herder, *Ideen*, 2: 3.

107. On this theme, see Owen Gingerich and James MacLachlan, *Nicolaus Copernicus: Making the Earth a Planet* (New York: Oxford University Press, 2005).

108. Dillenberger, *Protestant Thought and Natural Science.*

109. Herder, *Ideen*, 2: 26, 35, 75, 77.

110. Ibid., 61–79.

111. Manfred Kuehn, *Kant: A Biography* (New York: Cambridge University Press, 2001), 98–99.

112. Extraterrestrials should be a more prominent theme in the study of Kant's anthropology. For an example to be emulated, see David L. Clark, "Kant's Aliens: The Anthropology and Its Others," *New Centennial Review* 1, no. 2 (2001): 201–89.

113. Signe Toksvig, *Emmanuel Swedenborg: Scientist and Mystic* (New Haven, Conn.: Yale University Press, 1948), 71–73; Jaki, *Planets and Planetarians*, 111–54. Buffon had a copy of the work in which Swedenborg proposes the nebular hypothesis. Edward S. Holden, ed., *Essays in Astronomy* (New York: D. Appleton and Company, 1900), xvii.

114. Steven J. Dick, *Plurality of Worlds: The Origins of the Extraterrestrial Life Debate from Democritus to Kant* (Cambridge: Cambridge University Press, 1982); Michael J. Crowe, *The Extraterrestrial Life Debate, 1750–1900: The Idea of a Plurality of Worlds from Kant to Lowell* (Cambridge: Cambridge University Press, 1986); Patricia Fara, "Heavenly Bodies: Newtonianism, Natural Theology and the Plurality of Worlds Debate in the Eighteenth Century," *Journal for the History of Astronomy* 35 (2004): 143–60; Guthke, *Last Frontier*.

115. For a discussion of the theological issues that heliocentrism and the new science raised, see Dillenberger, *Protestant Thought and Natural Science*, 133–62.

116. Hilary Gatti, *Giordano Bruno and Renaissance Science* (Ithaca, N.Y.: Cornell University Press, 1999); Ernan McMullin, "Bruno and Copernicus," *Isis* 78, no. 1 (1987): 55–74; Paul-Henri Michel, *The Cosmology of Giordano Bruno* (Paris: Hermann, 1973); Christoph Lüthy, "Centre, Circle, Circumference: Giordano Bruno's Astronomical Woodcuts," *Journal for the History of Astronomy* 41, no. 3 (2010): 311–27; Miguel Ángel Granada, *Giordano Bruno: Universo infinito, unión con Dios, perfección del hombre* (Barcelona: Herder, 2002). On the role of homogeneity in Bruno's pluralism, see Michel, *The Cosmology of Giordano Bruno*, 182–86.

117. Christiaan Huygens, *Christiani Hugenii Kosmotheoros, sive de terris coelestibus, earum ornatu, conjecturae. Ad Constantinum Hugenium, fratrem*, ed. 1 ed. (Hagae-Comitum: Adrian Moetjens, 1698), 3–4.

118. Kant, *Allgemeine Naturgeschichte*, 171.

119. Blumenberg, *Genesis*, 103.

120. Liebersohn, "Anthropology Before Anthropology."

121. Immanuel Kant, "Träume eines Geistersehers, erläutert durch Träume der Metaphysik," in *Immanuel Kants Gesammelte Werke: Bd. II: Vorkritische Schriften II: 1757–1777* (Berlin: G. Reimer, 1905), 315–73.

122. Emmanuel Swedenborg, *De telluribus planetarum in mundo nostro solari quae vocantur planetae et de telluribus in coelo astrifero deque illarum incolis, tum de spiritibus et angelis ibi ex auditis et visis* (London, 1758).

123. Kant, "Träume," 328–31.

124. Ibid., 332.

125. Johann Philipp Andreae, *Neu Inventirtes, belehrendes und ergötzendes astronomisches Karten Spiel... der kunstliebenden Jugend vorgestellt durch Johann Philipp Andreae* (Nürnberg: Self, 1719).

126. *Nouvel Atlas des enfans, ou principes clairs pour apprendre facilement et en fort peu de tems la geographie, suivi d'un traité méthodique de la sphere, qui explique le mouvement des astres, les divers systemes du monde, & l'usage des globes* (Amsterdam: Chez B. Vlam, 1776). This text was probably based on another work that had been issued in Nuremberg: Homann, *Atlas compendiarius seu ita dictus scholasticus minor: In usum erudiendae juventutis adornatus* (Nürnberg: excudientibus Homanianis heredibus, 1753).

127. *Nouvel Atlas des enfans*, v.

128. In general, see Wolfgang-Hagen Hein, ed., *Alexander von Humboldt: Leben und Werk* (Frankfurt am Main: Weisbecker, 1985); Donald McCrory, *Nature's Interpreter: The Life and Times of Alexander von Humboldt* (Cambridge: Luttworth Press, 2010).

129. On Blumenbach and Göttingen, see John Zammito, "Policing Polygeneticism in Germany, 1775: (Kames,) Kant, and Blumenbach," in *The German Invention of Race*, ed. Sara Eigen and Mark Larrimore (Albany: State University of New York Press, 2006); Nicolaas A. Rupke, *Göttingen and the*

Development of the Natural Sciences (Göttingen: Wallstein, 2002); John Gascoigne, "Blumenbach, Banks, and the Beginning of Anthropology at Göttingen," in *Göttingen and the Development of the Natural Sciences*, ed. Nicolaas A. Rupke (Göttingen: Wallstein Verlag, 2002).

130. Alexander von Humboldt and Aimé Bonpland, *Voyage aux régions équinoxiales du nouveau continent, fait en 1799, 1800, 1801, 1802, 1803, et 1804* (Paris: Libraire grecque-latine-allemande, 1804); Miguel Ángel Puig-Samper Mulero, "El viajero científico: La visión de Humboldt sobre Nueva España," *Paraíso occidental: Norma y diversidad en el México Virreinal* (1998); Laura Dassow Walls, *The Passage to Cosmos: Alexander von Humboldt and the Shaping of America* (Chicago: University of Chicago Press, 2009); Gerard Helferich, *Humboldt's Cosmos: Alexander von Humboldt and the Latin American Journey That Changed the Way We See the World* (New York: Gotham, 2004).

131. Godlewska, *Geography Unbound*, 119–24; Carl B. Boyer and Uta C. Merzbach, *A History of Mathematics*, 2nd ed. (New York: John Wiley and Sons, 1991), 510–70; Howard Whitley Eves, *An Introduction to the History of Mathematics*, 5th ed. (Philadelphia: Saunders College Publishing, 1983), 422–54.

132. Alexander von Humboldt, *Kosmos: Entwurf einer physischen Weltbeschreibung*, 5 vols. (Stuttgart: J. G. Cotta'scher Verlag, 1845–62).

133. An overview of the translations: *Cosmos: Sketch of a Physical Description of the Universe*, 3rd ed., 4 vols. (London: Longman, Brown, Green J. Murray, 1847); *Cosmos: O saggio di una fisica descrizione del mondo* (Torino: G. Pomba E Compagni, 1847); *Cosmos: A Sketch of a Physical Description of the Universe*, trans. E. C. Otté, 5 vols. (New York: Harper and Brothers, 1850); *Cosmos, ó, ensayo de una descripción física del mundo*, trans. Francisco Díaz Quintero (México: V. García Torres, 1851); *Cosmos; Essai d'une description physique du monde*, trans. H. Faye and Ch. Galuski, 4 vols. (Paris: Gide et J. Baudry, Éditeurs, 1855).

134. On the richness of Humboldt's use of the term "cosmos," see Nicolaas Rupke's introduction to *Cosmos: A Sketch of a Physical Description of the Universe*, trans. E. C. Otté, vol. 1 (Baltimore: Johns Hopkins University Press, 1997), ix–x. This very useful version of Humboldt's great work is a reprint of an early English translation: *Cosmos: A Sketch of a Physical Description of the Universe*, trans. E. C. Otté, 5 vols., vol. 1 (London: H. G. Bohn, 1848).

135. Humboldt, *Kosmos*, 1: 80–81.

136. Ibid., 362–82. The categories used here appear in E. C. Otté's English translation as page headings within the text and accurately represent the content of Humboldt's discussion in the German version. *Cosmos: A Sketch of the Physical Description of the Universe*, trans. E. C. Otté, 5 vols., vol. 1 (Baltimore: Johns Hopkins University Press, 1997), 343–57.

137. Humboldt, *Kosmos*, 3: 3, 4: 3.

138. The dismissal of Iberian science has been a problem since the seventeenth century. See Jorge Cañizares-Esguerra, "Iberian Science in the Renaissance: Ignored How Much Longer?" *Perspectives on Science* 12, no. 1 (2004): 86–124.

139. Joseph-François Lafitau, *Mœurs des sauvages ameriquains, comparées aux mœurs des premiers temps*, 2 vols. (Paris: Saugrain l'aîné, 1724).

140. Wolfgang Pross's magnificent essay on Herder's relationship to European anthropology remains the best overview available—and it does not mention Herder's debt to astronomy. Pross, "Herder und die Anthropologie."

141. On the philosophical significance of this issue, see David Rapport Lachterman, *The Ethics of Geometry: A Genealogy of Modernity* (New York: Routledge, 1989), 187–205.

142. Claude Lévi-Strauss, *Anthropologie Structurale* (Paris: Plon, 1958).

143. Ibid., 415 (emphasis in the original).

Conclusion

1. Friedrich Nietzsche, *The Gay Science, with a Prelude in Rhymes and an Appendix of Songs*, trans. Walter Arnold Kaufmann (New York: Random House, 1974), 181.

2. Ibid.

3. Friedrich Schiller, "Spruch des Konfucius," in *Musen-Almanach für das Jahr 1800*, ed. Friedrich Schiller (Tübingen: J. G. Cotta'sche Buchhandlung, 1799), 210.

4. Fyodor Dostoyevsky, *The Brothers Karamazov*, trans. Constance Garnett (Mineola, N.Y.: Dover Publications, 2005), 212–13.

5. On the debate about Nietzsche, see Clayton Koelb, ed., *Nietzsche as Postmodernist: Essays Pro and Contra* (Albany: State University of New York Press, 1990). For exemplary studies of Nietzsche as the first postmodernist, see Johannes Willem Bertens and Joseph P. Natoli, *Postmodernism: The Key Figures* (Malden, Mass.: Blackwell Publishers, 2002); Allan Megill, *Prophets of Extremity: Nietzsche, Heidegger, Foucault, Derrida* (Berkeley: University of California Press, 1985); Gregory B. Smith, *Nietzsche, Heidegger, and the Transition to Postmodernity* (Chicago: University of Chicago Press, 1996). There is also a parallel literature that explores Dostoevsky's contributions to the rise of new literary forms. Much of this latter discussion goes back to Mikhail Bakhtin's imaginative interpretations of Dostoevsky's oeuvre. See M. M. Bakhtin, *Problems of Dostoevsky's Poetics* (Ann Arbor, Mich.: Ardis, 1973).

6. On the mathematical significance of Euclid's demotion, see Jeremy J. Gray, *Ideas of Space: Euclidean, Non-Euclidean, and Relativistic*, 2nd ed. (Oxford: Clarendon Press, 1989). I have taken much of my historical apparatus here from the forgotten nineteenth-century essay (which appeared in two parts): Paul Tannery, "La Géometrie imaginaire et la notion d'espace," *Revue Philosophique de la France et de l'Étranger* 2 (1876); "La Géometrie imaginaire, Part II (1877)."

7. See any of the following fine works: Perry Anderson, *The Origins of Postmodernity* (London: Verso, 1998); Johannes Willem Bertens, *The Idea of the Postmodern: A History* (London: Routledge, 1995); Bertens and Natoli, *Postmodernism: The Key Figures*; David Harvey, *The Condition of Postmodernity: An Enquiry into the Origins of Cultural Change* (Oxford: Blackwell, 1989); Jean François Lyotard, Robert Harvey, and Mark S. Roberts, eds., *Toward the Postmodern* (Atlantic Highlands, N.J.: Humanities Press, 1993); Steven Best and Douglas Kellner, *The Postmodern Turn* (New York: Guilford Press, 1997); John Docker, *Postmodernism and Popular Culture: A Cultural History* (Cambridge: Cambridge University Press, 1994).

8. Jean-François Lyotard, *The Postmodern Condition: A Report on Knowledge*, trans. Geoff Bennington and Brian Massumi (Minneapolis: University of Minnesota Press, 1984), xxiv. I have used a translation of the original French version: Jean-François Lyotard, *La Condition postmoderne: Rapport sur le savoir* (Paris: Éditions de Minuit, 1979).

9. Ibid., 3.

10. Carl B. Boyer, *The History of the Calculus and Its Conceptual Development (the Concepts of the Calculus)* (New York: Dover, 1959), 187–223.

11. Early modern thinkers well understood that something important had happened in the course of the seventeenth century. See, for example, the article "Géometrie," in *Encyclopédie*, ed. Denis Diderot and Jean-Baptiste le Rond D'Alembert, vol. 7 (Paris: Chez Briasson, 1754), 629–40.

12. Carl B. Boyer and Uta C. Merzbach, *A History of Mathematics*, 2nd ed. (New York: John Wiley & Sons, 1991), 519–22; Howard Whitley Eves, *An Introduction to the History of Mathematics*, 5th ed. (Philadelphia: Saunders College Publishing, 1983), 385–88; Morris Kline, *Mathematical Thought from Ancient to Modern Times*, 3 vols., vol. 3 (New York: Oxford University Press, 1990), 874. See also Paul Tannery's discussion of Descartes's effect on geometry in Tannery, "La Géometrie imaginaire, Part I (1876)," 436–38.

13. Detlef Laugwitz, *Bernhard Riemann, 1826–1866: Turning Points in the Conception of Mathematics* (Boston: Birkhäuser, 1999), 219–33.

14. Bernhard Riemann, *Ueber die Hypothesen, welche der Geometrie zu Grunde liegen*, vol. 13 (Göttingen, 1867).

15. G. Waldo Dunnington, Jeremy Gray, and Fritz-Egbert Dohse, *Carl Friedrich Gauss: Titan of Science* (Washington, D.C.: Mathematical Association of America, 2004), 85–99.

16. Ibid., 174–90.

17. Boyer and Merzbach, *History of Mathematics*, 547–50.

18. Pierre Cassou-Noguès, *Hilbert* (Paris: Les Belles Lettres, 2001), 56–61; Laugwitz, *Riemann*, 318–22.

19. David Hilbert, *Vorlesungen über Elemente der euklidischen Geometrie. Göttingen, W. S. 1898/99* (Göttingen, 1899).

20. *Gesammelte Abhandlungen*, 3 vols., vol. 3 (Berlin: Springer, 1932), 402–5. See also Constance Reid, *Hilbert* (New York: Springer-Verlag, 1996), 57–64.

21. Kline, *Mathematical Thought* 3:1092–93.

22. Tannery, "La Géometrie imaginaire, Part II (1877)," 557.

23. Edmund Husserl, "Über den Begriff der Zahl, psychologische Analysen" (Habilitationsschrift, University of Halle-Wittenberg, 1887).

24. Edmund Husserl, *Philosophie der Arithmetik. Psychologische und logische Untersuchungen* (Halle-Saale: C. E. M. Pfeffer, 1891).

25. Edmund Husserl, *Logische Untersuchungen*, 2., umgearb. aufl. ed., 2 vols. (Halle a.S.: Niemeyer, 1900).

26. Jeremy J. Gray, *Plato's Ghost: The Modernist Transformation of Mathematics* (Princeton, N.J.: Princeton University Press, 2008), 204.

27. Edmund Husserl, "Die Frage nach dem Ursprung der Geometrie als intentional-historisches Problem," *Revue Internationale de Philosophie* 1, no. 2 (1939): 203–25. This text originally included an appendix to Husserl's most important historical work, *Crisis of European Philosophy*, written around 1936 and published in 1954 as volume 6 of Husserl's collected works: Edmund Husserl, *Husserliana: Gesammelte Werke*, 42 vols., vol. 6: *Die Krisis der europäischen Wissenschaften und die transzendentale Phänomenologie*, ed. Walter Biemel (Den Haag: Martinus Nijhoff, 1950).

28. Edmund Husserl, *Cartesian Meditations: An Introduction to Phenomenology* (The Hague: M. Nijhoff, 1960), 7–8 (my emphasis).

29. Ibid., 27.

30. Edmund Husserl, *Die Krisis der europäischen Wissenschaften und die transzendentale Phänomenologie: Eine Einleitung in die phänomenologische Philosophie* (Den Haag: Martinus Nijhoff, 1954).

31. I have used the English version that is available in Jacques Derrida, *Edmund Husserl's Origin of Geometry: An Introduction*, trans. John P. Leavey (Lincoln: University of Nebraska Press, 1978), 157. The initial French translation is available in *L'Origine de la géometrie*, trans. Jacques Derrida (Paris: Presses Universitaires de France, 1962).

32. Thomas Heath's work on the history of Greek mathematics appeared in 1921 and takes a different view. In it Heath argued persuasively that the initial books of Euclid's *Elements* probably dated back to Pythagoras. Thomas Little Heath, *A History of Greek Mathematics*, 2 vols., vol. 1 (New York: Dover Publications, 1981; originally published, Oxford: Clarendon Press, 1921), 141–69.

33. Derrida, *Origin of Geometry*, 162.

34. See, for example, the classic work by Paul Johnson, *The Birth of the Modern: World Society, 1815–1830* (New York: HarperCollins, 1991).

35. The best work on this issue remains Megill, *Prophets of Extremity*. See also Lyotard, *Postmodern Condition*, 77, 81; Best and Kellner, *The Postmodern Turn*, 57–78; Smith, *Nietzsche, Heidegger, and the Transition to Postmodernity*. For a debate on Nietzsche's influence, see Koelb, *Nietzsche as Postmodernist: Essays Pro and Contra*.

36. Lyotard, *Postmodern Condition*, 3.

37. In general, see Ethan Kleinberg, *Generation Existential: Heidegger's Philosophy in France, 1927–1961* (Ithaca, N.Y.: Cornell University Press, 2005).

38. Emmanuel Levinas, *La Théorie de l'intuition dans la phénoménologie de Husserl* (Paris: Félix Alcan, 1930), and *En Découvrant l'Existence avec Husserl et Heidegger* (Paris: Librairie Philosophique, J. Vrin, 1949).

39. Edmund Husserl, *Méditations cartésiennes: Introduction á la phénoménologie* (Paris: A. Colin, 1931).

40. Derrida, *L'Origine de la géométrie*.

41. Derrida, *Origin of Geometry*.

42. Derrida, *Of Grammatology* (Baltimore: Johns Hopkins University Press, 1976).

43. Derrida, *Origin of Geometry*, 34.

44. Allan Megill, "Foucault, Structuralism, and the Ends of History," *Journal of Modern History* (1979); Alan Megill, "The Recepion of Foucault by Historians," *Journal of the History of Ideas* 48, no. 1 (1987).

45. Part of the inspiration for my approach comes from Anthony Close, "Centering the De-Centerers: Foucault and Las Meninas," *Philosophy and Literature* 11, no. 1 (1987).

46. Michel Foucault, *The Order of Things: An Archeology of the Human Sciences* (London: Routledge, 1970).

47. The original is available in Jorge Luis Borges, *Otras Inquisiciones* (Buenos Aires: Emec, 1960), 157.

48. Ibid., xv.

49. See, for example, Foucault, *Order of Things*; *The Birth of the Clinic: An Archaeology of Medical Perception*, trans. A. M. Sheridan Smith (New York: Vintage Books, 1975); *Discipline and Punish: The Birth of the Prison*, trans. Alan Sheridan (New York: Pantheon Books, 1977).

50. Euclid, *Euclid's Elements: All Thirteen Books Complete in One Volume*, trans. Thomas L. Heath (Santa Fe: Green Lion Press, 2002).

51. See Sartre's biting depiction of the "Self-Taught Man" in Jean-Paul Sartre, *Nausea* (Norfolk: New Directions, 1964).

52. On this great work's history, see Robert Darnton, *The Business of Enlightenment: A Publishing History of the Encyclopédie, 1775–1800* (Cambridge, Mass.: Harvard University Press, 1979).

53. For a handy modern translation, see René Descartes, *Discourse on Method*, trans. Lawrence Lafleur, 2nd rev. ed. (New York: Liberal Arts Press, 1956).

54. René Descartes, *The World and Other Writings*, trans. Stephen Gaukroger (New York: Cambridge University Press, 1998), 21–23.

55. For discussions of Descartes's relationship to European anthropology, see Murray Leaf, *Man, Mind, and Science: A History of Anthropology* (New York: Columbia University Press, 1979), 15–18; Wilhelm E. Mühlmann, *Geschichte der Anthropologie*, 2nd ed. (Frankfurt am Main: Athenäum Verlag, 1968), 39–40.

56. Another handy modern translation: René Descartes, *Meditations on First Philosophy*, trans. Laurence Lafleur (Indianapolis: Bobbs-Merrill, 1960).

57. Blaise Pascal, *Pensées*, trans. Roger Ariew (Indianapolis: Hackett, 2005), 61.

58. Ibid., 133.

59. For the text of the exchanges, see Samuel Clarke, Gottfried Wilhelm Leibniz, and Isaac Newton, *The Leibniz-Clarke Correspondence, Together with Extracts from Newton's Principia and Opticks*, ed. H. G. Alexander (Manchester: Manchester University Press, 1970). For analysis and discussion, see Alexandre Koyré, *From the Closed World to the Infinite Universe* (New York: Harper and Brothers, 1957), 235–72; Ezio Vailati, *Leibniz & Clarke: A Study of Their Correspondence* (New York: Oxford University Press, 1997).

60. Descartes, *Meditations on First Philosophy*, 26.

61. Friedrich Schiller, "Spruch des Konfucius," in *Musen-Almanach für das Jahr 1800*, ed. Friedrich Schiller (Tübingen: J. G. Cotta'sche Buchhandlung, 1799), 210. See Christoph Lüthy, "Where Logical Necessity Becomes Visual Persuasion: Descartes's Clear and Distinct Illustrations," in *Transmitting Knowledge: Words, Images, and Instruments in Early Modern Europe*, ed. Sachiko Kusukawa and Ian Maclean (Oxford: Oxford University Press, 2006). Although I read Descartes differently than does Lüthy, his stimulating essay has contributed to my own interpretation.

62. Foucault, *Discipline and Punish*, 30. See also Gary Gutting, "Michel Foucault: A User's Manual," in *The Cambridge Companion to Foucault*, ed. Gary Gutting (Cambridge: Cambridge University Press, 1994), 10; Michael S. Roth, "Foucault's 'History of the Present,'" *History and Theory* (1981): 32–46.

63. Keith Windschuttle, *The Killing of History: How Literary Critics and Social Theorists Are Murdering Our Past* (New York: Free Press, 1997); "Foucault as Historian," *Critical Review of International*

Social and Political Philosophy 1, no. 2 (1998): 5–35; Hans Ulrich Wehler, *Die Herausforderung der Kulturgeschichte* (München: C. H. Beck, 1998), 45–95.

64. Foucault, *Order of Things*, 3–16.

65. Foucault's idea of an underlying publicness in this painting cannot be sustained. See John F. Moffitt, "Velázquez in the Alcázar Palace in 1656: The Meaning of the Mise-En-Scène of Las Meninas," *Art History* 6, no. 3 (1983): 99.

66. Jonathan Brown and Carmen Garrido, *Velázquez: The Technique of Genius* (New Haven, Conn.: Yale University Press, 1998), 184.

67. Foucault, *Order of Things*, 16.

68. Ibid.

69. Samuel Y. Edgerton, *The Renaissance Rediscovery of Linear Perspective* (New York: Basic Books, 1975), 49–51; Martin Kemp, *The Science of Art: Optical Themes in Western Art from Brunelleschi to Seurat* (New Haven, Conn.: Yale University Press, 1990); Jonathan Brown, *Images and Ideas in Seventeenth-Century Spanish Painting* (Princeton, N.J.: Princeton University Press, 1978), 96.

70. John David North and John J. Roche, eds., *The Light of Nature: Essays in the History and Philosophy of Science Presented to A.C. Crombie* (Dordrecht: Martinus Nijhoff, 1985).

71. David C. Lindberg, *Theories of Vision from Al-Kindi to Kepler* (Chicago: University of Chicago Press, 1976), 11–14.

72. Richard Lorch, "Pseudo-Euclid on the Position of the Image in Reflection: Interpretations by an Anonymous Commentator, by Pena, and by Kepler," in North and Roche, *The Light of Nature*. See also Byron Ellsworth Hamann, "The Mirrors of Las Meninas: Cochineal, Silver, and Clay," *Art Bulletin* 92, no. 1–2 (2010): 6–35.

73. Brown, *Images and Ideas*, 97–98; Antonio Bonet Correa, "Velázquez, Arquitecto y Decorador," *Archivo español de arte* 33, no. 130 (1960): 215.

74. Jonathan M. Brown, "Painting in Seville from Pacheco to Murillo: A Study of Artistic Transition" (Ph.D. diss., Princeton University, 1964), 44–45; Zahira Véliz, "Francisco Pacheco's Comments on Painting in Oil," *Studies in Conservation* 27, no. 2 (1982): 49–57; Alfonso E. Pérez Sánchez, "The Artistic Milieu in Seville During the First Third of the Seventeenth Century," in *Zurbarán*, ed. Jeannine Baticle (New York: Metropolitan Museum of Art, 1987).

75. Foucault, *Order of Things*, 8.

76. Ibid.

77. I am not sympathetic to scholars' attempts to see Velázquez as "modern." José Antonio Maravall, *Velázquez y el espíritu de la modernidad*, 2nd ed. (Madrid: Alianza Editorial, 1987).

78. For a good discussion of this text, see also Brown, *Images and Ideas*, 44–62.

79. Francisco Pacheco, *Arte de la pintura, su antiguedad, y grandezas* (Seuilla: Simon Faxardo, 1649), 129.

80. Johannes Kepler and Ludwig Kepler, *Joh. Keppleri mathematici olim imperatorii somnium seu opus posthumum de astronomia lunari* ([S.l.]: haeredes authoris, 1634).

81. Karl Guthke, *The Last Frontier: Imagining Other Worlds, from the Copernican Revolution to Modern Science Fiction*, trans. Helen Atkins (Ithaca, N.Y.: Cornell University Press, 1990).

82. I suggest this as a complement to Hannah Arendt's arguments in "The Conquest of Space and the Stature of Man." *New Atlantis*, no. 18 (2007): 43–55.

83. Foucault, *Order of Things*, 17.

84. Ibid., 9.

85. On Kant's reading of homogeneous space, see Immanuel Kant, "Von dem ersten Grunde des Unterschiedes der Gegenden im Raume," in *Immanuel Kants Gesammelte Werke: Bd. II: Vorkritische Schriften II: 1757–1777* (Berlin).

86. Ibid., 377–78.

87. Stephen Kern, *The Culture of Time and Space, 1880–1918* (Cambridge, Mass.: Harvard University Press, 1983).

88. Ibid., 1.

Bibliography

In this work I have generally utilized published sources, which makes it difficult to divide this bibliography along clear lines of primary versus secondary sources. Nevertheless, to keep everything within one bibliography would yield only an extremely jumbled and effectively useless list of sources. I decided, therefore, to split this bibliography into two parts, using the year 1850—the end of the long early modern period—in order to approximate something of a shift from primary to secondary sources. My choice to use this year did not, however, work out perfectly, since the five volumes of Alexander von Humboldt's *Cosmos*, which I highlight in Chapter 6 as heliocentric anthropology's pinnacle, appeared between 1848 and 1862, whereas many primary source documents that I have consulted appeared in modern editions. I decided, thus, simply to use the year in which *Cosmos*' last volume appeared as the final point of separation. As a result, everything that appeared before 1863 can be treated as a primary source, although not everything that appeared after 1862 serves as a secondary source.

Manuscripts and Works Written or Published Through 1862

Acosta, José de. *Histoire naturelle et morale des Indes tant Occidentales qu'Orientales: Où il est traicté des choses remarquables du ciel, des élémens, métaux, plantes & animaux qui sont propres de ces païs. . . .* Translated by Robert Regnault Cauxois. Paris: M. Orry, 1598.

———. *Historia naturale e morale delle Indie*. Translated by Gio Paolo Galucci. Venetia: Pr. Bernardo Basa, 1596.

———. *Historia natvral y moral de las Indias, en qve se tratan las cosas notables del cielo, y elementos, metales, plantas y animales dellas: Y los ritos, y ceremonias, leyes, y gouierno, y guerras de los Indios.* Sevilla: de Leon, 1590.

———. *Historie naturael ende morael van de Westersche Indien: Waer inne ghehandelt wordt van de merckelijcktse Dinghen des Hemels, Elementen, Metalen, Planten ende ghedierten van dien.* Translated by Ian Huyghen van Linschoten. Enchuysen: Meyn, Rooman, 1598.

———. *The Naturall and Morall Historie of the East and West Indies: Intreating of the Remarkeable Things of Heaven, of the Elements, Mettalls, Plants and Beasts Which Are Proper to That Country: Together with the Manners, Ceremonies, Lawes, Governements, and Warres of the Indians.* Translated by Edward Grimeston. Val: Sims, 1604.

Agrippa, Heinrich Cornelius. *De occulta philosophia. Libri tres.* Louvain, 1531.

Andreae, Johann Philipp. *Neu inventirtes, belehrendes und ergötzendes astronomisches Karten Spiel: Das ist kunstrichtige Abbildung aller Gestirn des gantzen Firmamentes. Zu sonderbahren Nutzen der kunstliebenden Jugend vorgestellt durch Johann Philipp Andreae.* Nürnberg: Self, 1719.

Barrow, Isaac. *The Usefulness of Mathematical Learning Explained and Demonstrated: Being Mathematical Lectures . . . By Isaac Barrow, . . . to Which Is Prefixed, the Oratorical Preface of Our Learned Author, . . . Translated by the Revd. Mr. John Kirkby.* Translated by John Kirkby. London: printed for Stephen Austen, 1734.

Bauer, Carl. "Die Erde und ihre Bewohner." Collection of the Utrecht University Museum, inv. no. UM-373, 1825.

Biblia Latina. Mainz: Johann Gutenberg and Johannes Fust, 1454.

Biblia: Das ist: Die gantze Heilige Schrifft: Deudsch, Auffs new zugericht. Translated by Martin Luther. 2 vols. Vol. 1. Gedrückt zu Wittemberg: Durch Hans Lufft, 1545.

Buffon, Georges-Louis Leclerc, Comte de. *Histoire naturelle, générale et particuliére avec la description du cabinet du roy.* 29 vols. Paris: L'Imprimerie Royale, 1749–88.

Bürja, Abel. *Lehrbuch der Astronomie.* 5 vols. Vol. 1. Berlin: Schöne, 1794.

Chappuzeau, Samuel. *Idée du monde, ou introduction facile & méthodique à la cosmographie et à l'histoire.* Celle: André Holven, 1690.

Clavius, Christoph. *Christophori Clavii bambergensis in sphaeram Ioannis de Sacro Bosco commentarius.* Romae: Apud Victorium Helianum, 1570.

Cluverius, Philippus. *Philippi Cluuerii introductionis in universam geographiam, tam veterem quàm novam, libri VI., cui adjuncta est D. Heinsii oratio in obitum eiusdem P. Cluuerii, ed. ultima.* Edited by Josephus Vorstius. Amsterdam, 1629.

Copernicus, Nicolaus. *De lateribvs et angvlis triangulorum, tum planorum rectilineorum, tum sphæricorum, libellus eruditissimus & utilissimus, cum ad plerasque prolemæi demonstrationes intelligendas, tum uero ad alia multa, scriptus à clarissimo & doctissimo uiro D. Nicolao Copernico toronensi. Additus est canon semissium subtensarum rectarum linearum in circulo.* Vittembergae: Johannes Lufft, 1542.

———. *De revolutionibus orbium coelestium, libri VI.* Norimbergae: Ioh. Petreium, 1543.

Cusa, Nicholas of. *Opera.* Paris: in Aidibus Ascensianis, 1514.

Descartes, René. *Discours de la méthode pour bien conduire sa raison et chercher la vérité dans les sciences, plus la dioptrique, les météores et la géométrie qui sont des essais de cette méthode.* Leyde: L'Imprimerie de Jan Maire, 1637.

———. *L'Homme de René Descartes et un traitté de la formation du foetus du mesme auteur.* Paris: Charles Angot, 1664.

———. *Le Monde de Mr. Descartes, ou, le traité de la lumiere et des avtres principavx objets des sens.* Paris: Michel Bobin & Nicolas le Gras, 1664.

Doppelmayr, Johann Gabriel. *Kurze Einleitung zur edlen Astronomie, in welcher zum Fundament derselben das Systema Solare & Planetarium nach des weitberühmten Herrn Chrisiani Hugenii copernicanischen Grund-Sätzen erkläret/ Und in unterschiedlichen neu-verfertigten Homännischen Charten allen wahren Freunden der Sternen-Wissenschaft deutlich vor Augen gelegt wird.* Nürnberg: Joh. Baptistae Homann, 1708.

Dürer, Albrecht. *Underweysung der Messung/ mit dem Zirkel und Richtscheyt/ in Linien Ebnen unnd gantzen Corporen.* Norimbergae, 1525.

Duval, Pierre. *La Connoissance et l'usage des globes; et des cartes de géographie.* Paris: Chez l'Autheur, 1654.

Euclid. *Analyseis geo-||metricæ sex librorum|| Euclidis.|| Primi et qvinti factae à|| Christiano Herlino: reliqvae unà cvm com-||mentariis, & scholiis perbreuibus in eosdem sex libros geo-||metricos: à Cunrado Dasypodio.|| cvm indice . . . || Pro schola Argentinensi.||.* Edited by Christian Herlin. Straßburg: Josias Rihel, 1566.

———. *Das sibend, acht und neunt Buch, des hoch berümbten Mathematici Euclidis Megarensis, in welchen der Operationen unnd Regulen aller gemainer Rechnung, Ursach Grund und Fundament, angezaigt wirt zu Gefallen allen den/ so die Kunst der Rechnung liebhaben/ durch Magistrum Johann Scheybl/ der löblichen Universitet zu Tübingen des Euclidis und Arithmetic Ordinarien/ auß dem Latein ins Teütsch gebracht. . . .* Translated by Johann Scheubel. Augspurg: Valentin Otmar, 1555.

———. *Die sechs erste Bücher Euclidis, vom Anfang oder Grund der Geometrj. Jn welchen der rechte Grund, nitt allain der Geometrj (versteh alles kunstlichen/ gwisen/ und vortailigen Gebrauchs des Zirkels/ Linials oder Richtscheittes und andrer Werckzeüge/ so zu allerlaj Abmessen dienstlich) sonder auch der fürnemsten Stuck und Vortail der Rechenkunst/ furgeschrieben und dargethan ist./ Auß griechischer Sprach in die Teütsch gebracht, aigentlich erklärt, auch mit verstentlichen Exempeln, gründlichen Figuren/ und allerlaj den Nutz für Augen stellenden Anhängen geziert/ der Massen vormals in teutscher Sprach nie gesehen worden . . . durch Wilhelm Holtzman, genant Xylander/ von Augspurg.* Translated by Guillelmus Xylander. Basel: Ioanns Oporini, 1562.

———. *Elementa geometriae.* Venetia: Erhard Ratdolt, 1482.

———. *Elementa geometriae ex Evclide singulari prudentia collecta a Ioanne Vogelin . . . Arithmeticae practicae per Georgium Peurbachium mathematicum. Cum praefatione Philippi Melanthonis*. Vitebergae, 1536.

———. *Elementorvm libri XV. Breviter demonstrati, operâ Is. Barrow, cantabrigensis*. Translated by Isaac Barrow. Cantabrigiae: Guilielmo Nealand, 1655.

———. *Elementos geometricos de Evclides, los seis primeros libros de los planos, y los onzeno, y dozeno de los solidos: Con algvnos selectos theoremas de Archimedes/ traducidos, y explicados por P. Jacobo Kresa de La Compañia de Jesus, Cathedratico de Mathematicas en los Estudios Reales del Colegio Imperial de Madrid; y en interin en la Armada Real en Cadiz*. Translated by Jacobo Kresa. Brvsselas: Francisco Foppens, 1689.

———. *Euclide megarense philosopho solo introduttore delle scientie mathematice; diligentemente reassetto, et alla integrità ridotto per il degno professore di tal scientie Nicolò Tartalea, brisciano, secondo le due tradottioni e per commune commodo & vtilità di Latino in volgar tradotto; con una ampla espositione dello istesso tradottore di nouo aggionta; talmente chiara, che ogni mediocre ingegno, senza la notitia, ouer suffragio di alcun'altra scientia con facilità, sera capace à poterlo intendere*. Translated by Gabriele Tadino, Niccolò Tartaglia, Pietro Facolo, Guilielmo de Monferra, Venturino Ruffinelli, and Lodovico Vittorio Fachino. Stampato in Vinegia: Per Venturino Ruffinelli ad instantia e requisitione de Guilielmo de Monferra, & de Pietro di Facolo da Vinegia libraro, & de Nicolò Tartalea brisciano tradottore, 1543.

———. *Euclidis elementorum, Libri XIV*. Roma: Vincentius Accoltus, 1574.

———. *Euclidis megarensis philosophi acutissimi mathematicorumque omnium sine controuersia principis o[per]a / Lucas Paciolus theologus insignis, altissima mathematica[rum] disciplinarum scientia rarissimus iudicio castigatissimo detersit emendauit*. Venetia: A. Paganius Paganinus characteribus elegantissimis accuratissime imprimebat, 1509.

———. *Euclid's Elements of Geometry, from the Latin Translation of Commandine. To Which Is Added, a Treatise of the Nature and Arithmetick of Logarithms; Likewise Another of the Elements of Plain and Spherical Trigonometry; with a Preface, Shewing the Usefulness and Excellency of This Work by Doctor John Keil, F.R.S. And Late Professor of Astronomy in Oxford. Now Done into English*. London: Thomas Woodward, 1723.

———. *Evclidis elementorvm libri XV: Accessit liber XVI. De qvinqve solidorvm regularivm inter se comparatione; ad exemplaria R.P. Christophori Clauijè Societ. Iesv, & aliorum collati, emendati & aucti*. Edited by Christoph Clavius. Coloniae: Gosuinus Cholinus, 1607.

———. *Evclidis elementorvm libri XV: Acceßit XVI. De solidorvm regvlarivm cuiuslibet intra quodlibet comparatione; omnes perspicuis demonstrationibvs, accuratisq'ue scholiis illustrati, ac multarum rerum accessione locupletati*. Edited by Christoph Clavius. Nunc tertio ed., summaq', diligentia recogniti, atque emendati ed. Colonia: Ioh. Baptistae Ciottus, 1591.

———. *Les Six Premiers Livres des Élémens d'Euclide, traduicts et commentez par J. Errard*. Translated by J. Errard. Paris, 1598.

———. *Les Six Premiers Livres des Éléments d'Euclide*. Translated by Pierre Forçadel Béziers. Paris, 1564.

———. *Los Seis Libros primeros de la geometria de Evclides. Traduzidos en lengua Española por Rodrigo Çamorano astrólogo y mathemàtico, y cathedrático de cosmographia por su Magestad en la Casa dela Contratacion de Seuilla dirigidos al Jllustre Señor Luciano Negron, canonigo dela Sancta Yglesia de Seuilla*. Translated by Rodrigo Zamorano. Seuilla: Alonso de la Barrera, 1576.

———. *Teutsch-Redender Euclides, oder acht Bücher von denen Anfängen der Mesz-Kunst: Auff eine neue end gantz leichte Art, zu Nutzen allen Generalen, Ingeniern, Natur- und Warheit-Kündigern, Bau-Meistern, Künstlern und Handwerckern / in teutscher Sprach eingerichtet und bewiesen durch A. E. B. V. P.* Translated by Anton Ernst Burckhard von Birckenstein. Wienn: Philipp Fievers, Buch- und Kunst-Händler, 1694.

———. *The Elements of Geometrie of the Most Auncient Philosopher Euclide of Megara: Faithfully (Now First) Translated into the English Toung, by H. Billingsley, Citizen of London. Whereunto Are Annexed*

Certaine Scholies, Annotations, and Inventiones, of the Best Mathematiciens, Both of Time Past, and in This Our Age. Translated by Henry Billingsley. London: John Daye, 1570.

Euclid, and Proclus. *Eukleidou stoicheion bibl. 15. Ekton theonos synousion: eis tou autou to proton, exegematon Proklou bibl. 4. Adiecta præfatiuncula in qua de disciplinis mathematicis nonnihil*. Translated by Simon Grynäus and Johann Herwagen. Basileae: Apud Ioan. Heruagium, 1533.

Fiorentino, Mauro. *Sphera volgare novamente tradotta con molte notande additioni di geometria, cosmographia, arte navicatoria, et stereometria, proportioni, et qvantità delli ellementi, distanze, grandeze, et movimienti di tutti li corpi celesti, cose certamente rade et maravigliose*. Venetia: Bartholomeo Zanetti, 1537.

Fontenelle, Bernard de. *Entretiens sur la pluralité des mondes*. Paris: Chez la veuve C. Blageart, 1686.

Frisius, Gemma. *De Principiis astronomiae et cosmographiae, deque usu globi ab eodem editi. Item de orbis divisione, & insulis, rebusque nuper inventis*. Antwerp: Gregorius Bontius, 1530.

"Géometrie." In *Encyclopédie*, edited by Denis Diderot and Jean-Baptiste le Rond D'Alembert. 28 vols. Vol. 7. 629–40. Paris: Chez Briasson, 1754.

Gigli, Ottavio, ed. *Studi sulla divina commedia, di Galileo Galilei, Vincenzo Borghini ed altri*. Florence: Felice Le Monnier, 1855.

Hegel, Georg Wilhelm Friedrich. "Disseratio philosophica de orbitis planetarum." Habilitation, University of Jena, 1801.

———. *System der Wissenschaft: Erster Theil, Die Phänomenologie des Geistes*. Bamberg: Joseph Anton Goebhardt, 1807.

Herder, Johann Gottfried. *Ideen zur Philosophie der Geschichte der Menschheit*. 4 vols. Riga: Johann Friedrich Hartknoch, 1784–91.

Hobbes, Thomas. *Behemoth; or an Epitome of the Civil Wars of England, from 1640 to 1660*. London: s.n., 1679.

———. *Elementorum philosophiae sectio prima de corpore*. Londini: Andrea Crook, 1655.

———. *Elementorum philosophiae sectio secunda de homine*. Londini: Typis T. Childe. Sumptibus Andr. Crooke, & væneunt sub insigni viridis draconis in Cæmetirio Paulino, 1658.

———. *Hobbs's Tripos, in Three Discourses: The First, Humane Nature . . . The Second, De corpore Politico . . . The Third, of Liberty and Necessity*. 3rd ed. London, 1684.

———. *Humane Nature: Or, the Fundamental Elements of Policie. Being a Discoverie of the Faculties, Acts, and Passions of the Soul of Man, from Their Original Causes, According to Such Philosophical Principles as Are Not Commonly Known or Asserted. By Tho. Hobbs of Malmsbury*. London: printed by T. Newcomb, for Fra: Bowman of Oxon, 1650.

———. *Leviathan, or the Matter, Forme, & Power of a Common-Wealth Ecclesiasticall and Civill*. London: Printed for Andrew Crooke by William Wilson, at the Green Dragon in St. Pauls Church-Yard, 1651.

———. *Religio elementorvm philosophiæ sectio tertia de cive*. Parisis: s.n., 1642.

———. *Six Lessons to the Professors of the Mathematiques One of Geometry the Other of Astronomy, in the Chaires Set up by the Noble and Learned Sir Henry Savile in the University of Oxford*. London: Printed by J.M. for Andrew Crook, 1656.

Hocker, Johann Ludwig. *Einleitung zur Kenntnis und Gebrauch der Erd- und Himmels-Kugel auf das deutlichste und Leichteste in Frag und Antwort*. Nürnberg: Peter Conrad Monath, 1734.

The Holy Bible Conteyning the Old Testament, and the New: Newly Translated out of the Originall Tongues: & with the Former Translations Diligently Compared and Reuised, by His Maiesties Speciall Co[M]Mandement. Appointed to Be Read in Churches. Edited by Cornelius Bol. London: By Robert Barker, printer to the Kings most excellent Maiestie, 1611.

Homann. *Atlas compendiarius seu ita dictus scholasticus minor: in usum erudiendae juventutis adornatus*. Norimbergae: excudientibus Homanianis heredibus, 1753.

Hues, Robert. *Learned Treatise of Globes: Both Celestiall and Terestriall: With Their Several Uses. Written First in Latine, by Mr. Robert Hues: And by Him So Published*. Translated by John Chilmead. London: Andrew Kemb, 1659.

———. *Tractatus de globis coelesti et terrestri eorumque usu*. Amstelodami: Henricus Hondius, 1627.

Humboldt, Alexander von. *Cosmos, ó, ensayo de una descripción física del mundo*. Translated by Francisco Díaz Quintero. Biblioteca Mexicana Popular y Económica. México: V. García Torres, 1851.

———. *Cosmos; Essai d'une description physique du monde*. Translated by H. Faye and Ch. Galuski. 4 vols. Paris: Gide et J. Baudry, Éditeurs, 1855.

———. *Cosmos: O saggio di una fisica descrizione del mondo*. Torino: G. Pomba e Compagni, 1847.

———. *Cosmos: A Sketch of a Physical Description of the Universe*. Translated by E. C. Otté. 5 vols. New York: Harper and Brothers, 1850.

———. *Cosmos: Sketch of a Physical Description of the Universe*. 3rd ed. 4 vols. London: Longman, Brown, Green J. Murray, 1847.

———. *Kosmos: Entwurf einer physischen Weltbeschreibung*. 5 vols. Stuttgart: J. G. Cotta'scher Verlag, 1845–62.

———. *Kritische Untersuchungen über die historische Entwickelung der geographischen Kenntnisse von der Neuen Welt und die Fortschritte der nautischen Astronomie in dem 15ten und 16ten Jahrhundert*. 2 vols. Vol. 1. Berlin: Nicolai'sche Buchhandlung, 1836.

Humboldt, Alexander von, and Aimé Bonpland. *Voyage aux régions équinoxiales du nouveau continent, fait en 1799, 1800, 1801, 1802, 1803, et 1804*. Paris: Libraire grecque-latine-allemande, 1804.

Huygens, Christiaan. *Christiani Hugenii Kosmotheoros, sive de terris coelestibus, earum ornatu, conjecturae. Ad Constantinum Hugenium, fratrem*. Hagae-Comitum: Adrian Moetjens, 1698.

Ibn Athari, Masha'allah. *De scientia motus orbis*. Norimbergae: Johannes Stabius, 1505.

Kant, Immanuel. *Allgemeine Naturgeschichte und Theorie des Himmels: Oder Versuch von der Verfassung und dem Mechanischen Ursprunge des ganzen Weltgebäudes nach Newtonischen Grundsätzen abgehandelt*. Königsberg: Petersen, 1755.

———. *Jmmanuel Kants physische Geographie*. Mainz: Vollmer, 1801.

Kepler, Johannes. *Prodromus dissertationvm cosmographicarvm, continens mysterivm cosmographicvm, de admirabili proportione orbivm coelestivm, deqve cavsis coelorum numeri, magnitudinis, motuumq'ue periodicorum genuinis & proprijs, demonstratvm, per qvinqve regularia corpora geometrica / a M. Ioanne Keplero, . . . Addita est erudita narratio M. Georgii Ioachimi Rhetici, de libris reuolutionum, atq[ue] admirandis de numero, ordine, & distantijs sphærarum mundi hypothesibus, excellentissimi mathematici, totiusq[ue] astronomiæ restauratoris D. Nicolai Copernici*. Tvbingae: Georg Gruppenbach, 1596.

Kepler, Johannes, and Ludwig Kepler. *Joh. Keppleri mathematici olim imperatorii somnium seu opus posthumum de astronomia lunari*. [s.l.]: haeredes authoris, 1634.

Kästner, Abraham G. "Das Lob der Sternkunst." *Hamburgisches Magazin, oder gesammlete Schriften, zum Unterricht und Vergnügen* 1, no. 2 (1747): 206–22.

Lamy, Bernard. *Les Élémens de géométrie ou de la mesure du corps. Qui comprennent les Élémens d'Euclide, les plus belles propositions d'Archimède & l'analise*. Paris: André Pralard, 1685.

Lafitau, Joseph-François. *Moeurs des sauvages ameriquains, comparées aux moeurs des premiers temps*. 2 vols. Paris: Saugrain l'aîné, 1724.

Lambert, Johann Heinrich. *Cosmologische Briefe über die Einrichtung des Weltbaues*. Augsburg: Eberhard Kletts Wittib, 1761.

Lefèvre d'Étaples, Jaques. *Quatuor libri meteororum Aristotelis paraphrasi Iacobi Stapulensis explanati*. [s.l.]: Valentius Schumann, 1516.

Locke, John. *An Essay Concerning Humane Understanding*. London: T. Basset, 1690.

———. *Some Thoughts Concerning Education*. London: Printed for A. and J. Churchill, 1693.

———. *Two Treatises of Government: In the Former, the False Principles, and Foundation of Sir Robert Filmer, and His Followers, Are Detected and Overthrown. The Latter Is an Essay Concerning the True Original, Extent, and the End of Civil Government*. London: A. Churchill, 1690.

Manesson-Mallet, Allain. *Description de l'univers: Contenant les differents systèmes du monde, les cartes generales & particulières de la géographie ancienne & moderne: Les Plans & les profils des principales villes & des autres lieux plus considerables de la Terre; avec les portraits des souverains qui y commandent, leurs blasons, titres & livrées: Et les moeurs, religions, gouvernemens & divers habillemens de chaque nation. . . .* 5 vols. Paris: Denys Thierry, 1685.

Melanchthon, Philip. "Praefatio in arithmeticen (1536)." In *Corpus Reformatorum Philippi Melanthonis operae quae supersunt omnia*, edited by Carolus Gottlieb Bretschneider and Henricus Ernestus Bindseil, 284–92. Halle: Apud C. A. Schwetschke et filium, 1834–60.

Mersenne, Marin. *Euclidis elementorum libri, Apollonii Pergae conica, sereni de sectione*. Paris, 1626.

———. *L'Impieté des deistes, athées, et libertins de ce temps, combatuë, & renuersée de point en point par raisons tirées de la philosophie, & de la théologie*. Pierre Bilaine, 1624.

———. *Universae geometriae mixtaeque mathematicae synopsis, et bini refractionum demonstratarum tractatus*. Parisiis: Antonium Bertier, 1644.

Müller, Johann Wolfgang. *Anweisung zur Kenntnis und dem Gebrauch der künstlichen Himmels- und Erdkugeln besonders in Rücksicht auf die neuesten Nürnberger Globen, für die Höhern Classen der Schulen und Liebhaber der Sphaerologie*. 2 vols. Nürnberg: Johann Georg Klinger, 1791–92.

Münster, Sebastian. *Canones super novum instrumentum luminarium, docentes quo pacto per illud inveniantur Solis et Lunae medii et veri motus, lunationes, coniunctiones, oppositiones, Caput Draconis, . . .* Baslieae: Cratander, 1534.

———. *Cosmographia: Beschreibung aller Lender durch Sebastianium Munsterum in welcher begriffen/ aller Völcker/ Herrschafften/ Stetten/ und namhafftiger Flecken/ herkommen: Sitten/ Gebreüch/ Ordnung/ Glauben/ Secten. Und Hantierung/ durch die ganze Welt/ und fürnemmlich Teütsche Nation. Was auch besunder in jedem Landt gefunden/ unnd darin beschene Sey. Alles mit Figuren und schönen Landt Tafeln erklert/ und füre Augen gestelt*. Basel: Heinrich Petri, 1544.

———. *Cosmographei. Mappa Evropae, eygentlich fürgebildet/ außgelegt vnnd beschribenn. Vonn aller Land vnd Stett Ankunfft/ Gelegenheyt/ Sitten/ ietzi-ger Handtierung vnd Wesen. Wie weit Stett vnnd Länder inn Europa von einander Gelegen/ leichtlich zufinden. Des Polus in ieglicher Statt Erhebung/ daher vil Nutzbarkeyt/ als die Son[n]uhr/ Compast/ Chilinder [et]c. zumachen. . . . künstlich vnnd gewisse Anleytung/ einen um[m]Kreiß einer Statt oder Landschafft zuerzeichnen/ Mappen vnd Landtaffeln zumachen/ durch Sebastianum Munsterum|| an Tag geben*. Franckfurt am Main: Egenolff, 1537.

———. *Erklerung des newen Instruments der Sunnen, nach allen seinen Scheyben und Circkeln; item eyn Vermanung . . . an alle Liebhaber der Künstenn im Hilff zu thun zu warer unnd rechter Beschreybung Teutscher Nation*. Oppenheym: Kobel, 1528.

Newton, Isaac. *The Mathematical Principles of Natural Philosophy by Sir Isaac Newton*. Translated by Andrew Motte. 2 vols. London: Benjamin Motte, 1729.

———. *Philosophiae naturalis principia mathematica*. Londini: J. Societatis Regiae ac Typis J. Streater, 1687.

Newton, John. *Mathematical Elements, in III Parts, the First, Being a Discourse of Practical Geometry, the Three Parts of Continued Quantity, Lines, Planes, and Solides. The Second, a Description and Use of Coelestial and Terrestrial Globes. The Third, the Delineation of the Globe Upon the Plain of Any Great Circle, According to the Stereographick, or Circular Projection*. London: R. and W. Leybourn, 1660.

Nouvel Atlas des enfans, ou principes clairs pour apprendre facilement et en fort peu de tems la géographie, suivi d'un traité méthodique de la sphère, qui explique le mouvement des astres, les divers systèmes du monde, & l'usage des globes. Amsterdam: Chez B. Vlam, 1776.

Oresme, Nicole. "Traité de la sphère: Aristote, Du Ciel et Du Monde [De Caelo Et De Mundo], traduction française par Nicole Oresme." In *Département des Manuscrits*: Bibliothéque Nationale de France, 1410–1420.

Pacheco, Francisco. *Arte de la pintura, su antiguedad, y grandezas*. Seuilla: Simon Faxardo, 1649.

Pacioli, Luca. "De viribus quantitatis." Biblioteca Universitaria di Bologna, 1508.

———. *Summa de arithmetica geometria proportioni & proportionalità: continentia de tutta l'opera*. Venetijs: Paganino de Paganini, 1494.

———. *Summa de arithmetica geometria: proportioni: Et proportionalità: Nouamente impressa*. Toscolano, 1523.

Pope, Alexander. *An Essay on Man. Address'd to a Friend. Part I*. 2nd ed. Dublin: Printed by S. Powell, for George Risk . . . George Ewing . . . and William Smith . . . 1734.

Ptolemy, Claudius. *Claudii Ptolemaei Alexandrini geographicae enarrationis libri octo; adiecta insuper ab eodem scholia, quibus exoleta urbium nomina . . . exponuntur / Ex Bilibaldi Pirckeymheri translationem sed ad Graeca et prisca exemplaria à Michaele Villanovano iam primum recogniti . . .* Translated by Willibald Pirckheimer. Edited by Michael Servetus. Lugduni: Trechsel, 1535.

———. *Claudii Ptolemaei geographicae enarrationis libri octo*. Strasbourg: Koberger, 1525.

———. "Geographia." In Manuscripts and Archives Division: New York Public Library, 1460–70.

———. *Geographia universalis, vetvs et nova, complectens Clavdii Ptolemæi Alexandrini enarrationis libros VIII. quorum primus noua translatione Pirckheimeri et accessione commentarioli illustrior quàm hactenus fuerit, redditus est. reliqu. castigatiores facti sunt. Addita sunt insuper scholia . . . succedunt tabulæ Ptolemaic[a]e, opera Sebastiani Munsteri nouo paratæ modo. His adiectæ sunt plurim[a]e nouæ tabulæ, moderna[e] orbis faciem. explicantes . . . vltimo annexum est compendium geographic[a]e descriptionis, in quo uarij gentium & regionum ritus & mores explicantur. Pr[a]efixus est . . . Index . . .* Basileae: Heinrich Petri, 1540.

———. *Klaudiou Ptolemaiou Alexandreōs philosophou en tis malisa te paideumenou. Peri tēs geōgraphikēs biblia oktō, meta pasēs akribeias entiptothenta*. Basel: Frobenius, 1533.

———. *Tabvlae geographicae Cl: Ptolomei: Ad mentem autoris restitutae & emendate*. Cologne: Godefridus Kempen, 1578.

Ramus, Peter. *Euclides*. Paris: Thomas Richard, 1549.

Recorde, Robert. *The Pathway to Knowledge, Containing the First Principles of Geometrie, as They May Moste Aptly Be Applied unto Practise, Bothe for Use of Instrumentes Geometricall, and Astronomicall and Also for Proiection of Plattes in Everye Kinde, and Therefore Much Necessary for All Sortes of Men*. London: Reynold Wolfe, 1551.

Regiomontanus, Johannes. *Epytoma Joannis de Monte Regio in Almagestum Ptolomei*. Venice, 1496.

Reisch, Gregorius. *Margarita philosophica*. Argentorati [Strasbourg]: Schottus, 1504.

Robertson, William. *The History of America*. 3 vols. Vol. 1. Dublin: Price, 1777.

Rost, Johann Leonard. *Atlas Portatilis Coelestis, oder Compendiöse Vorstellung des gantzen Welt-Gebäudes/ in den Anfangs-Gründen der wahren Astronomie. Dadurch man nicht nur zur Erlernung dieser unentbehrlichen Wissenschaft/ auf eine sehr leichte Art gelangen; sondern auch zugleich daraus, sich einen bessern Begriff von dem wahren Fundament/ so wol der Geographie als Schiffahrt/ zueignen kan. Den Liebhabern zu Gefallen; absonderlich aber ser studirenden Jugend zum Unterrichte, in möglichster Deutlichkeit abgefasset: Und durch mehr als anderthalb Hundert Figuren erkläret*. Nürnberg: Johann Ernst Adelbulner, 1723.

Rousseau, Jean-Jacques. *Du contrat social, ou, principes du droit politique*. Ed. sans cartons, à laquelle on a ajoûté une lettre de l'auteur au seul ami qui-lui reste dans le monde ed. Amsterdam: Chez Marc-Michel Rey, 1762.

Sanson, Nicholas. *Description de tout l'universe, en plusieurs cartes, & en divers traitez de geographie et d'histoire; où sont décrits succinctement & avec une methode belle & facile ses empires, ses peuples, ses colonies, leur moeurs, langues, religions, richesses, &c, et ce qu'il y a de plus beau & plus rares dans toutes ses parties & dans ses isles*. Amsterdam: Chez François Halma, 1700.

Schickard, Wilhelm. *Kurze Anweisung wie künstliche Land-Tafeln auss rechtem Grund zu Machen/ und die biss her begangne Irrthumb zu Verbessern/ sampt etlich new erfundenen Vörtheln/ Die Polus Höhin auffs leichtest/ und doch scharpff gnug zu forschen*. Tübingen: Johann Georg Cotta, 1669.

Schiller, Friedrich. "Spruch des Konfucius." In *Musen-Almanach für das Jahr 1800*, edited by Friedrich Schiller, 209–10. Tübingen: J. G. Cotta'sche Buchhandlung, 1799.

Schöner, Johannes. *Globi stelliferi, sive sphaerae stellarum fixarum usus, & explicationes, quibus quicquid de primo mobili demonstrari solet, id uniuersum prope continetur, directionum autem ipsarum quas uocant, ratio accuratis, est exposita. Autore Ioanne Schonero Carolostadio, atque haec omnia multò quâm ante emendatiora & copiosiora singulari ac studio in lucem edita fuere anno Christi M. D. XXXIII*. Norimbergae: Petreius, 1533.

———. *Luculentissima quaedā Terrae totius descriptio: Cū multis vtilissimis cosmographiæ iniciis. Nouaq, & q ante fuit verior Europæ nostræ formatio*. Norimbergae: Impressum ī excusoria officina I. Stuchssen, 1515.

———. *Opera mathematica Ioannis Schoneri in vnvm volvmen congesta.* Norimbergae: Impressa in officina I. Montani & V. Neuberi, 1551.

———. *Opusculum geographicum ex diversorum libris ac cartis summa cura & diligentia collectum, accomodatum ad recenter elaboratum ab eodem globum descriptionis Terrenae.* Norimbergae: Johann Petreius, 1533.

———. *Solidi ac sphaerici corporis sive globi astronomici canones, usum et expeditam praxim ejus dem expromentes.* Norimbergae: Ioannes Stuchs, 1516.

Swedenborg, Emmanuel. *De telluribus planetarum in mundo nostro solari quae vocantur planetae et de telluribus in coelo astrifero deque illarum incolis, tum de spiritibus et angelis ibi ex auditis et visis.* London, 1758.

Tartaglia, Niccolò. *La Nova Scientia, con una gionta al terzo libro.* Vinegia: Camillo Castelli, 1583.

———. *Nova Scientia inventa da Nicolò Tartalea.* Vinegia: Stephano da Sabio, ad instantia di Nicolò Tartalea, 1537.

Thevet, André. *La Cosmographie vniverselle d'André Thevet Cosmographe dv Roy: Illvstrée de diverses figvres des choses plvs remarqvables vevës par l'auteur, & incogneuës de noz anciens & modernes*, 2 vols. Paris: Pierre l'Huillier, 1575.

Thucydides. *Eight Bookes of the Peloponnesian Warre.* Translated by Thomas Hobbes. London: Imprinted for Richard Mynne, 1634.

Varenius, Bernhardus. *Bernhardi Varenii geographia generalis in qua affectiones generales Telluris explicantur. Adjecta est appendix, . . . a Jacobo Jurin.* Cantabrigiæ: typis academicis. Impensis Cornelii Crownfield, 1712.

———. *Bernhardi Varenj Med. D. geographia generalis in qua affectiones generales Telluris explicantur, summâ curâ quam plurimis in locis emendata, & XXXIII schematibus novis, aere incisis, unà cum tabb. aliquot quæ desiderabantur aucta & illustrata.* Edited by Isaac Newton. Editio secunda auctior & emendatior ed. Cantabrigiæ: ex officina Joann Hayes . . . , sumptibus Henrici Dickinson . . . , 1681.

———. *Descriptio Regni Iaponiae: Cum quibusdam Affinis Materiæ, ex Variis Auctoribus Collecta et in Ordinem Redacta.* 2 vols. Amstelodami Amsterdam: apud Ludovicum Elzevirium, 1649.

———. *Geographia Generalis in qua affectiones generales Telluris explicantur.* Amstelodami: Apud Ludovicum Elzevirium, 1650.

———. *Bernhardi Vareni geographia generalis in qua affectiones generales Telluris explicantur, summâ curâ quam plurimis in locis emendata, & XXXiii schematibus novis, aere incisis, unà cum tabb. aliquot quae desiderabantur.* Edited by Isaac Newton. Cantabrigiae: Ex officina Joann Hayes . . . sumptibus Henrici Dickinson, 1672.

Voltaire. *Élémens de la philosophie de Neuton: Mis à la portée de tout le monde.* Amsterdam: Desbordes, 1738.

———. *Le Micromégas de Mr. de Voltaire avec une histoire des croisades & un nouveau plan de l'ésprit humain.* London: J. Robinson, 1752.

Walker, Joseph. *Astronomy's Advancement, or, News for the Curious; Being a Treatise of Telescopes: And an Account of the Marvelous Astronomical Discoveries of Late Years Made Throughout Europe; with the Figures of the Sun, Moon, and Planets; with Copernicus His System, in Twelve Copper Plates. Also an Abstract Touching the Distance, Faces, Bulks, and Orbs of the Heavenly Bodies, the Best Way of Using Instruments for Satisfaction, &C. Out of the Best Astronomers, Ancient and Modern, Viz. Mr. Hooke, Mr. Boilleau, Mr. Hevelius, Father Kircher, &C.* London: Philip Lea, 1684.

Wallis, John. *De sectionibus conicis, nova methodo expositis, tractatus.* Oxonii: Thomas Robinson, 1655.

———. *Due Correction for Mr Hobbes: Or Schoole Discipline, for Not Saying His Lessons Right: In Answer to His Six Lessons, Directed to the Professors of Mathematicks.* Oxford: Printed by Leonard Lichfield printer to the University for Tho: Robinson, 1656.

———. *Elenchus geometriæ Hobbianæ, sive, geometricorum, quæ in ipsius elementis philosophiæ [Sect. 1, De corpore] à Thomas Hobbes . . . proferuntur, refutatio.* Oxoniensis, 1655.

———. *Johannis Wallisii, SS. Th. D. Geometriæ Professoris Saviliani in celeberrimâ Academia Oxoniensi, operum mathematicorum pars altera: Qua continentur de angulo contactus & semicirculi, disquisitio geometrica. De sectionibus conicis tractatus. Arithmetica infinitorum: Sive de curvilineorum quadra-*

turâ, &C. ecclipseos solaris observatio. Oxonii: Typis Leon: Lichfield Academiae typographi, impensis Tho. Robinson, 1656.

Wolff, Christian. *Vernünfftige Gedancken von der Würckungen der Natur*. Halle: Rengerischen Buchhandlung, 1725.

Young, Edward. *The Complaint: Or, Night Thoughts on Life, Death, and Immortality*. London: A. Millar, 1750.

Works Published from 1863 Onward

Ackerknecht, Erwin. "George Forster, Alexander von Humboldt, and Ethnology." *Isis* 46, no. 2 (1955): 83–95.

Acosta, José de. *Historia Natural y Moral de las Indias: En que Se Tratan de las Cosas Notables del Cielo, Elementos, Metales, Plantas y Animales Dellas, y Los Ritos, y Ceremonías, Leyes y Gobierno de los Indios*. Edited by Edmundo O'Gorman. Biblioteca Americana. 2nd ed. Mexico City: Fondo de Cultura Económica, 1962.

———. *Historia Natural y Moral de las Indias (Sevilla, Juan de Léon, 1590)*. Edited by Barbara G. Bedall. Hispaniae Scientia. Repr. facs. ed. Valencia: Albatros Ed., 1977.

Agassi, Joseph. *The Very Idea of Modern Science: Francis Bacon and Robert Boyle*. Boston Studies in the Philosophy and History of Science. Dordrecht: Springer, 2013.

Albert the Great. *The Commentary of Albertus Magnus on Book 1 of Euclid's Elements of Geometry*. Translated by Anthony Lo Bello. Ancient Mediterranean and Medieval Texts and Contexts. Boston: Brill Academic Publishers, 2003.

Alberti, Leon Battista. *The Mathematical Works of Leon Battista Alberti*. Basel: Springer Science and Business Media, 2010.

Albertson, David. *Mathematical Theologies: Nicholas of Cusa and the Legacy of Thierry of Chartres*. Oxford Studies in Historical Theology. New York: Oxford University Press, 2014.

Alexander, Amir R. *Geometrical Landscapes: The Voyages of Discovery and the Transformation of Mathematical Practice*. Stanford, Calif.: Stanford University Press, 2002.

———. *Infinitesimal: How a Dangerous Mathematical Theory Shaped the Modern World*. New York: Scientific American, 2014.

Ammon, Laura. "Bernardino de Sahagún, Jose de Acosta and the Sixteenth-Century Theology of Sacrifice in New Spain." *Journal of Colonialism and Colonial History* 12, no. 2 (2011).

Andersen, Kirsti. *The Geometry of an Art: The History of the Mathematical Theory of Perspective from Alberti to Monge*. Sources and Studies in the History of Mathematics and Physical Sciences. New York: Springer, 2007.

Anderson, Perry. *The Origins of Postmodernity*. London: Verso, 1998.

Aquinas, St. Thomas. *The Summa Theologica of St. Thomas Aquinas*. Translated by Fathers of the English Dominican Province. 25 vols. Vol. 1. London: Burns, Oates and Washbourne, Ltd., 1911.

Arageorgis, Aristidis. "Aristotle and the Atomists vis-à-vis the Mathematicians." *Philosophical Inquiry* 39, no. 1 (2015): 164–80.

Arendt, Hannah. "The Conquest of Space and the Stature of Man." *New Atlantis*, no. 18 (2007): 43–55.

Aristotle. *The Complete Works of Aristotle: The Revised Oxford Translation*. Bollingen Series 71. 2 vols. Vol. 1. Princeton, N.J.: Princeton University Press, 1984.

Arnaud, Pascal. "L'Image du globe dans le monde romain: Science, iconographie, symbolique." *Mélanges de l'École Française de Rome. Antiquité* 96, no. 1 (1984): 53–116.

Arnim, Hans Friedrich August von. *Stoicorum veterum fragmenta*. 4 vols. Lipsiae: in aedibus B. G. Teubneri, 1903.

Artmann, Benno. *Euclid: The Creation of Mathematics*. New York: Springer, 1999.

Ashcraft, Richard. *Revolutionary Politics & Locke's Two Treatises of Government*. Princeton, N.J.: Princeton University Press, 1986.

Ashworth, William B., Jr. "The Calculating Eye: Baily, Herschel, Babbage and the Business of Astronomy." *British Journal for the History of Science* 27, no. 4 (1994): 409–41.

Aubrey, John. *Brief Lives: A Modern English Version*. Woodbridge: Boydell Press, 1982.

Aujac, Germaine. *Claude Ptolémée, astronome, astrologue, géographe: Connaissance et représentation du monde habité*. Paris: Editions du CTHS, 1993.

———. "Greek Cartography in the Early Roman World." In *The History of Cartography: Cartography in Prehistoric, Ancient and Medieval Europe and the Mediterranean*, edited by J. B. Harley and David Woodward, 161–76. Chicago: University of Chicago Press, 1987.

Avotins, Ivars. "On Some Epicurean and Lucretian Arguments for the Infinity of the Universe." *Classical Quarterly* 33, no. 2 (1983): 421–27.

Axtell, James L. "Locke, Newton, and 'The Elements of Natural Philosophy.'" *Paedagogica Europaea* (1965): 235–45.

———. "Locke's Review of the 'Principia.'" *Notes and Records of the Royal Society of London* (1965): 152–61.

Baasner, Rainer. *Das Lob der Sternkunst: Astronomie in der deutschen Aufklärung*. Göttingen: Vandenhoeck and Ruprecht, 1987.

Babicz, Józef. "The Celestial and Terrestrial Globes of the Vatican Library, Dating from 1477, and Their Maker Donnus Nicolaus Germanus (ca. 1420–ca. 1490)." *Der Globusfreund: Wissenschaftliche Zeitschrift für Globenkunde* 35/37 (1987): 155–66.

———. "Donnus Nicolaus Germanus—Probleme seiner Biographie und sein Platz in der Rezeption der Ptolemäischen Geographie." *Wolfenbütteler Forschungen* 7 (1980): 9–42.

Bagrow, Leo. "The Origin of Ptolemy's Geographia." *Geografiska Annaler* 27 (1945): 318–67.

Bailey, Martin. "Dürer's Comet." *Apollo-London-Apollo Magazine Limited* 141 (1995): 19–32.

Baker, J. N. L. "The Geography of Bernhard Varenius." *Transactions and Papers (Institute of British Geographers)* 21 (1955): 51–60.

Bakhtin, M. M. *Problems of Dostoevsky's Poetics*. Ann Arbor, Mich.: Ardis, 1973.

Baldasso, Renzo. "Portrait of Luca Pacioli and Disciple: A New, Mathematical Look." *Art Bulletin* 92, no. 1/2 (2010): 83–102.

Barnard, Alan. *History and Theory in Anthropology*. Cambridge: Cambridge University Press, 2000.

Barnes, Jonathan, ed. *The Complete Works of Aristotle*. 2 vols. Bollingen Series, vol. 71, 2. Princeton, N.J.: Princeton University Press, 1984.

Bartrum, Giulia, and Günter Grass. *Albrecht Dürer and His Legacy: The Graphic Work of a Renaissance Artist*. London: British Museum Press, 2002.

Bataillon, Marcel. "L'Unité du genre humaine du P. Acosta au P. Clavigero." In *Mélanges à la mémoire de Jean Sarrailh*, edited by Marcel Bataillon, 75–95. Paris: Centre de Recherches de l'Institut d'Etudes Hispaniques, 1966.

Baumgold, Deborah. *Contract Theory in Historical Context: Essays on Grotius, Hobbes, and Locke*. Brill's Studies in Intellectual History. Leiden: Brill, 2010.

Beck, Lewis White. *Early German Philosophy: Kant and His Predecessors*. Cambridge, Mass.: Belknap Press, Harvard University Press, 1969.

Bell, David A. *Lawyers and Citizens: The Making of a Political Elite in Old Regime France*. New York: Oxford University Press, 1994.

Bell, Matthew. *Goethe's Naturalistic Anthropology: Man and Other Plants*. Oxford Modern Languages and Literature Monographs. Oxford: Clarendon Press, 1994.

Bellitto, Christopher M., Thomas M. Izbicki, and Gerald Christianson. *Introducing Nicholas of Cusa: A Guide to a Renaissance Man*. New York: Paulist Press, 2004.

Bénatouïl, Thomas. "Les Critiques épicuriennes de la géométrie." In *Construction: Festschrift for Gerhard Heinzmann*, edited by Gerhard Heinzmann, 151–62. London: College Publications, 2010.

Benmakhlouf, Ali. "Nature et cosmos: incursions en philosophie ancienne et médiévale." *Revue de métaphysique et de morale*, no. 3 (2004): 343–52.

Bennett, J. A. *The Divided Circle: A History of Instruments for Astronomy, Navigation, and Surveying*. Oxford: Phaidon, 1987.

———. "The English Quadrant in Europe: Instruments and the Growth of Consensus in Practical Astronomy." *Journal for the History of Astronomy* 23 (1992): 1–14.

Bennett, J. A., and Domenico Bertoloni Meli. *Sphaera Mundi: Astronomy Books in the Whipple Museum, 1478–1600*. Cambridge: Whipple Museum of the History of Science, 1994.

Bennett, Jim. "Practical Geometry and Operative Knowledge." *Configurations* 6, no. 2 (1998): 195–222.

Bennett, John W. "Comments on 'the Renaissance Foundations of Anthropology.' " *American Anthropologist* 68 (1966): 215–20.

Bentham, Jeremy. "To Sir Samuel Bentham, 20–26 August 1773." In *Electronic Enlightenment*, edited by Robert McNamee. Oxford: Oxford University Press, 2011. http://www.e-enlightenment.com.

Berg, Eberhard. *Zwischen den Welten: Über die Anthropologie der Aufklärung und ihr Verhältnis zu Entdeckungs-Reise und Welt-Erfahrung mit besonderem Blick auf das Werk Georg Forster*. Beiträge Zur Kulturanthropologie. Berlin: Reimer, 1982.

Berggren, J. L., and Alexander Jones, eds. *Ptolemy's Geography: An Annotated Translation of the Theoretical Chapters*. Princeton, N.J.: Princeton University Press, 2000.

Bernard, Alain. "The Significance of Ptolemy's *Almagest* for Its Early Readers." *Revue de synthèse* 6, no. 4 (2010): 495–521.

Bertens, Johannes Willem. *The Idea of the Postmodern: A History*. London Routledge, 1995.

Bertens, Johannes Willem, and Joseph P. Natoli. *Postmodernism: The Key Figures*. Malden, Mass.: Blackwell Publishers, 2002.

Besse, Jean-Marc. *Les Grandeurs de la Terre: Aspects du savoir géographique à la Renaissance*. Collection Sociétés, Espaces, Temps. Lyon: ENS, 2003.

Best, Steven, and Douglas Kellner. *The Postmodern Turn*. Critical Perspectives. New York: Guilford Press, 1997.

Blanckaert, Claude. *Naissance de l'ethnologie?: Anthropologie et missions en Amérique XVIe– XVIIIe siècle*. Paris: Ed. du Cerf, 1985.

Blasjö, Viktor. "A Critique of the Arguments for Maragha Influence on Copernicus." *Journal for the History of Astronomy* 45, no. 2 (2014): 183–95.

Blumenberg, Hans. *Die Genesis der kopernikanischen Welt*. Frankfurt am Main: Suhrkamp, 1975.

———. *The Genesis of the Copernican World*. Translated by Robert M. Wallace. Cambridge, Mass.: MIT Press, 1987.

———. *The Legitimacy of the Modern Age*. Translated by Robert M. Wallace. Cambridge, Mass.: MIT Press, 1985.

———. *Die Legitimität der Neuzeit*. Frankfurt am Main: Suhrkamp, 1966.

Bödeker, Hans Erich. "Menschheit, Humanität, Humanismus." In *Geschichtliche Grundbegriffe*, edited by Otto Brunner, Werner Conze and Reinhart Koselleck, 1063–1128. Stuttgart: Klett-Cotta, 1982.

Böhme, Hartmut. *Albrecht Dürer: Melencolia I: Im Labyrinth der Deutung*. Fischer-Taschenbücher. Orig.-Ausg. ed. Frankfurt a.M.: Fischer Taschenbuch Verl., 1989.

Borges, Jorge Luis. *Otras Inquisiciones*. Obras Completas. Buenos Aires: Emec, 1960.

Bott, Gerhard, and Johannes Karl Wilhelm Willers. *Focus Behaim Globus: Germanisches Nationalmuseum, Nürnberg, 2. Dezember 1992 Bis 28. Februar 1993*. 2 vols. Nürnberg: Verlag des Germanischen Nationalmuseums, 1992.

Bouwsma, William J. *The Waning of the Renaissance, 1550–1640*. New Haven, Conn.: Yale University Press, 2000.

Bowen, Margarita. *Empiricism and Geographical Thought: From Francis Bacon to Alexander von Humboldt*. Cambridge Geographical Studies. Cambridge: University Press, 1981.

Boyer, Carl B. *The History of the Calculus and Its Conceptual Development (the Concepts of the Calculus)*. New York: Dover, 1959.

———. *The Rainbow: From Myth to Mathematics*. Princeton, N.J.: Princeton University Press, 1987.

Boyer, Carl B., and Uta C. Merzbach. *A History of Mathematics*. 2nd ed. New York: John Wiley and Sons, 1991.

Brading, David A. *The First America: The Spanish Monarchy, Creole Patriots, and the Liberal State, 1492–1867*. Cambridge: Cambridge University Press, 1991.

Braudel, Fernand. *The Mediterranean and the Mediterranean World in the Age of Philip II*. 2 vols. New York: Harper and Row, 1972.

Brient, Elizabeth. *The Immanence of the Infinite: Hans Blumenberg and the Threshold to Modernity*. Washington, D.C.: Catholic University of America Press, 2002.

———. "Transitions to a Modern Cosmology: Meister Eckhart and Nicholas of Cusa on the Intensive Infinite." *Journal of the History of Philosophy* 37, no. 4 (1999): 575–600.

Broc, Numa. *La Géographie de la Renaissance (1420–1620)*. Mémoires de la Section de Géographie / Ministère des Universités, Comité des Travaux Historiques et Scientifiques. Paris: Bibliothèque Nationale, 1980.

Brockliss, L. W. B. *The University of Oxford: A History*. Oxford: Oxford University Press, 2016.

Brown, J. V. "Abstraction and the Object of the Human Intellect According to Henry of Ghent." *Vivarium* 11, no. 1 (1973): 80–104.

Brown, John N. "Voltaire and Astronomy." *Journal of the British Astronomical Association* 99, no. 1 (1989): 29–31.

Brown, Jonathan. *Images and Ideas in Seventeenth-Century Spanish Painting*. Princeton, N.J.: Princeton University Press 1978.

Brown, Jonathan, and Carmen Garrido. *Velázquez: The Technique of Genius*. New Haven, Conn.: Yale University Press, 2003.

Brown, Jonathan M. "Painting in Seville from Pacheco to Murillo: A Study of Artistic Transition." Ph.D. dissertation, Princeton University, 1964.

Brown, Lloyd A. *The Story of Maps*. Boston: Little, Brown, 1949.

Brown, Stephen. "The Intellectual Context of Later Medieval Philosophy: Universities, Aristotle, Arts, Theology." In *Medieval Philosophy*, edited by John Marenbon, 188–203. Routledge History of Philosophy. London: Routledge, 1998.

Bruce, Steve. *God Is Dead: Secularization in the West*. Oxford: Blackwell, 2002.

Brummelen, Glen van. *The Mathematics of the Heavens and the Earth: The Early History of Trigonometry*. Princeton, N.J.: Princeton University Press, 2009.

Bunce, Robin. "Thomas Hobbes' Relationship with Francis Bacon—An Introduction." *Hobbes Studies* 16, no. 1 (2003): 41–83.

Burckhardt, Jacob. *Die Kultur der Renaissance in Italien*. Berlin: Deutsche Buch-Gemeinshaft, 1961.

Burgaleta, Claudio M. *José de Acosta, S.J. (1540–1600): His Life and Thought*. Chicago: Jesuit Way, 1999.

Burmeister, Karl Heinz. *Johannes Campensis und Sebastian Münster: Ihre Stellung in der Geschichte der hebräischen Sprachstudien*. Louvain: Bibliothèque de l'Université, 1970.

———. *Neue Forschungen zu Sebastian Münster: Mit einem Anhang von Ernst Emmerling: Graphische Bildnisse Sebastian Münsters*. Beiträge zur Ingelheimer Geschichte. Ingelheim: Historischer Verein, 1971.

———. *Sebastian Münster: Versuch eines Biographischen Gesamtbildes*. Basler Beiträge zur Geschichtswissenschaft. Basel: Helbing und Lichtenhahn, 1963.

Burnett, Charles. "Scientific Speculations." In *A History of Twelfth-Century Western Philosophy*, edited by Peter Dronke, 151–76. Cambridge: Cambridge University Press, 1988.

Busard, H. L. L., ed. *The First Latin Translation of Euclid's Elements Commonly Ascribed to Adelard of Bath: Books I–VIII and Books X.36–XV.2*. Toronto: Pontifical Institute of Mediaeval Studies, 1983.

Büttner, Frank. *Sammeln, Ordnen, Veranschaulichen: Zur Wissenskompilatorik in der Frühen Neuzeit*. Pluralisierung & Autorität. Münster: Lit, 2003.

Büttner, Manfred. *Die Geographia Generalis vor Varenius: Geograph. Weltbild u. Providentiallehre*. Erdwissenschaftliche Forschung. Wiesbaden: Franz Steiner Verlag, 1973.

Büttner, Manfred, and Józef Babicz. *Zur Entwicklung der Geographie vom Mittelalter bis zu Carl Ritter*. Abhandlungen und Quellen zur Geschichte der Geographie und Kosmologie. Paderborn: F. Schöningh, 1982.

Button, Mark E. *Contract, Culture, and Citizenship: Transformative Liberalism from Hobbes to Rawls*. University Park: Pennsylvania State University Press, 2008.

Butzer, Karl W. "From Columbus to Acosta: Science, Geography, and the New World." *Annals of the Association of American Geographers* 82, no. 3 (1992): 543–65.

Campbell, Gordon. *Bible: The Story of the King James Version, 1611–2011*. Oxford: Oxford University Press, 2011.

Campbell, Tony. "A Descriptive Census of Willem Blaeu's Sixty-Eight Centimeter Globes." *Imago Mundi* 28 (1976): 21–50.

Cañizares-Esguerra, Jorge. "Iberian Science in the Renaissance: Ignored How Much Longer?" *Perspectives on Science* 12, no. 1 (2004): 86–124.

Carracido, José R. *El P. José de Acosta y su importancia en la literatura científica española*. Madrid: Est. Tipográfico "Sucesores de Rivadeneyra," 1899.

Casarella, Peter J. *Cusanus: The Legacy of Learned Ignorance*. Washington, D.C.: Catholic University of America Press, 2006.

Casini, P. "Newton's 'Principia' and the Philosophers of the Enlightenment." *Notes and Records of the Royal Society of London* 42, no. 1 (1988): 35–52.

Cassirer, Ernst. *Die Platonische Renaissance in England und die Schule von Cambridge*. Leipzig: B. G. Teubner, 1932.

———. *The Individual and the Cosmos in Renaissance Philosophy*. Oxford: Blackwell, 1963.

———. *The Philosophy of the Enlightenment*. Princeton, N.J.: Princeton University Press, 1968.

Cassou-Noguès, Pierre. *Hilbert*. Paris: Les Belles Lettres, 2001.

Certeau, Michel de, and James Hovde. "Writing vs. Time: History and Anthropology in the Works of Lafitau." *Yale French Studies* 59 (1980): 37–64.

Chadwick, Owen. *The Secularization of the European Mind in the Nineteenth Century*. Cambridge: Cambridge University Press, 1990.

Clagett, Marshall. "The Medieval Latin Translations from the Arabic of the Elements of Euclid, with Special Emphasis on the Versions of Adelard of Bath." *Isis* 44, no. 1/2 (1953): 16–42.

———, ed. *Nicole Oresme and the Medieval Geometry of Qualities and Motions: A Treatise on the Uniformity and Difformity of Intensities Known as Tractatus de configurationibus qualitatum et motuum*. Madison: University of Wisconsin Press, 1968.

———. "Nicole Oresme and Medieval Scientific Thought." *Proceedings of the American Philosophical Society* 108, no. 4 (1964): 298–309.

———. "The Use of Points in Medieval Natural Philosophy and Most Particularly in the 'Questiones de spera' of Nicole Oresme." In *Studies in Medieval Physics and Mathematics*, 216–21. London: Variorum Reprints, 1979.

Clark, David L. "Kant's Aliens: The Anthropology and Its Others." *New Centennial Review* 1, no. 2 (2001): 201–89.

Clarke, Katherine. *Between Geography and History: Hellenistic Constructions of the Roman World*. Oxford Classical Monographs. Oxford: Clarendon Press, 1999.

Clarke, Samuel, Gottfried Wilhelm Leibniz, and Isaac Newton. *The Leibniz-Clarke Correspondence, Together with Extracts from Newton's Principia and Opticks*. Edited by H. G. Alexander. Philosophical Classics. Manchester: Manchester University Press, 1970.

Clay, Diskin. *Lucretius and Epicurus*. Ithaca, N.Y.: Cornell University Press, 1983.

Close, Anthony. "Centering the De-Centerers: Foucault and Las Meninas." *Philosophy and Literature* 11, no. 1 (1987): 21–36.

Clucas, Stephen. "The Atomism of the Cavendish Circle: A Reappraisal." *Seventeenth Century* 9, no. 2 (1994): 247–73.

Coffin, Charles M., ed. *The Complete Poetry and Selected Prose of John Donne*. New York: Random House, 1952.

Colish, Marcia L. "Early Scholastic Angelology." *Recherches de Théologie Ancienne et Médiévale* 62 (1980): 80–109.

———. *Medieval Foundations of the Western Intellectual Tradition, 400–1400*. Yale Intellectual History of the West. New Haven, Conn.: Yale University Press, 1997.

Copenhaver, Brian P. "Hermes Trismegistus, Proclus, and the Question of a Philosophy of Magic in the Renaissance." In *Hermeticism and the Renaissance: Intellectual History and the Occult in Early Modern Europe*, edited by Ingrid Merkel and Allen George Debus. Folger Institute Symposia, 79–110. Washington, D.C.: Folger Books, 1988.

———, ed. *Hermetica: The Greek Corpus Hermeticum and the Latin Asclepius in a New English Translation.* Cambridge: Cambridge University Press, 1995.

Copenhaver, Brian P., and Charles B. Schmitt. *Renaissance Philosophy.* A History of Western Philosophy. Oxford: Oxford University Press, 1992.

Copernicus, Nicolaus, and Georg Joachim Rhäticus. *Three Copernican Treatises: The Commentariolus of Copernicus, the Letter Against Werner, the Narratio Prima of Rheticus.* Edited by Edward Rosen. Records of Civilization, Sources and Studies. New York: Columbia University Press, 1939.

Cormack, Lesley B. "The World at Your Fingertips: English Renaissance Globes as Cosmographical, Mathematical and Pedagogical Instruments." *Archives Internationales d'Histoire des Sciences* 59, no. 163 (2009): 485–97.

Correa, Antonio Bonet. "Velázquez, arquitecto y decorador." *Archivo Español de Arte* 33, no. 130 (1960): 215.

Cosgrove, Dennis. *Apollo's Eye: A Cartographic Genealogy of the Earth in the Western Imagination.* Baltimore: Johns Hopkins University Press, 2001.

———. "Images of Renaissance Cosmography, 1450–1650." In *The History of Cartography,* edited by J. B. Harley and David Woodward, 55–98. Chicago: University of Chicago Press, 1987.

Cotter, Charles H. *The Astronomical & Mathematical Foundations of Geography.* London: Hollis and Carter, 1966.

Cranston, Maurice William, and R. S. Peters. *Hobbes and Rousseau: A Collection of Critical Essays.* Modern Studies in Philosophy. Garden City, N.Y.: Anchor Books, 1972.

Cranz, F. Edward, Thomas M. Izbicki, and Gerald Christianson. *Nicholas of Cusa and the Renaissance.* Variorum Collected Studies Series. Aldershot: Ashgate, 2000.

Crombie, A. C. "Mathematics and Platonism in the Sixteenth-Century Italian Universities and in Jesuit Educational Policy." In *Prismata: Naturwissenschaftsgeschichtliche Studien: Festschrift für Willy Hartner,* edited by Willy Hartner, Yasukatsu Maeyama, and Walter G. Saltzer, 63–94. Wiesbaden: Franz Steiner Verlag, 1977.

Crosby, Alfred W. *The Measure of Reality: Quantification and Western Society, 1250–1600.* Cambridge: Cambridge University Press, 1998.

Crowe, Michael J. *The Extraterrestrial Life Debate, 1750–1900: The Idea of a Plurality of Worlds from Kant to Lowell.* Cambridge: Cambridge University Press, 1986.

Cusa, Nicholas of. *De Ludo Globi = The Game of Spheres.* Janus Series. New York: Abaris Books, 1986.

———. *Der Laie über den Geist: Lateinisch-Deutsch.* Translated by Renate Steiger. Schriften Des Nikolaus von Kues in deutscher Übersetzung. Edited by Ernst Hoffmann. Hamburg: Felix Meiner Verlag, 1995.

———. *Nicholas of Cusa on Learned Ignorance: A Translation and an Appraisal of De Docta Ignorantia.* Translated by Jasper Hopkins. 2nd ed. Minneapolis: A. J. Banning Press, 1990.

———. *Opera omnia: Iussu et auctoritate Academiae Litterarum Heidelbergensis ad codicum fidem edita.* 28 vols. Vol. 3. Hamburg: Felix Meiner Verlag, 1959.

———. *The Vision of God.* Translated by Emma Gurney Salter. New York: Ungar Publishing Company, 1960.

Da Vinci, Leonardo. *The Notebooks of Leonardo Da Vinci / Arranged and Rendered into English with Introductions by Edward MacCurdy.* Edited by Edward MacMurdy. New York: Reynal and Hitchcock, 1938.

Dahl, Edward H., and Jean-François Gauvin. *Sphaerae Mundi: Early Globes at the Stewart Museum.* Sillery, Que.: McGill-Queen's University Press, 2000.

Dainville, François de, and Marie-Madeleine Compère. *L'Éducation des Jésuites: XVIe–XVIIIe siècles.* Paris: Les Éditions de Minuit, 1978.

Darnton, Robert. *The Business of Enlightenment: A Publishing History of the Encyclopédie, 1775–1800.* Cambridge, Mass.: Harvard University Press, 1979.

Dear, Peter. *Discipline & Experience: The Mathematical Way in the Scientific Revolution.* Science and Its Conceptual Foundations. Chicago: University of Chicago Press, 1995

———. *Mersenne and the Learning of the Schools.* Cornell History of Science Series. Ithaca, N.Y.: Cornell University Press, 1988.

———. "Mersenne's Suggestion: Cartesian Meditation and the Mathematical Model of Knowledge in the Seventeenth Century." In *Descartes and His Contemporaries: Meditations, Objections, and Replies*, edited by Roger Ariew and Marjorie Grene, 44–62. Chicago University of Chicago Press, 1995.

———. *Revolutionizing the Sciences: European Knowledge and Its Ambitions, 1500–1700.* Princeton, N.J.: Princeton University Press, 2001.

Dekker, Elly. "Globes in Renaissance Europe." In *The History of Cartography*, edited by J. B. Harley and David Woodward, 135–59. Chicago: University of Chicago Press, 1987.

———. *Illustrating the Phaenomena Celestial Cartography in Antiquity and the Middle Ages.* Oxford: Oxford University Press, 2013.

Dekker, Elly, Silke Ackermann, and Kristen Lippincott. *Globes at Greenwich: A Catalogue of the Globes and Armillary Spheres in the National Maritime Museum, Greenwich.* Oxford: Oxford University Press and the National Maritime Museum, 1999.

Dekker, Elly, and P. C. J. van der Krogt. *Globes from the Western World.* London: Zwemmer, 1993.

Demerath, Nicholas J., III. "Secularization and Sacralization: Deconstructed and Reconstructed." In *The Sage Handbook of the Sociology of Religion*, edited by James A. Beckford and Nicholas J. Demerath III, 57–80. Los Angeles: SAGE Publications, 2007.

Denby, David. "Herder: Culture, Anthropology and the Enlightenment." *History of the Human Sciences* 18, no. 1 (2005): 55–76.

Depré, Olivier. "The Ontological Foundations of Hegel's Dissertation of 1801." In *Hegel and the Philosophy of Nature*, edited by Stephen Houlgate, 257–82. Albany: State University of New York Press, 1998.

Derrida, Jacques. *Edmund Husserl's Origin of Geometry: An Introduction.* Translated by John P. Leavey. Lincoln: University of Nebraska Press, 1978.

———. *L'Origine de la géométrie.* Translated by Jacques Derrida. Épiméthée Essais Philosophiques. Paris: Presses universitaires de France, 1962.

———. *Of Grammatology.* Translated by Gayatri Chakravorty Spivak. Baltimore: Johns Hopkins University Press, 1976.

Descartes, René. *Discourse on Method.* Translated by Lawrence Lafleur. The Library of Liberal Arts. 2nd (rev.) ed. New York: Liberal Arts Press, 1956.

———. *Meditations on First Philosophy.* Translated by Laurence Lafleur. Library of Liberal Arts. Indianapolis: Bobbs-Merrill, 1960.

———. "To Frans Burman, 16 April 1648." In *Electronic Enlightenment*, edited by Robert McNamee. Oxford: Oxford University Press, 2011. http://www.e-enlightenment.com.

———. *The World and Other Writings.* Translated by Stephen Gaukroger. Cambridge Texts in the History of Philosophy. New York: Cambridge University Press, 1998.

Devine, James. "The Positive Political Economy of Individualism and Collectivism: Hobbes, Locke, and Rousseau." *Politics and Society* 28, no. 2 (2000): 265–304.

Dick, Steven J. *Plurality of Worlds: The Origins of the Extraterrestrial Life Debate from Democritus to Kant.* Cambridge: Cambridge University Press, 1982.

Dickey, Laurence W. *Hegel: Religion, Economics, and the Politics of Spirit, 1770–1807.* Ideas in Context. Cambridge: Cambridge University Press, 1987.

Dietz, Mary G., ed. *Thomas Hobbes and Political Theory.* Lawrence: University Press of Kansas, 1990.

Dilke, O. A. W. "Cartography in the Byzantine Empire." In *The History of Cartography: Cartography in Prehistoric, Ancient and Medieval Europe and the Mediterranean*, edited by J. B. Harley and David Woodward, 256–75. Chicago: University of Chicago Press, 1987.

———. *The Roman Land Surveyors: An Introduction to the Agrimensores*. Newton Abbot: David and Charles, 1971.

Dillenberger, John. *Protestant Thought and Natural Science: A Historical Interpretation*. Notre Dame: University of Notre Dame Press, 1988.

Dilthey, Wilhelm. "Auffassung und Analyse des Menschen im 15. und 16. Jahrhundert." In *Wilhelm Diltheys Gesammelte Schriften*, 1–89. 14 vols. Vol. 2 Leipzig: B. G. Teubner, 1921.

———. "Die Funktion der Anthropologie in der Kultur des 16. und 17. Jahrhunderts." In *Wilhelm Diltheys Gesammelte Schriften*, 416–92. 14 vols. Vol. 2. Leipzig: B. G. Teubner, 1921.

Dobbs, Betty Jo Teeter, and Margaret C. Jacob. *Newton and the Culture of Newtonianism*. Atlantic Highlands, N.J.: Humanities Press, 1995.

Docker, John. *Postmodernism and Popular Culture: A Cultural History*. Cambridge: Cambridge University Press, 1994.

Dohrn-van Rossum, Gerhard. *History of the Hour: Clocks and Modern Temporal Orders*. Chicago: University of Chicago Press, 1996.

Dolnikowski, Edith Wilks. *Thomas Bradwardine: A View of Time and a Vision of Eternity in Fourteenth-Century Thought*. Studies in the History of Christian Thought. Leiden: Brill, 1995.

Donne, John. *Complete Poetry and Selected Prose of John Donne*. Modern Library. New York: Random House, 1952.

Doorly, Patrick. "Dürer's "Melencolia I": Plato's Abandoned Search for the Beautiful." *Art Bulletin* 86, no. 2 (2004): 255–76.

Dostoyevsky, Fyodor. *The Brothers Karamazov*. Translated by Constance Garnett. Mineola, N.Y.: Dover Publications, 2005.

Douglas, Mary. *Implicit Meanings: Essays in Anthropology*. London: Routledge, 1993.

Drake, Stillman. *Galileo: A Very Short Introduction*. Oxford: Oxford University Press, 2001.

———. *Galileo at Work: His Scientific Biography*. Chicago: University of Chicago Press, 1978.

Droixhe, Daniel, and Pol-Pierre Gossiaux, eds. *Anthropologie et histoire au siècle des lumières: Buffon, Voltaire, Rousseau, Helvétius, Diderot*. Paris: Albin Michel, 1995. Éditions Albin Michel, S.A. 1971.

———. *L'Homme des lumières et la découverte de l'autre*. Études sur le XVIIIe siècle. Volume Hors Série, 3. Bruxelles: Éditions de l'Université de Bruxelles, 1985.

Duhem, Pierre. *Le Système du monde; histoire des doctrines cosmologiques de Platon à Copernic*. Paris: A. Hermann, 1913.

Dunnington, G. Waldo, Jeremy Gray, and Fritz-Egbert Dohse. *Carl Friedrich Gauss: Titan of Science*. Washington, D.C.: Mathematical Association of America, 2004.

Durand, Dana B. "Nicole Oresme and the Medieval Origins of Modern Science." *Speculum* 16, no. 2 (1941): 167–85.

Dussler, Luitpold. *Raphael: Kritisches Verzeichnis der Gemälde, Wandbilder und Bildteppiche*. Munich: Bruckmann, 1966.

Duzer, Chet A. van. *Johann Schoner's Globe of 1515: Transcription and Study*. Transactions of the American Philosophical Society Held at Philadelphia for Promoting Useful Knowledge. Philadelphia: American Philosophical Society, 2010.

Eastwood, Bruce. *Ordering the Heavens: Roman Astronomy and Cosmology in the Carolingian Renaissance*. History of Science and Medicine Library. Leiden: Brill, 2007.

Edgerton, Samuel Y. "Alberti's Perspective: A New Discovery and a New Evaluation." *Art Bulletin* 48, no. 3–4 (1966): 367–78.

———. *The Heritage of Giotto's Geometry: Art and Science on the Eve of the Scientific Revolution*. Ithaca, N.Y.: Cornell University Press, 1991.

———. *The Renaissance Rediscovery of Linear Perspective*. New York: Basic Books, 1975.

Edson, Evelyn. *Mapping Time and Space: How Medieval Mapmakers Viewed Their World*. British Library Studies in Map History. London: British Library, 1999.

———. *The World Map, 1300–1492: The Persistence of Tradition and Transformation*. Baltimore: Johns Hopkins University Press, 2007.

Eichberger, Dagmar, and Charles Zika, eds. *Dürer and His Culture*. Cambridge: Cambridge University Press, 1998.

Eisenstein, Elizabeth. *The Printing Press as an Agent of Change: Communications and Cultural Transformations in Early Modern Europe*. 2 vols. Cambridge: Cambridge University Press, 1979.

Eisler, Colin T. *Dürer's Animals*. Washington, D.C.: Smithsonian Institution Press, 1991.

Elliott, John H. *The Old World and the New, 1492–1650*. Cambridge: Cambridge University Press, 1970.

Engelfriet, Peter M. *Euclid in China: The Genesis of the First Chinese Translation of Euclid's Elements, Books I–VI (Jihe Yuanben, Beijing, 1607) and Its Reception up to 1723*. Edited by W. L. Idema. Sinica Leidensia. Leiden: Brill, 1998.

Erdheim, Mario. "Anthropologische Modelle des 16. Jahrhunderts: Oviedo (1478–1557), Las Casas (1475–1566), Sahagún (1499–1540), Montaigne (1533–1592)." In *Klassiker der Kulturanthropologie: Von Montaigne bis Margaret Mead*, edited by Wolfgang Marschall, 19–50. München: C. H. Beck, 1990.

Erigena, Johannes Scotus. *Eriugena's Commentary on the Dionysian Celestial Hierarchy*. Translated by Paul Rorem. Studies and Texts. Toronto: Pontifical Institute of Mediaeval Studies, 2005.

Eriksen, Thomas Hyllans, and Finn Sivert Nielsen. *A History of Anthropology*. London: Pluto Press, 2001.

Euclid. *Euclid's Elements: All Thirteen Books Complete in One Volume*. Translated by Thomas L. Heath. Santa Fe: Green Lion Press, 2002.

———. *The Thirteen Books of Euclid's Elements*. Translated by Thomas Little Heath. 2nd ed. 3 vols. Vol. 1. New York: Dover Publications, 1956.

Evans, Gillian R. "The 'Sub-Euclidean' Geometry of the Earlier Middle Ages, up to the Mid-Twelfth Century." *Archive for History of Exact Sciences* 16, no. 2 (1976): 105–18.

Evans-Pritchard, E. E. *History of Anthropological Thought*. London: Faber and Faber, 1981.

Eves, Howard Whitley. *An Introduction to the History of Mathematics*. 5th ed. Philadelphia: Saunders College Publishing, 1983.

Fara, Patricia. "Heavenly Bodies: Newtonianism, Natural Theology and the Plurality of Worlds Debate in the Eighteenth Century." *Journal for the History of Astronomy* 35 (2004): 143–60.

Faull, Katherine M., ed. *Anthropology and the German Enlightenment: Perspectives on Humanity*. Lewisburg, Pa.: Bucknell University Press, 1995.

Fauser, Alois. *Die Welt in Händen: Kurze Kulturgeschichte des Globus*. Stuttgart: Schuler, 1967.

———. *Kulturgeschichte des Globus*. Reprint ed. Vienna: Vollmer, 1967.

Feingold, Mordechai, ed. *Before Newton: The Life and Times of Isaac Barrow*. Cambridge: Cambridge University Press, 1990.

———. "Isaac Barrow and the Foundation of the Lucasian Professorship." In *From Newton to Hawking: A History of Cambridge University's Lucasian Professors of Mathematics*, edited by Kevin C. Knox and Richard Noakes, 45–67. Cambridge: Cambridge University Press, 2003.

———. *The Newtonian Moment: Isaac Newton and the Making of Modern Culture*. New York: Oxford University Press, 2004.

Fenton, William N., and Elizabeth L. Moore. "J.-F. Lafitau (1681–1746), Precursor of Scientific Anthropology." *Southwestern Journal of Anthropology* 25, no. 2 (1969): 173–87.

Ficino, Marsilio. *Three Books on Life: A Critical Edition and Translation*. Translated by Carol V. Kaske and John R. Clark. Medieval & Renaissance Texts & Studies. Binghamton: Center for Medieval and Early Renaissance Studies, 1989.

Fiorini, Matteo. *Sfere terrestri e celesti di autore italiano, oppure fatte o conservate in Italia*. Roma: La Società Geografica Italiana, 1899.

Flood, John L. "Luther and Tyndale as Bible Translators: Achievement and Legacy." In *Landmarks in the History of the German Language*, edited by Geraldine Horan, Nils Langer and Sheila Watts, 35–56. Bern: Peter Lang, 2009.

Folkerts, Menso. *Boethius Geometrie II: Ein mathematisches Lehrbuch des Mittelalters*. Wiesbaden: Franz Steiner Verlag, 1970.

———, ed. *The Development of Mathematics in Medieval Europe: The Arabs, Euclid, Regiomontanus.* Variorum Collected Studies Series; 811. Aldershot: Ashgate, 2006.
———. "Euclid in Medieval Europe." Chap. 3 in *The Development of Mathematics in Medieval Europe: The Arabs, Euclid, Regiomontanus,* edited by Menso Folkerts. Variorum Collected Studies Series; 811, 1–64. Aldershot: Ashgate, 2006.
———. "The Importance of the Pseudo-Boethian Geometria During the Middle Ages." In *Boethius and the Liberal Arts: A Collection of Essays,* edited by Michael Masi, 187–209. Utah Studies in Literature and Linguistics. Bern: Peter Lang, 1981.
Ford, Thayne R. "Stranger in a Foreign Land: Jose de Acosta's Scientific Realizations in Sixteenth-Century Peru." *Sixteenth Century Journal* 29, no. 1 (1998): 19–33.
Foucault, Michel. *The Birth of the Clinic: An Archaeology of Medical Perception.* Translated by A. M. Sheridan Smith. New York: Vintage Books, 1975.
———. *Discipline and Punish: The Birth of the Prison.* Translated by Alan Sheridan. New York: Pantheon Books, 1977.
———. *The Order of Things: An Archeology of the Human Sciences.* Routledge Classics. London: Routledge, 1970.
Frame, Donald M. *Montaigne's Discovery of Man: The Humanization of Humanist.* New York: Columbia University Press, 1955.
Freely, John. *Before Galileo: The Birth of Modern Science in Medieval Europe.* New York: Overlook Duckworth, 2012.
Funkenstein, Amos. *Theology and the Scientific Imagination from the Middle Ages to the Seventeenth Century.* Princeton, N.J.: Princeton University Press, 1986.
Garber, Daniel. "On the Frontlines of the Scientific Revolution: How Mersenne Learned to Love Galileo." *Perspectives on Science* 12, no. 2 (2004): 135–63.
———. "Philosophia, Historia, Mathematica: Shifting Sands in the Disciplinary Geography of the Seventeenth Century." In *Scientia in Early Modern Philosophy,* edited by Tom Sorell, G. A. J. Rogers, and Jill Kraye, 1–17. Dordrecht: Springer, 2010.
Garber, Jörn, and Heinz Thoma. *Zwischen Empirisierung und Konstruktionsleistung: Anthropologie im 18. Jahrhundert.* Tübingen: Niemeyer, 2004.
Gascoigne, John. "Blumenbach, Banks, and the Beginning of Anthropology at Göttingen." In *Göttingen and the Development of the Natural Sciences,* edited by Nicolaas A. Rupke, 86–98. Göttingen: Wallstein Verlag, 2002.
Gatti, Hilary. *Giordano Bruno and Renaissance Science.* Ithaca, N.Y.: Cornell University Press, 1999.
Gatto, Romano. "Christoph Clavius' 'Ordo servandus in addiscendis disciplinis mathematicis' and the Teaching of Mathematics in Jesuit Colleges at the Beginning of the Modern Era." *Science & Education* 15 (2006): 235–58.
Gaukroger, Stephen. *Descartes: An Intellectual Biography.* Oxford: Clarendon Press, 1995.
———. *Francis Bacon and the Transformation of Early-Modern Philosophy.* Cambridge: Cambridge University Press, 2001.
Gautier Dalché, Patrick. "Avant Behaim: Les Globes Terrestres au XVe siècle." *Médiévales* 58, no. 1 (2010): 43–61.
———. *La Géographie de Ptolémée en Occident (IVe–XVIe siècle).* Terrarum Orbis. Turnhout: Brepols, 2009.
———. "The Reception of Ptolemy's *Geography.*" In *The History of Cartography,* edited by J. B. Harley and David Woodward, 285–364. Chicago: University of Chicago Press, 1987.
Gay, Peter. *The Enlightenment: An Interpretation.* 2 vols. New York: Vintage Books, 1968.
Geertz, Clifford. *Local Knowledge: Further Essays in Interpretive Anthropology.* New York: Basic Books, 1983.
Ghent, Henry of. *Henrici de Gandavo opera omnia.* Ancient and Medieval Philosophy Series 2. 38 vols. Vol. 6. Louvain: Louvain University Press, 1978.
Ghillany, Friedrich Wilhelm. *Geschichte des Seefahrers Ritter Martin Behaim.* Nürnberg: Bauer and Raspe, J. Merz, 1853.

Giehlow, Karl. "Dürers Stich 'Melencolia' und der Maximilianische Humanistenkreis." *Mitteilungen der Gesellschaft für verfielfaltigende Kunst*, no. 2, 3, 4 (1903–1904): 29–41, 6–21, 57–78.

Gilreath, James, and Douglas L. Wilson. *Thomas Jefferson's Library: A Catalog with the Entries in His Own Order.* Washington, D.C.: Library of Congress 1989.

Gingerich, Owen. "From Copernicus to Kepler: Heliocentrism as Model and as Reality." *Proceedings of the American Philosophical Society,* 117, no. 6 (1973): 513–22.

Gingerich, Owen, and James MacLachlan. *Nicolaus Copernicus: Making the Earth a Planet.* New York: Oxford University Press, 2005.

Glacken, Clarence J. *Traces on the Rhodian Shore: Nature and Culture in Western Thought from Ancient Times to the End of the Eighteenth Century.* Berkeley: University of California Press, 1976.

Glat, Mark. "John Locke's Historical Sense." *Review of Politics* 43, no. 1 (1981): 3–21.

Goddu, André. "The Impact of Ockham's Reading of the Physics on the Mertonians and Parisian Terminists." *Early Science and Medicine* 6, no. 3 (2001): 204–36.

———. *The Physics of William of Ockham.* Leiden: Brill, 1984.

Godlewska, Anne Marie. *Geography Unbound: French Geographic Science from Cassini to Humboldt.* Chicago: University of Chicago Press, 1999.

Goeury, Marianne. "L'Atomisme Épicurien du temps à la lumière de la physique d'Aristote." *Les Études Philosophiques,* no. 4 (2013): 535–52.

Goldstein, Thomas. "The Renaissance Concept of the Earth in Its Influence upon Copernicus." *Terrae Incognitae* 4, no. 1 (1972): 19–51.

Goldwin, Robert A. "Locke's State of Nature in Political Society." *Western Political Quarterly* 29, no. 1 (1976): 126–35.

Goodman, Dena. *The Republic of Letters: A Cultural History of the French Enlightenment.* Ithaca, N.Y.: Cornell University Press, 1994.

Gorski, Philip S. "Historicizing the Secularization Debate: Church, State, and Society in Late Medieval and Early Modern Europe, ca. 1300 to 1700." *American Sociological Review* (2000): 138–67.

Gossiaux, Pol-Pierre, ed. *L'Homme et la nature: Genéses de l'anthropologie à l'age classique, 1580–1750: Anthologie.* Bruxelles: DeBoeck, 1993.

Goulding, Robert. "Henry Savile Reads His Euclid." In *For the Sake of Learning: Essays in Honor of Anthony Grafton,* edited by Ann Blair, Anja-Silvia Goeing, and Anthony Grafton, 780–97. Scientific and Learned Cultures and Their Institutions. Leiden: Brill, 2016.

Gourevitch, Victor. "Rousseau's Pure State of Nature." *Interpretation* 16, no. 1 (1988): 23–59.

Gowland, Angus. "The Problem of Early Modern Melancholy." *Past & Present* 191 (2006): 77–120.

Grafton, Anthony. *Commerce with the Classics: Ancient Books and Renaissance Readers.* Jerome Lectures. Ann Arbor: University of Michigan Press, 1997.

———. *Defenders of the Text: The Traditions of Scholarship in an Age of Science, 1450–1800.* Cambridge, Mass.: Harvard University Press, 1991.

Grafton, Anthony, April Shelford, and Nancy G. Siraisi. *New Worlds, Ancient Texts: The Power of Tradition and the Shock of Discovery.* Cambridge, Mass.: Harvard University Press, 1992.

Granada, Miguel A. *Giordano Bruno: Universo infinito, unión con dios, perfección del hombre.* Barcelona: Herder, 2002.

———. "New Visions of the Cosmos." In *The Cambridge Companion to Renaissance Philosophy,* edited by James Hankins, 270–86. Cambridge: Cambridge University Press, 2007.

Grant, Edward. "Aristotelianism and the Longevity of the Medieval World View." *History of Science* 16, no. 2 (1978): 93–106.

———. "Cosmology." In *Science in the Middle Ages,* edited by David C. Lindberg, 265–302. Chicago: University of Chicago Press, 1978.

———. "Medieval and Seventeenth-Century Conceptions of an Infinite Void Space Beyond the Cosmos." *Isis* 60, no. 1 (1969): 39–60.

———. "Motion in the Void and the Principle of Inertia in the Middle Ages." *Isis* 55, no. 3 (1964): 265–92.

———. "Science and Theology in the Middle Ages." In *God and Nature: Historical Essays on the Encounter Between Christianity and Science*, edited by David C. Lindberg and Ronald L. Numbers, 49–75. Berkeley: University of California Press, 1986.

———. "Scientific Imagination in the Middle Ages." *Perspectives on Science* 12, no. 4 (2004): 394–423.

———. "Scientific Thought in Fourteenth-Century Paris: Jean Buridan and Nicole Oresme." *Annals of the New York Academy of Sciences* 314, no. 1 (1978): 105–26.

———. "The Medieval Cosmos: Its Structure and Operation." *Journal for the History of Astronomy* 28 (1997): 147–67.

———. *Much Ado About Nothing: Theories of Space and Vacuum from the Middle Ages to the Scientific Revolution*. Cambridge: Cambridge University Press, 1981.

———. *Planets, Stars, and Orbs: The Medieval Cosmos, 1200–1687*. Cambridge: Cambridge University Press, 1996.

———. *Science and Religion, 400 B.C. to A.D. 1550: From Aristotle to Copernicus*. Greenwood Guides to Science and Religion. Westport, Conn.: Greenwood Press, 2004.

———. *The Nature of Natural Philosophy in the Late Middle Ages*. Studies in Philosophy and the History of Philosophy. Washington, D.C.: Catholic University of America Press, 2010.

Grant, Hardy. "Geometry and Politics: Mathematics in the Thought of Thomas Hobbes." *Mathematics Magazine* 63, no. 3 (1990): 147–54.

Grasshoff, Gerd. *The History of Ptolemy's Star Catalogue*. Studies in the History of Mathematics and Physical Sciences. New York: Springer-Verlag, 1990.

Gray, Jeremy J. *Ideas of Space: Euclidean, Non-Euclidean, and Relativistic*. Oxford Science Publications. 2nd ed. Oxford: Clarendon Press, 1989.

———. *Plato's Ghost: The Modernist Transformation of Mathematics*. Princeton, N.J.: Princeton University Press, 2008.

———. *Worlds out of Nothing: A Course in the History of Geometry in the 19th Century*. London: Spinger-Verlag, 2007.

Greenblatt, Stephen. *Renaissance Self-Fashioning: From More to Shakespeare*. Chicago: University of Chicago Press, 1980.

———. *The Swerve: How the World Became Modern*. New York: W. W. Norton and Company, 2011.

Gregory, Tullio. "The Platonic Inheritance." In *A History of Twelfth-Century Western Philosophy*, edited by Peter Dronke, 54–80. Cambridge: Cambridge University Press, 1988.

Grendler, Paul F. *The Universities of the Italian Renaissance*. Baltimore: Johns Hopkins University Press, 2002.

———. "The Universities of the Renaissance and Reformation." *Renaissance Quarterly* 57, no. 1 (2004): 1–12.

Griffiths, Richard. *The Bible in the Renaissance: Essays on Biblical Commentary and Translation in the Fifteenth and Sixteenth Centuries*. Aldershot: Ashgate, 2001.

Grotans, Anna A. *Reading in Medieval St. Gall*. Cambridge Studies in Palaeography and Codicology. Cambridge: Cambridge University Press, 2006.

Groten, Manfred. "Nikolaus von Kues: Vom Studenten zum Kardinal—Lebensweg und Lebenswelt eines Spätmittelalterlichen Intellektuellen." In *Nicholas of Cusa: A Medieval Thinker for the Modern Age*, edited by Kazuhiko Yamaki, 112–24. Richmond, Surrey: Curzon Press, 2002.

Guarino, Sergio. "La Formazione Veneziana di Jacopo De'Barbari." In *Giorgione e la cultura Venetra tra '400 e '500: Mito, allegoria, analisi iconologica*, edited by Maurizio Calvesi, 186–98. Roma: De Luca, 1981.

Guerlac, Henry. "Copernicus and Aristotle's Cosmos." *Journal of the History of Ideas* 29, no. 1 (1968): 109–13.

Günther, Siegmund. *Martin Behaim. Zeichnungen von Otto E. Lau*. Bamberg: Buchnersche Verlagsbuchhandlung, 1890.

———. *Varenius*. Amsterdam: Meridan, 1970.

Guthke, Karl. *The Last Frontier: Imagining Other Worlds, from the Copernican Revolution to Modern Science Fiction*. Translated by Helen Atkins. Ithaca, N.Y.: Cornell University Press, 1990.

Gutting, Gary. "Michel Foucault: A User's Manual." In *The Cambridge Companion to Foucault*, edited by Gary Gutting, 1–28. Cambridge Cambridge University Press, 1994.
Guttridge, George H., ed. *The Correspondence of Edmund Burke, III.* Cambridge: Cambridge University Press, 1961.
Hahm, David E. *The Origins of Stoic Cosmology.* Columbus: Ohio State University Press, 1977.
Hall, A. Rupert. *From Galileo to Newton.* New York: Dover Publications, 1981.
———. *Henry More and the Scientific Revolution.* Cambridge Science Biographies Series. Cambridge: Cambridge University Press, 1996.
———. *Isaac Newton, Adventurer in Thought.* Blackwell Science Biographies. Oxford: Blackwell, 1992.
———. *The Revolution in Science, 1500–1750.* 3rd ed. London: Longman, 1983.
———. *The Scientific Revolution, 1500–1800: The Formation of the Modern Scientific Attitude.* London: Longman, 1962.
Halley, Edmond. "To John Wallis, 25 February 1687." In *Electronic Enlightenment*, edited by Robert McNamee. Oxford: University of Oxford, 2011. http://www.e-enlightenment.com.
Hamann, Byron Ellsworth. "The Mirrors of Las Meninas: Cochineal, Silver, and Clay." *Art Bulletin* 92, no. 1–2 (2010): 6–35.
Hamann, Günther. "Der Behaim-Globus als Vorbild der Stabius Dürer Karte von 1515." *Der Globusfreund* 25–27 (1977–79): 135–47.
Hamel, Jürgen. *Nicolaus Copernicus: Leben, Werk und Wirkung.* Heidelberg: Spektrum, 1994.
Hamilton, James J. "Hobbes's Study and the Hardwick Library." *Journal of the History of Philosophy* 16, no. 4 (1978): 445–53.
Hampson, Norman. *The Enlightenment.* Harmondsworth: Penguin, 1968.
Hankins, James, ed. *The Cambridge Companion to Renaissance Philosophy.* Cambridge: Cambridge University Press, 2007.
———. "The Myth of the Platonic Academy of Florence." *Renaissance Quarterly* 44, no. 3 (1991): 429–75.
———. *Plato in the Italian Renaissance.* Columbia Studies in the Classical Tradition. 2 vols. Leiden: Brill, 1990.
———. "The Study of the *Timaeus* in Early Renaissance Italy." In *Natural Particulars: Nature and the Disciplines in Renaissance Europe*, edited by Anthony Grafton and Nancy G. Siraisi, 77–120. Dibner Institute Studies in the History of Science and Technology. Cambridge, Mass.: MIT Press, 1999.
Hankins, James, and Ada Palmer. *The Recovery of Ancient Philosophy in the Renaissance: A Brief Guide.* Quaderni di Rinascimento / Istituto Nazionale di Studi sul Rinascimento. Florence: L. S. Olschki, 2008.
Hankinson, R. J. "Philosophy of Science." In *The Cambridge Companion to Aristotle*, edited by Jonathan Barnes, 140–67. Cambridge: Cambridge University Press, 1995.
Hannam, James. *God's Philosophers: How the Medieval World Laid the Foundations of Modern Science.* London: Icon, 2010.
Harley, J. B. *The New Nature of Maps: Essays in the History of Cartography.* Baltimore: Johns Hopkins University Press, 2001.
Harris, Marvin. *The Rise of Anthropological Theory: A History of Theories of Culture.* London: Routledge & Kegan Paul, 1968.
Hartner, Willy. "The Astronomical Instruments of Cha-Ma-Lu-Ting, Their Identification, and Their Relations to the Instruments of the Observatory of Marāgha." *Isis* 41, no. 2 (1950): 184–94.
Harvey, David. *The Condition of Postmodernity: An Enquiry into the Origins of Cultural Change.* Oxford: Blackwell, 1989.
Harvey, P. D. A. *The Hereford World Map: Medieval World Maps and Their Context.* London: British Library, 2006.
Hayes, Kevin J. "How Thomas Jefferson Read the Qur'ān." *Early American Literature* 39, no. 2 (2004): 247–61.
Hazard, Paul. *La Crise de la conscience européene (1680–1715).* Paris: Boivin, 1935.
———. *La Pensée Européenne au XVIIIe siècle: De Montesquieu à Lessing.* Paris: Boivin, 1946.

Headley, John M. "The Universalizing Principle and Process: On the West's Intrinsic Commitment to a Global Context." *Journal of World History* 13, no. 2 (2002): 291–321.

Heath, Terrence. "Logical Grammar, Grammatical Logic, and Humanism in Three German Universities." *Studies in the Renaissance* 18 (1971): 9–64.

Heath, Thomas Little. *A History of Greek Mathematics*. 2 vols. New York: Dover Publications, 1981. Originally published, Oxford: Clarendon Press, 1921.

Heckscher, William S. "Melancholia (1541) an Essay in the Rhetoric of Description by Joachim Camerarius." In *Joachim Camerarius: (1500–1574); Beiträge zur Geschichte des Humanismus im Zeitalter der Reformation*, edited by Frank Baron, 32–120. Humanistische Bibliothek Reihe 1, Abhandlungen. München: Wilhelm Fink Verlag, 1978.

Hedeman, Anne D. "Gothic Manuscript Illustration: The Case of France." In *A Companion to Medieval Art: Romanesque and Gothic in Northern Europe*, edited by Conrad Rudolph, 421–42. Blackwell Companions to Art History. Malden, Mass.: Blackwell, 2006.

Hegel, Georg Wilhelm Friedrich. "Geometrical Studies." Translated by Alan L. Paterson. *Hegel Bulletin* 29, no. 1–2 (2008): 132–53.

———. *Lectures on the Philosophy of World History, Introduction: Reason in World History*. Translated by H. B. Nisbet. Cambridge: Cambridge University Press, 1975.

———. *Phenomenology of Spirit*. Translated by Arnold V. Miller and J. N. Findlay. Oxford: Oxford University Press, 1977.

Heilbron, John L. *Elements of Early Modern Physics*. Berkeley: University of California Press, 1981.

Hein, Wolfgang-Hagen, ed. *Alexander von Humboldt: Leben und Werk*. Frankfurt am Main: Weisbecker, 1985.

Helferich, Gerard. *Humboldt's Cosmos: Alexander von Humboldt and the Latin American Journey That Changed the Way We See the World*. New York: Gotham, 2004.

Herder, Johann Gottfried. *Herders Werke in Fünf Bänden*. Edited by Regine Otto. 5th ed. 5 vols. Vol. 4. Berlin: Hermann Duncker, 1978.

Heyd, Michael. *Be Sober and Reasonable: The Critique of Enthusiasm in the Seventeenth and Early Eighteenth Centuries*. Brill's Studies in Intellectual History. Leiden: Brill, 1995.

Hilbert, David. *Gesammelte Abhandlungen*. 3 vols. Vol. 3. Berlin: Springer, 1932.

———. *Vorlesungen über Elemente der Euklidischen Geometrie. Göttingen, W. S. 1898/99*. Göttingen, 1899.

Hindess, Barry. "Locke's State of Nature." *History of the Human Sciences* 20, no. 3 (2007): 1–20.

Hobbes, Thomas. *Leviathan*. The Penguin Classics. Harmondsworth: Penguin Books, 1985.

Hodgen, Margaret T. *Early Anthropology in the Sixteenth and Seventeenth Centuries*. Philadelphia: University of Pennsylvania Press, 1964.

Hoebel, E. A. "William Robertson: An 18th-Century Anthropologist-Historian." *American Anthropologist* 62 (1960): 648–55.

Hoffman, Piotr. *Freedom, Equality, Power: The Ontological Consequences of the Political Philosophies of Hobbes, Locke, and Rousseau*. Studies in European Thought. New York: Peter Lang, 1999.

Hofmann, Catherine. *Le Globe & son image*. Paris: Bibliothèque nationale de France, 1995.

Hofmann, Joseph E. "Dürers Verhältnis zur Mathematik." In *Albrecht Dürers Umwelt: Festschrift zum 500. Geburtstag Albrecht Dürers am 21. Mai 1971*, edited by Otto Herding, 132–51. Nürnberg: Selbstverl. des Vereins für Geschichte der Stadt Nürnberg, 1971.

Holden, Edward S., ed. *Essays in Astronomy*. New York: D. Appleton and Company, 1900.

Homann, Frederick A. "Christopher Clavius and the Renaissance of Euclidean Geometry." *Archivum Historicum Societatis Iesu Roma* 52, no. 4 (1983): 233–46.

Hoogvliet, Margriet. "The Medieval Texts of the 1486 Ptolemy Edition by Johann Reger of Ulm." *Imago Mundi* 54 (2002): 7–18.

Hoorn, Tanja van. *Dem Leibe Abgelesen: Georg Forster im Kontext der physischen Anthropologie des 18. Jahrhunderts*. Tübingen: M. Niemeyer, 2004.

Hopkins, Jasper, and Nicholas. *A Concise Introduction to the Philosophy of Nicholas of Cusa*. Minneapolis: University of Minnesota Press, 1978.

Horst, Thomas. "Der Niederschlag von Entdeckungsreisen auf Globen des frühen 16. Jahrhunderts." *Der Globusfreund: Wissenschaftliche Zeitschrift für Globenkunde* 55/56 (2009): 23–38.

Hoskin, Michael A., ed. *The Cambridge Concise History of Astronomy*. Cambridge: Cambridge University Press, 1999.

Howard, Dick. *The Primacy of the Political: A History of Political Thought from the Greeks to the French & American Revolutions*. Columbia Studies in Political Thought / Political History. New York: Columbia University Press, 2010.

Howse, Derek, ed. *The Greenwich List of Observatories: A World List of Astronomical Observatories, Instruments and Clocks, 1670–1850*. Chalfont St. Giles: Science History Publications, 1986.

Hudry, Françoise. *Le Livre des XXIV philosophes*. Collection Krisis. Grenoble: J. Millon, 1989.

Hudry, Françoise, and Marius Victorinus. *Le Livre des XXIV philosophes: Résurgence d'un texte du IV*[e] *siècle*. Histoiredes Doctrines de l'Antiquité Classique. Paris: J. Vrin, 2009.

Huff, Toby E. *The Rise of Early Modern Science: Islam, China, and the West*. Cambridge: Cambridge University Press, 1993.

Hull, Gordon. *Cosmos: A Sketch of the Physical Description of the Universe*. Translated by E. C. Otté. 5 vols. Vol. 1. Baltimore: Johns Hopkins University Press, 1997.

———. *Hobbes and the Making of Modern Political Thought*. Continuum Studies in British Philosophy. London: Continuum, 2009.

Husserl, Edmund. *Cartesian Meditations: An Introduction to Phenomenology*. The Hague: M. Nijhoff, 1960.

———. "Die Frage Nach dem Ursprung der Geometrie als intentional-historisches Problem." *Revue Internationale de Philosophie* 1, no. 2 (1939): 203–25.

———. *Die Krisis der europäischen Wissenschaften und die transzendentale Phänomenologie: Eine Einleitung in die phänomenologische Philosophie*. Husserliana. Den Haag: M. Nijhoff, 1954.

———. *Husserliana: Gesammelte Werke*. 42 vols. Vol. 6: Die Krisis der europäischen Wissenschaften und die transzendentale Phänomenologie. Den Haag: M. Nijhoff, 1950.

———. *Logische Untersuchungen*. 2., umgearb. aufl. ed. 2 vols. Halle an der Saale: Niemeyer, 1900.

———. *Méditations Cartésiennes: introduction á la phénoménologie*. Paris: A. Colin, 1931.

———. *Philosophie der Arithmetik. Psychologische und logische Untersuchungen*. Halle-Saale: C. E. M. Pfeffer, 1891.

———. "Über den Begriff der Zahl, Psychologische Analysen." Habilitationsschrift, University of Halle-Wittenberg, 1887.

Hutchison, Jane Campbell. *Albrecht Dürer: A Biography*. Princeton, N.J.: Princeton University Press, 1990.

———. *Albrecht Dürer: A Guide to Research*. Artist Resource Manuals vol. 3. New York: Garland, 2000.

Hutton, Sarah. "Émilie du Châtelet's 'Institutions de Physique' as a Document in the History of French Newtonianism." *Studies in the History of the Philosophy of Science* 35, no. 3 (2004): 515–31.

Ishihara, Aeka. "Goethe und die Astronomie seiner Zeit. Eine Astronomisch-Literarische Landschaft um Goethe." *Goethe-Jahrbuch*, no. 117 (2000): 103–17.

Israel, Jonathan I. *The Dutch Republic: Its Rise, Greatness, and Fall, 1477–1806*. Oxford History of Early Modern Europe. Oxford: Oxford University Press, 1995.

———. *Enlightenment Contested: Philosophy, Modernity, and the Emancipation of Man, 1670–1752*. Oxford: Oxford University Press, 2006.

———. *Radical Enlightenment: Philosophy and the Making of Modernity, 1650–1750*. Oxford: Oxford University Press, 2001.

Iversen, Margaret. "Retrieving Warburg's Tradition." *Art History* 16, no. 4 (1993): 541–53.

Jack, Malcolm. "One State of Nature: Mandeville and Rousseau." *Journal of the History of Ideas* 39, no. 1 (1978): 119–24.

Jacob, Margaret C. *The Cultural Meaning of the Scientific Revolution*. New Perspectives on European History. New York: A. A. Knopf, 1988.

———. "The Enlightenment Redefined: The Formation of Modern Civil Society." *Social Research* 58, no. 2 (1991): 475–95.

———. *The Newtonians and the English Revolution, 1689–1720.* Ithaca, N.Y.: Cornell University Press, 1976.

———. *The Radical Enlightenment: Pantheists, Freemasons, and Republicans.* London: Allen and Unwin, 1981.

———. *Strangers Nowhere in the World: The Rise of Cosmopolitanism in Early Modern Europe.* Philadelphia: University of Pennsylvania Press, 2006.

Jaki, Stanley L. "Goethe and the Physicists." *American Journal of Physics* 37, no. 2 (1969): 195–203.

———. *Planets and Planetarians: A History of Theories of the Origin of Planetary Systems.* Edinburgh: Scottish Academic Press, 1978.

Jammer, Max. *Concepts of Space: The History of Theories of Space in Physics.* Cambridge, Mass.: Harvard University Press, 1954.

Jardine, Nicholas. "The Places of Astronomy in Early-Modern Culture." *Journal for the History of Astronomy* 29 (1998): 49–62.

Jesseph, Douglas Michael. *Squaring the Circle: The War Between Hobbes and Wallis.* Science and Its Conceptual Foundations. Chicago: University of Chicago Press, 1999.

Johnson, Christine R. *The German Discovery of the World: Renaissance Encounters with the Strange and Marvelous.* Studies in Early Modern German History. Charlottesville: University of Virginia Press, 2008.

Johnson, Paul. *The Birth of the Modern: World Society, 1815–1830.* New York: HarperCollins, 1991.

Jones, Alexander. "Ptolemy's Mathematical Models and Their Meaning." In *Mathematics and the Historian's Craft: The Kenneth O. May Lectures,* edited by Glen van Brummelen and Michael Kinyon, 23–42. CMS Books in Mathematics. New York: Springer, 2005.

———. "The Stoics and the Astronomical Sciences." In *The Cambridge Companion to the Stoics,* edited by Brad Inwood, 328–44. Cambridge: Cambridge University Press, 2003.

Jones, Roger, and Nicholas Penny. *Raphael.* New Haven, Conn.: Yale University Press, 1983.

Joost-Gaugier, Christiane L. *Italian Renaissance Art: Understanding Its Meaning.* Chichester: Wiley-Blackwell, 2013.

———. "Ptolemy and Strabo and Their Conversation with Appelles and Protogenes: Cosmography and Painting in Raphael's School of Athens." *Renaissance Quarterly* 51, no. 3 (1998): 761–87.

———. *Raphael's Stanza della Segnatura: Meaning and Invention.* Cambridge: Cambridge University Press, 2002.

Jullien, Vincent. *Philosophie naturelle et géométrie au XVIIe siècle.* Sciences, Techniques et Civilisations du Moyen Age à l'Aube des Lumières. Paris: Honoré Champion Éditeur, 2006.

Jürgens, Klaus H. "Neue Forschungen zu dem Münchener Selbstbildnis des Jahres 1500 von Albrecht Dürer." *Kunsthistorisches Jahrbuch Graz,* no. 19–20; 21 (1983–84; 1985): 167–90; 34–64.

Jurgensen, Ray C. *Modern Geometry: Structure and Method.* Boston: Houghton Mifflin, 1963.

Jurgensen, Ray C., Richard G. Brown, and John W. Jurgensen. *Geometry.* Evanston, Ill.: McDougal Littell, 2000.

Jurgensen, Ray C., Richard G. Brown, and Alice M. King. *Geometry.* Teacher's ed. Boston: Houghton Mifflin Harcourt, 1982.

Kahn, Joel. "Culture: Demise or Resurrection?" *Critique of Anthropology* 9, no. 2 (1989): 5–25.

Kandler, Karl-Hermann. *Nikolaus von Kues: Denker zwischen Mittelalter und Neuzeit.* Göttingen: Vandenhoeck and Ruprecht, 1995.

Kant, Immanuel. "Von dem ersten Grunde des Unterschiedes der Gegenden im Raume." In *Immanuel Kants Gesammelte Werke: Bd. II: Vorkritische Schriften II: 1757–1777,* 375–84. Berlin: G. Reimer und Deutsche Akademie der Wissenschaften zu Berlin, 1905.

Kant, Immanuel. *Critique of Pure Reason.* The Cambridge Edition of the Works of Immanuel Kant. Cambridge: Cambridge University Press, 1998.

———. *Immanuel Kant Werkausgabe.* Edited by Wilhelm Weischedel. 12 vols. Vol. 7. Frankfurt am Main: Suhrkamp Verlag, 1974.

———. *Immanuel Kant Werkausgabe*. Edited by Wilhelm Weischedel. 12 vols. Vol. 1. Frankfurt am Main: Suhrkamp Verlag, 1974.

———. "Träume eines Geistersehers, erläutert durch Träume der Metaphysik." In *Immanuel Kants Gesammelte Werke: Bd. II: Vorkritische Schriften II: 1757–1777*, 315–73. Berlin: G. Reimer und Deutsche Akademie der Wissenschaften zu Berlin, 1905.

Kargon, Robert H. *Atomism in England from Hariot to Newton*. Oxford: Clarendon Press, 1966.

Kemp, Martin. *The Science of Art: Optical Themes in Western Art from Brunelleschi to Seurat*. New Haven, Conn.: Yale University Press, 1990.

Kennedy, Roger. "Jefferson and the Indians." *Wintherthur Portfolio* 27, no. 2/3 (1992): 105–21.

Kern, Stephen. *The Culture of Time and Space, 1880–1918*. Cambridge, Mass.: Harvard University Press, 1983.

Kleinberg, Ethan. *Generation Existential: Heidegger's Philosophy in France, 1927–1961*. Ithaca, N.Y.: Cornell University Press, 2005.

Kleinknecht, Thomas. "'Reise Der Aufklärung': Selbstverortung, Empirie und epistemologischer Diskurs bei Herder, Lessing, Lichtenberg und Anderen." *Berichte zur Wissenschaftsgeschichte* 22 (1999): 95–111.

Klibansky, Raymond, Erwin Panofsky, and Fritz Saxl. *Saturn and Melancholy: Studies in the History of Natural Philosophy, Religion, and Art*. London: Nelson, 1964.

———. *Saturn und Melancholie: Studien zur Geschichte der Naturphilosophie und Medizin, der Religion und der Kunst*. Translated by Christa Buschendorf. Frankfurt am Main: Suhrkamp, 1990.

Klima, Gyula. *John Buridan*. Great Medieval Thinkers. Oxford: Oxford University Press, 2009.

Kline, Morris. *Mathematical Thought from Ancient to Modern Times*. 3 vols. Vol. 3. New York: Oxford University Press, 1990.

———. *Mathematics in Western Culture*. New York: Oxford University Press, 1953.

Kline, Naomi Reed. *Maps of Medieval Thought: The Hereford Paradigm*. Woodbridge: Boydell Press, 2001.

Klingenberg, Wilhelm. *Mathematik und Melancholie: Von Albrecht Dürer bis Robert Musil*. Abhandlungen der Mathematisch-Naturwissenschaftlichen Klasse / Akademie der Wissenschaften und der Literatur. Stuttgart: Franz Steiner Verlag, 1997.

Kluckhohn, Clyde. *Anthropology and the Classics*. Providence: Brown University Press, 1961.

Knight, Isabel F. *The Geometric Spirit: The Abbé de Condillac and the French Enlightenment*. Yale Historical Publications. New Haven, Conn.: Yale University Press, 1968.

Knorr, Wilbur R. *The Ancient Tradition of Geometric Problems*. Boston: Birkhäuser, 1986.

Koelb, Clayton, ed. *Nietzsche as Postmodernist: Essays Pro and Contra*. SUNY Series in Contemporary Continental Philosophy. Albany: State University of New York Press, 1990.

Koerner, Joseph L. *The Moment of Self-Portraiture in German Renaissance Art*. Chicago: University of Chicago Press, 1993.

Kohler, Alfred "Die Entwicklung der Darstellung Afrikas auf deutschen Globen des 15. und 16. Jahrhunderts." *Der Globusfreund: Wissenschaftliche Zeitschrift für Globenkunde* 18/20 (1970): 85–96.

Konstan, David. "Problems in Epicurean Physics." *Isis* (1979): 394–418.

Kooiman, Willem Jan. *Luther and the Bible*. Philadelphia: Muhlenberg Press, 1961.

Koyré, Alexandre. *From the Closed World to the Infinite Universe*. New York: Harper and Brothers, 1957.

———. *Galileo Studies*. European Philosophy and the Human Sciences. Atlantic Highlands, N.J.: Humanities Press, 1978.

———. *Newtonian Studies*. Cambridge, Mass.: Harvard University Press, 1965.

Krauss, Werner. *Zur Anthropologie des 18. Jahrhunderts: Die Frühgeschichte der Menschheit im Blickpunkt der Aufklärung*. Berlin: Akademie-Verlag, 1978.

Kraye, Jill. "The Revival of Hellenistic Philosophies." In *The Cambridge Companion to Renaissance Philosophy*, edited by James Hankins, 97–112. Cambridge: Cambridge University Press, 2007.

Kreuzer, Johann. *Gestalten mittelalterlicher Philosophie: Augustinus, Eriugena, Eckhart, Tauler, Nikolaus v. Kues*. München: Fink, 2000.

Krogt, P. C. J. van der. *Old Globes in the Netherlands: A Catalogue of Terrestrial and Celestial Globes Made Prior to 1850 and Preserved in Dutch Collections.* Utrecht: HES, 1984.

———. "Gerard Mercator and His Cosmography: How the *Atlas* Became an Atlas." *Archives Internationales d'Histoire des Sciences* 59, no. 163 (2009): 465–83.

Kuehn, Manfred. *Kant: A Biography.* New York: Cambridge University Press, 2001.

Kundert, Werner. "Hermann Conring als Professor der Universität Helmstedt." In *Hermann Conring (1606–1681): Beiträge zu Leben und Werk,* edited by Michael Stolleis, 399–412. Berlin: Duncker and Humblot, 1983.

Kusukawa, Sachiko. *The Transformation of Natural Philosophy: The Case of Philip Melanchthon.* Ideas in Context. Cambridge: Cambridge University Press, 1995.

Lachterman, David Rapport. *The Ethics of Geometry: A Genealogy of Modernity.* New York: Routledge, 1989.

Laird, W. R. "Archimedes Among the Humanists." *Isis* 82, no. 4 (1991): 628–38.

Landauer, Carl. "Erwin Panofsky and the Renascence of the Renaissance." *Renaissance Quarterly* 47, no. 2 (1994): 255–81.

Lattis, James M. *Between Copernicus and Galileo: Christoph Clavius and the Collapse of Ptolemaic Cosmology.* Chicago: University of Chicago Press, 1995.

Laugwitz, Detlef. *Bernhard Riemann, 1826–1866: Turning Points in the Conception of Mathematics.* Boston: Birkhäuser, 1999.

Leaf, Murray. *Man, Mind, and Science: A History of Anthropology.* New York: Columbia University Press, 1979.

Leff, Gordon. *The Dissolution of the Medieval Outlook: An Essay on Intellectual and Spiritual Change in the Fourteenth Century.* New York: Harper and Row, 1976.

———. *Medieval Thought: St. Augustine to Ockham.* Harmondsworth: Penguin Books, 1958.

———. *William of Ockham: The Metamorphosis of Scholastic Discourse.* Manchester: Manchester University Press, 1975.

Lehmann, Klaus. "Der Bildungsweg des jungen Bernhard Varenius." In *Bernhard Varenius (1622–1650),* edited by Margret Schuchard, 59–90. Brill's Studies in Intellectual History. Leiden: Brill, 2007.

Leibniz, Gottfried Wilhelm. *Writings on China.* Translated by Daniel J. Cook and Henry Rosemont. Chicago: Open Court, 1994.

Leinkauf, Thomas, and Carlos G. Steel. *Platons Timaios als Grundtext der Kosmologie in Spätantike, Mittelalter und Renaissance = Plato's Timaeus and the Foundations of Cosmology in Late Antiquity, the Middle Ages and Renaissance.* Ancient and Medieval Philosophy Series 1. Louvain: Louvain University Press, 2005.

León Portilla, Miguel. *Bernardino de Sahagún, First Anthropologist.* Translated by Mauricio J. Mixco. Norman: University of Oklahoma Press, 2002.

Lestringant, Frank. *L'Atelier du cosmographe: Ou l'image du monde à la Renaissance.* Bibliothèque De Synthèse. Paris: A. Michel, 1991.

Levao, Ronald. *Renaissance Minds and Their Fictions: Cusanus, Sidney, Shakespeare.* Berkeley: University of California Press, 1985.

Levi, Anthony. *Renaissance and Reformation: The Intellectual Genesis.* New Haven, Conn.: Yale University Press, 2002.

Lévi-Strauss, Claude. *Anthropologie structurale.* Paris: Plon, 1958.

Levinas, Emmanuel. *En Découvrant l'existence avec Husserl et Heidegger.* Librairie Philosophique. Paris: J. Vrin, 1949.

———. *La Théorie de l'intuition dans la phénoménologie de Husserl.* Paris: Félix Alcan, 1930.

Liebersohn, Harry. "Anthropology Before Anthropology." In *A New History of Anthropology,* edited by Henrika Kucklick, 17–32. Malden, Mass.: Blackwell Publishing, 2008.

———. *Aristocratic Encounters: European Travelers and North American Indians.* Cambridge: Cambridge University Press, 1998.

———. *The Travelers' World: Europe to the Pacific.* Cambridge, Mass.: Harvard University Press, 2006.

Lindberg, David C. *Theories of Vision from Al-Kindi to Kepler.* University of Chicago History of Science and Medicine. Chicago: University of Chicago Press, 1976.

Lindberg, David C., and Robert S. Westman. *Reappraisals of the Scientific Revolution.* Cambridge: Cambridge University Press, 1990.

Linden, Mareta. *Untersuchungen zum Anthropologie Begriff des 18. Jahrhunderts.* Bern: Herbert Lang/ Peter Lang, 1976.

Little, Becky. "NASA's 'Blue Marbles': Pictures of Earth from 1972 to Today: To Celebrate NASA's Newest Photo of Earth, Here's a Look at Some of the Ones That Came Before It." *National Geographic,* July 21, 2015, https://news.nationalgeographic.com/2015/07/150721-pictures-earth-nasa-dscovr-spacex-space-science/.

Locke, John. "Instructions for the Education of Edward Clarke's Children, C. 8 February 1686." In *Electronic Enlightenment,* edited by Robert McNamee. Oxford: Oxford University Press, 2011. http://www.e-enlightenment.com.

Long, A. A. *Hellenistic Philosophy: Stoics, Epicureans, Sceptics.* Berkeley: University of California Press, 1986.

Lorch, Richard. "Pseudo-Euclid on the Position of the Image in Reflection: Interpretations by an Anonymous Commentator, by Pena, and by Kepler." In *The Light of Nature: Essays in the History and Philosophy of Science Presented to A. C. Crombie,* edited by John David North and John J. Roche. Archives Internationales d'Histoire des Idées, 135–44. Dordrecht: M. Nijhoff, 1985.

Lowden, John. *The Making of the Bibles Moralisées.* 2 vols. Vol. 1. University Park: Pennsylvania State University Press, 2000.

Lüthy, Christoph. "Where Logical Necessity Becomes Visual Persuasion: Descartes's Clear and Distinct Illustrations." In *Transmitting Knowledge: Words, Images, and Instruments in Early Modern Europe,* edited by Sachiko Kusukawa and Ian Maclean, 97–133. Oxford-Warburg Studies. Oxford: Oxford University Press, 2006.

Lüthy, Christoph. "Centre, Circle, Circumference: Giordano Bruno's Astronomical Woodcuts." *Journal for the History of Astronomy* 41 (2010): 311–27.

Lynch, Terence. "The Geometric Body in Dürer's Engraving Melencolia I." *Journal of the Warburg and Courtauld Institutes* 45 (1982): 226–32.

Lyotard, Jean François, Robert Harvey, and Mark S. Roberts, eds. *Toward the Postmodern.* Philosophy and Literary Theory. Atlantic Highlands, N.J.: Humanities Press, 1993.

Lyotard, Jean-François. *La Condition Postmoderne: Rapport sur le savoir.* Collection Critique. Paris: Éditions de Minuit, 1979.

———. *The Postmodern Condition: A Report on Knowledge.* Translated by Geoff Bennington and Brian Massumi. Theory and History of Literature. Minneapolis: University of Minnesota Press, 1984.

Mackinnon, Nick. "The Portrait of Fra Luca Pacioli." *Mathematical Gazette* 77, no. 479 (1993): 130–219.

Malet, Antoni. "Isaac Barrow on the Mathematization of Nature: Theological Voluntarism and the Rise of Geometrical Optics." *Journal of the History of Ideas* 58, no. 2 (1997): 265–87.

———. "Renaissance Notions of Number and Magnitude." *Historia Mathematica* 33 (2006): 63–81.

Malet, Antoni, and Cozzoli Daniele. "Mersenne and Mixed Mathematics." *Perspectives on Science* 18, no. 1 (2010): 1–8.

Malherbe, Michel, and Jean-Marie Pousseur. *Francis Bacon, Science et Méthode: Actes du Colloque du Nantes.* De Pétrarque à Descartes. Paris: J. Vrin, 1985.

Maravall, José Antonio. *Velázquez y el espíritu de la modernidad.* 2nd ed. Madrid: Alianza Editorial, 1987.

Marenbon, John. *Medieval Philosophy: An Historical and Philosophical Introduction.* Routledge History of Philosophy. London: Routledge, 1998.

Marks, Jonathan. "Who Lost Nature? Rousseau and Rousseauism." *Polity* 34, no. 4 (2002): 479–502.

Marquard, Odo. "Zur Geschichte des philosophischen Begriffs 'Anthropologie' seit dem Ende des 18. Jahrhunderts." In *Collegium Philosophicum: Studien Joachim Ritter zum 60. Geburtstag,* edited by Ernst-Wolfgang Böckenförde, 209–39. Basel: Schwabe and Co., 1965.

Marrone, Steven P. "Henry of Ghent and Duns Scotus on the Knowledge of Being." *Speculum* 63, no. 1 (1988): 22–57.

———. *Truth and Scientific Knowledge in the Thought of Henry of Ghent.* Speculum Anniversary Monographs. Cambridge: Medieval Academy of America, 1985.

Martin, Craig. *Renaissance Meteorology: Pomponazzi to Descartes.* Baltimore: Johns Hopkins University Press, 2011.

Mastnak, Tomasz. *Hobbes's Behemoth: Religion and Democracy.* Exeter: Imprint Academic, 2009.

May, J. A. "The Geographical Interpretation of Ptolemy in the Renaissance." *Tijdschrift voor Economische en Sociale Geografie* 73, no. 6 (1982): 350–61.

Mazzotti, Massimo. "Newton for Ladies: Gentility, Gender and Radical Culture." *British Journal for the History of Science* 32, no. 2 (2004): 119–46.

McCluskey, Stephen C. *Astronomies and Cultures in Early Medieval Europe.* Cambridge: Cambridge University Press, 1998.

McCrory, Donald. *Nature's Interpreter: The Life and Times of Alexander von Humboldt.* Cambridge: Luttworth Press, 2010.

McCrossen, Alexis. *Marking Modern Times: A History of Clocks, Watches, and Other Timekeepers in American Life.* Chicago: University of Chicago Press, 2013.

McLean, Matthew. *The "Cosmographia" of Sebastian Münster: Describing the World in the Reformation.* St. Andrews Studies in Reformation History. Aldershot: Ashgate, 2007.

McMullin, Ernan. "Bruno and Copernicus." *Isis* 78, no. 1 (1987): 55–74.

Megill, Allan. "Foucault, Structuralism, and the Ends of History." *Journal of Modern History* (1979): 451–503.

———. *Prophets of Extremity: Nietzsche, Heidegger, Foucault, Derrida.* Berkeley: University of California Press, 1985.

———. "The Recepion of Foucault by Historians." *Journal of the History of Ideas* 48, no. 1 (1987): 117–41.

Meinel, Christoph. "Early Seventeenth-Century Atomism: Theory, Epistemology, and the Insufficiency of Experiment." *Isis* (1988): 68–103.

———. *In Physicis futurum saeculum respicio: Joachim Jungius und die naturwissenschaftliche Revolution des 17. Jahrhunderts.* Veröffentlichung der Joachim Jungius-Gesellschaft der Wissenschaften Hamburg. Göttingen: Vandenhoeck and Ruprecht, 1984.

Mendham, Matthew D. "Gentle Savages and Fierce Citizens against Civilization: Unraveling Rousseau's Paradoxes." *American Journal of Political Science* 55, no. 1 (2011): 170–87.

Mett, Rudolf. *Regiomontanus: Wegbereiter des neuen Weltbildes.* Stuttgart: B. G. Teubner, 1996.

Michel, Paul-Henri. *The Cosmology of Giordano Bruno.* Ithaca, N.Y.: Cornell University Press, 1973.

Miert, Dirk van. *Humanism in an Age of Science: The Amsterdam Athenaeum in the Golden Age, 1632–1704.* Brill's Studies in Intellectual History. Leiden: Brill, 2009.

Mignolo, Walter D. *The Darker Side of the Renaissance: Literacy, Territoriality, and Colonization.* Ann Arbor: University of Michigan Press, 1995.

———. *The Darker Side of Western Modernity: Global Futures, Decolonial Options.* Latin America Otherwise: Languages, Empires, Nations. Durham, N.C.: Duke University Press, 2011.

———. "José de Acosta's Historia Natural y Moral de las Indias: Occidentalism, the Modern/Colonial World, and the Colonial Difference." Translated by Frances López Morillas. In *Natural and Moral History of the Indies,* edited by Jane E. Mangan. Durham, N.C.: Duke University Press, 2002.

Miller, P. N. "Citizenship and Culture in Early Modern Europe." *Journal of the History of Ideas* 57, no. 4 (1996): 725–42.

Misner, Charles W. "Cosmology and Theology." In *Cosmology, History, and Theology,* edited by Wolfgang Yourgrau, Allen duPont Breck, and Hannes Alfvén, 75–100. New York: Plenum Press, 1977.

Mitrović, Branko. "Leon Battista Alberti and the Homogeneity of Space." *Journal of the Society of Architectural Historians* (2004): 424–39.

Moffitt, John F. "Velázquez in the Alcázar Palace in 1656: The Meaning of the Mise-En-Scène of Las Meninas." *Art History* 6, no. 3 (1983): 271–300.

Molland, George A. "The Geometrical Background to the 'Merton School.'" *British Journal for the History of Science* 4, no. 2 (1968): 108–25.

———. *Mathematics and the Medieval Ancestry of Physics*. Vol. 481. Aldershot: Variorum Publishing, 1995.

Monfasani, John. *Byzantine Scholars in Renaissance Italy: Cardinal Bessarion and Other Émigrés: Selected Essays*. Aldershot: Variorum, 1995.

Moore, Jerry D. *Visions of Culture: An Introduction to Anthropological Theories and Theorists*. Walnut Creek, Calif.: AltaMira Press, 1997.

Moran, Dermot. "Nicholas of Cusa and Modern Philosophy." In *The Cambridge Companion to Renaissance Philosophy*, edited by James Hankins, 173–92. Cambridge: Cambridge University Press, 2007.

Moravia, Sergio. *Beobachtende Vernunft: Philosophie und Anthropologie in der Aufklärung*. Translated by Elisabeth Piras. Frankfurt am Main: Fischer Taschenbuch Verlag, 1989.

Moretto, Antonio. "Hegel on Greek Mathematics and the Modern Calculus." In *Hegel and Newtonianism*, edited by M. J. Petry. Archives Internationales d'Histoire des Idées / International Archives of the History of Ideas, 149–65. Dordrecht: Springer, 1993.

Morgan, Augustus de. *The Globes, Celestial and Terrestrial*. London: Malby and Co., 1845.

Morris, Christopher W. *The Social Contract Theorists: Critical Essays on Hobbes, Locke, and Rousseau*. Critical Essays on the Classics. Lanham, Md.: Rowman and Littlefield, 1999.

Morrow, John. *A History of Political Thought: A Thematic Introduction*. New York: New York University Press, 1998.

Mosley, Adam. *Bearing the Heavens: Tycho Brahe and the Astronomical Community of the Late Sixteenth Century*. Cambridge: Cambridge University Press, 2007.

———. "The Cosmographer's Role in the Sixteenth Century: A Preliminary Study." *Archives Internationales d'Histoire des Sciences* 59, no. 163 (2009): 423–39.

Mueller, Ian. "Aristotle on Geometrical Objects." *Archiv für Geschichte der Philosophie* 52 (1970): 156–71.

Mühlmann, Wilhelm E. *Geschichte der Anthropologie*. 2nd ed. Frankfurt am Main: Athenäum Verlag, 1968.

Müller, Klaus E. *Geschichte der antiken Ethnologie*. Rowohlts Enzyklopädie. Reinbek bei Hamburg: Rowohlt Taschenbuch Verlag, 1997.

Murdoch, John. "Euclid: Transmission of the Elements." In *Dictionary of Scientific Biography*, edited by Charles Coulston Gillispie, 437–59. New York: Charles Scribner's Sons, 1971.

———. "The Medieval Euclid: Salient Aspects of the Translations of the *Elements* by Adelard of Bath and Campanus of Novara." *Revue de Synthèse* 89, no. 2 (1968): 67–94.

Muris, Oswald. "Der Globus des Martin Behaim." *Mitteilungen der Geographischen Gessellschaft Wien* 97 (1955): 169–82.

Muris, Oswald, and Gert Saarmann. *Der Globus im Wandel der Zeiten: Eine Geschichte der Globen*. Berlin: Columbus Verlag, 1961.

Murphy, Trevor. *Pliny the Elder's Natural History: The Empire in the Encyclopedia*. Oxford: Oxford University Press, 2004.

Nauert, Charles G. "The Clash of Humanists and Scholastics: An Approach to Pre-Reformation Controversies." *Sixteenth Century Journal* 4, no. 1 (1973): 1–18.

Navarro Brotons, Victor. "The Reception of Copernicus in Sixteenth-Century Spain: The Case of Diego de Zúñiga." *Isis* 86, no. 1 (1995): 52–78.

Neal, Katherine. *From Discrete to Continuous: The Broadening of Number Concepts in Early Modern England*. Dordrecht: Kluwer, 2002.

———. "Mathematics and Empire, Navigation and Exploration: Henry Briggs and the Northwest Passage Voyages of 1631." *Isis* 93, no. 3 (2002): 435–53.

New American Standard Bible. Carol Stream, Ill.: Creation House, 1971.

Newman, William R., and Anthony Grafton. *Secrets of Nature: Astrology and Alchemy in Early Modern Europe*. Transformations: Studies in the History of Science and Technology. Cambridge, Mass.: MIT Press, 2001.

Nietzsche, Friedrich Wilhelm. *The Dawn of Day.* Translated by J. M. Kennedy. Dover Philosophical Classics. Mineola, N.Y.: Dover Publications, 2007.

———. *The Gay Science, with a Prelude in Rhymes and an Appendix of Songs.* Translated by Walter A. Kaufmann. New York: Random House, 1974.

Nisbet, Hugh Barr. *Herder and the Philosophy and History of Science.* Cambridge, Mass.: Modern Humanities Research Association, 1970.

———. *Herder and Scientific Thought.* Cambridge, Mass.: Modern Humanities Research Association, 1970.

North, John D. *Cosmos: An Illustrated History of Astronomy and Cosmology.* Chicago: University of Chicago Press, 2008.

North, John David, and John J. Roche, eds. *The Light of Nature: Essays in the History and Philosophy of Science Presented to A.C. Crombie.* Archives Internationales d'Histoire des Idées, vol. 110. Dordrecht: M. Nijhoff, 1985.

Nyhus, Paul L. "The Franciscans in South Germany, 1400–1530: Reform and Revolution." *Transactions of the American Philosophical Society* (1975): 1–47.

Oakeshott, Michael. *Lectures in the History of Political Thought.* Selected Writings / Michael Oakeshott. Exeter: Imprint Academic, 2006.

Oberman, Heiko Augustinus. *The Harvest of Medieval Theology: Gabriel Biel and Late Medieval Nominalism.* The Robert Troup Paine Prize Treatise. Cambridge, Mass.: Harvard University Press, 1963.

———. *Spätscholastik und Reformation.* Zürich: EVZ-Verlag, 1965.

Obhof, Ute. "Der Erdglobus, der Amerika benannte: Die Überlieferung der Globensegmente von Martin Waldseemüller." *Der Globusfreund: Wissenschaftliche Zeitschrift für Globenkunde* 55/56 (2009): 13–22.

Occam, William of. *Quodlibetal Questions.* 2 vols. New Haven, Conn.: Yale University Press, 1991.

Oestmann, Günther. "Johannes Stoefflers Himmelsglobus." *Der Globusfreund: Wissenschaftliche Zeitschrift für Globenkunde* 43/44 (1995): 71–74.

Oestreich, Gerhard. *Antiker Geist und moderner Staat bei Justus Lipsius (1547–1606): Der Neustoizismus als Politische Bewegung.* Schriftenreihe der Historischen Kommission bei der Bayerischen Akademie der Wissenschaften. Göttingen: Vandenhoeck and Ruprecht, 1989.

———. *Neostoicism and the Early Modern State.* Edited by Brigitta Oestreich and H. G. Koenigsberger. Cambridge: Cambridge University Press, 1982.

Osler, Margaret J., ed. *Rethinking the Scientific Revolution.* Cambridge: Cambridge University Press, 2000.

Overfield, James H. *Humanism and Scholasticism in Late Medieval Germany.* Princeton, N.J.: Princeton University Press, 1984.

Ozment, Steven E. *The Age of Reform, 1250–1550: An Intellectual and Religious History of Late Medieval and Reformation Europe.* New Haven, Conn.: Yale University Press, 1980.

Pacioli, Luca. *De viribus quantitatis. Facsimile ad uso professionale.* Sansepolcro: Aboca Edizioni, 2009.

Paganini, Gianni. "How Did Hobbes Think of the Existence and Nature of God?: *De Motu, Loco et Tempore* as a Turning Point in Hobbes's Philosophical Career." In *The Bloomsbury Companion to Hobbes,* edited by S. A. Lloyd, 285–303. London: Bloomsbury, 2013.

Pagden, Anthony. *European Encounters with the New World: From Renaissance to Romanticism.* New Haven, Conn.: Yale University Press, 1993.

———., ed. *Facing Each Other: The World's Perception of Europe and Europe's Perception of the World.* 2 vols. Vol. 31. An Expanding World. Aldershot: Ashgate/Variorum, 2000.

———. *The Fall of Natural Man: The American Indian and the Origins of Comparative Ethnology.* Cambridge: Cambridge University Press, 1986.

———. *Lords of All the World: Ideologies of Empire in Spain, Britain and France, c. 1500–c. 1800.* New Haven, Conn.: Yale University Press, 1995.

———. *Peoples and Empires: A Short History of European Migration, Exploration, and Conquest, from Greece to the Present.* Modern Library Chronicles. Vol. 6. New York: Modern Library, 2003.

———. *Spanish Imperialism and the Political Imagination: Studies in European and Spanish-American Social and Political Theory, 1513–1830*. New Haven, Conn.: Yale University Press, 1990.

———. "Stoicism, Cosmopolitanism, and the Legacy of European Imperialism." *Constellations* 7, no. 1 (2000): 3–22.

———. *The Uncertainties of Empire: Essays in Iberian and Ibero-American Intellectual History*. Aldershot: Ashgate/Variorum, 1994.

Pagels, Elaine H. *Adam, Eve, and the Serpent*. New York: Random House, 1988.

Palladini, Fiammetta. "Pufendorf, Disciple of Hobbes: The Nature of Man and the State of Nature: The Doctrine of *Socialitas*." *History of European Ideas* 34 (2008): 26–60.

Pannenberg, Wolfhart. *Anthropologie in theologischer Perspektive*. Göttingen: Vandenhoeck and Ruprecht, 1983.

Panofsky, Erwin. *Albrecht Dürer*. 3rd ed. London: Oxford University Press, 1948.

———. "Artist, Scientist, Genius: Notes on the 'Renaissance-Dämmerung.'" In *The Renaissance: Six Essays*, edited by Wallace K. Ferguson, Robert S. Lopez, George Sarton, Roland H. Bainton, Leicester Bradner, and Erwin Panofsky, 123–82. New York: Harper and Row, 1962.

Panofsky, Erwin, and Fritz Saxl. *Dürers 'Melencolia I': Eine quellen- und typengeschichtliche Untersuchung*. Leipzig: B. G. Teubner, 1923.

Pantin, Isabelle. "Kepler's *Epitome*: New Images for an Innovative Book." In *Transmitting Knowledge: Words, Images, and Instruments in Early Modern Europe*, edited by Sachiko Kusukawa and Ian Maclean, 217–37. Oxford-Warburg Studies. Oxford: Oxford University Press, 2006.

Pascal, Blaise. *Pensées*. Translated by Roger Ariew. Indianapolis: Hackett Publishing Company, 2005.

Paterson, Alan L. "GWF Hegel: Geometrical Studies Introduction." *Hegel Bulletin* 29, no. 1–2 (2008): 118–31.

Patrides, C. A. *The Cambridge Platonists*. The Stratford-upon-Avon Library. London: Edward Arnold, 1969.

Paulus, Jean. *Henri de Gand: essai sur les tendances de sa métaphysique*. Études de Philosophie Médiévale. Paris: J. Vrin, 1938.

Paviot, Jacques. "Ung Mapmonde Rond, en Guise de Pom(M)E: Ein Erdglobus von 1440–44, Hergestellet für Philipp den Guten, Herzog von Burgund." *Der Globusfreund: Wissenschaftliche Zeitschrift für Globenkunde* 43/44 (1995): 19–29.

Peck, Linda Levy. "Hobbes on the Grand Tour: Paris, Venice, or London?" *Journal of the History of Ideas* 57, no. 1 (1996): 177–83.

Pedersen, Olaf. "Astronomy." In *Science in the Middle Ages*, edited by David C. Lindberg, 303–37. Chicago: University of Chicago Press, 1978.

———. *Early Physics and Astronomy: A Historical Introduction*. Rev. ed. Cambridge: Cambridge University Press, 1993.

———. "In Quest of Sacrobosco." *Journal for the History of Astronomy* 16, no. 3 (1985): 175–220.

Pedersen, Olaf, and Alexander Jones. *A Survey of the Almagest: With Annotation and New Commentary by Alexander Jones*. Sources and Studies in the History of Mathematics and Physical Sciences. New York: Springer, 2011.

Pérez Sánchez, Alfonso E. "The Artistic Milieu in Seville During the First Third of the Seventeenth Century." In *Zurbarán*, edited by Jeannine Baticle, 37–52. New York: Metropolitan Museum of Art, 1987.

Perry, Marvin. *An Intellectual History of Modern Europe*. Boston: Houghton Mifflin, 1993.

Pesic, Peter. "Secrets, Symbols, and Systems: Parallels Between Cryptanalysis and Algebra, 1580–1700." *Isis* (1997): 674–92.

Peterson, Mark A. *Galileo's Muse: Renaissance Mathematics and the Arts*. Cambridge, Mass.: Harvard University Press, 2011.

Pfotenhauer, Helmut. *Literarische Anthropologie: Selbstbiographien und ihre Geschichte, am Leitfaden des Leibes*. Stuttgart: Metzler, 1987.

Pickles, William. "The Notion of Time in Rousseau's Political Thought." In *Hobbes and Rousseau: A Collection of Critical Essays*, edited by Maurice William Cranston and R. S. Peters, 366–400. Modern Studies in Philosophy. Garden City, N.Y.: Anchor Books, 1972.

Pinkard, Terry P. *Hegel: A Biography*. Cambridge: Cambridge University Press, 2000.

Pino, Fermín del. "Contribución del Padre Acosta a la constitución de la etnología y su evolucionismo." *Revista de Indias* (1978): 507–43.

———. "Culturas clásicas y americanas en la obra del Padre Acosta." In *America y la España del siglo XVI*, edited by Francisco de Solano and Fermín del Pino, 327–62. Madrid: C.S.I.C., 1982.

———. "Edición de crónicas de Indias e historia intellectual, o la distancia entre José de Acosta y José Alcina (1)." *Revista de Indias* L, no. 190 (1990): 861–78.

———. "La Historia Indiana del P. Acosta y su ponderación 'científica' del Perú." In *Dos Mundos, dos culturas, o de la Historia (Natural y Moral) entre España y el Perú*, edited by Fermín del Pino Díaz, 23–38. Textos y Estudios Coloniales y de la Independencia. Frankfurt am Main: Vervuert; Iberoamericana, 2004.

———. "Los Reinos de Méjico y Cuzco en la obra del P. Acosta." In *Economia y Sociedad en Los Andes y Mesoamerica*, edited by José Alcina Franch, 13–43. Madrid: Universidad Complutense de Madrid, 1979.

Plato. *The Collected Dialogues of Plato, Including the Letters*. Edited by Edith Hamilton and Huntington Cairns. Bollingen Series. Princeton, N.J.: Princeton University Press, 1961.

Plessner, Helmuth. "Der Mensch als Lebewesen." In *Philosophische Anthropologie*, edited by Werner Schüssler, 71–83. München: Verlag Karl Alber Freiburg, 2000.

Pocock, J. G. A. "Enthusiasm: The Antiself of Enlightenment." *Huntington Library Quarterly* 60, no. 1–2 (1999): 7–28.

———. *Virtue, Commerce, and History: Essays on Political Thought and History, Chiefly in the Eighteenth Century*. Cambridge: Cambridge University Press, 1985.

Polaschek, Erich. "Ptolemy's 'Geography' in a New Light." *Imago Mundi* 14 (1959): 17–37.

Pomata, Gianna, and Nancy G. Siraisi. "Introduction." In *Historia: Empiricism and Erudition in Early Modern Europe*, edited by Gianna Pomata and Nancy G. Siraisi, 1–38. Transformations: Studies in the History of Science and Technology. Cambridge, Mass.: MIT Press, 2005.

Pope, Alexander. *Essay on Man and Other Poems*. New York: Dover Publications, 1994.

Popkin, Richard H. *The History of Scepticism: From Savonarola to Bayle*. Rev. and expanded ed. Oxford: Oxford University Press, 2003.

Porter, Roy. *The Creation of the Modern World: The Untold Story of the British Enlightenment*. New York: W. W. Norton and Company, 2000.

Pratt, Mary Louise. *Imperial Eyes: Travel Writing and Transculturation*. London: Routledge, 1992.

Proclus. *A Commentary on the First Book of Euclid's Elements*. Translated by Glenn R. Morrow. Princeton, N.J.: Princeton University Press, 1970.

Pross, Wolfgang. "Herder und die Anthropologie der Aufklärung." In *Johann Gottfried Herder: Werke*, edited by Wolfgang Pross, 1128–1216. Munich: Carl Hanser Verlag, 1987.

———. *Ptolemy's Almagest*. Translated by G. J. Toomer. Princeton, N.J.: Princeton University Press, 1998.

Puig-Samper Mulero, Miguel Ángel. "El viajero científico: La visión de Humboldt sobre Nueva España." *Paraíso occidental: Norma y Diversidad en el México Virreinal* (1998): 197–211.

Purinton, Jeffrey S. "Magnifying Epicurean Minima." *Ancient Philosophy* 14, no. 1 (1994): 115–46.

Raeff, Marc. *The Well-Ordered Police State: Social and Institutional Change Through Law in the Germanies and Russia, 1600–1800*. New Haven, Conn.: Yale University Press, 1983.

———. "The Well-Ordered Police State and the Development of Modernity in Seventeenth- and Eighteenth-Century Europe: An Attempt at a Comparative Approach." *American Historical Review* 80, no. 5 (1975): 1221–43.

Ragep, F. Jamil. "Copernicus and His Islamic Predecessors: Some Historical Remarks." *History of Science* 45, no. 1 (2007): 65–81.

———. "Tusi and Copernicus: The Earth's Motion in Context." *Science in Context* 14, no. 1/2 (2001): 145–63.

Randles, W. G. L. *De La Terre Plate au globe terrestre: Une mutation épistémologique rapide (1480–1520)*. Cahiers des Annales. Paris: Librairie Armand Colin, 1980.

———. *The Unmaking of the Medieval Christian Cosmos, 1500–1700: From Solid Heavens to Boundless Aether.* Aldershot: Ashgate, 1999.

Ravenstein, Ernest George. *Martin Behaim: His Life and His Globe.* London: George Philip and Son, 1908.

Rebel, Ernst. *Albrecht Dürer: Maler und Humanist.* München: Bertelsmann, 1996.

Reichenbach, Andreas. *Martin Behaim. Ein Deutscher Seefahrer aus dem Fünfzehnten Jahrhundert.* C. Kiesler's Jugend-Bibliothek. Wurzen: C. Kiesler, 1889.

Reid, Constance. *Hilbert.* New York: Springer-Verlag, 1996.

Reid, Thomas. "To James Gregory, 1786." In *Electronic Enlightenment,* edited by Robert McNamee. Oxford: University of Oxford, 2011. http://www.e-enlightenment.com.

Riemann, Bernhard. *Ueber die Hypothesen, welche der Geometrie zu Grunde liegen.* Abhandlungen der Königlichen Gesellschaft der Wissenschaften zu Göttingen. Vol. 13. Göttingen, 1867.

Riley, Patrick. *Will and Political Legitimacy: A Critical Exposition of Social Contract Theory in Hobbes, Locke, Rousseau, Kant, and Hegel.* Cambridge, Mass.: Harvard University Press, 1982.

Rist, John M. *Epicurus: An Introduction.* Cambridge: CUP Archive, 1972.

Ritter, Gerhard. *Studien zur Spätscholastik.* Sitzungsberichte der Heidelberger Akademie der Wissenschaften Philosophisch-Historische Klasse. 3 vols. Heidelberg: C. Winter, 1921.Roger, Jacques. *Buffon: A Life in Natural History.* Translated by Sara Lucille Bonnefoi. Cornell History of Science Series. Ithaca, N.Y.: Cornell University Press, 1997.

———. *The Life Sciences in Eighteenth-Century French Thought.* Translated by Robert Ellrich. Stanford, Calif.: Stanford University Press, 1997.

Rogers, G. A. J., Jean-Michel Vienne, and Yves Charles Zarka. *The Cambridge Platonists in Philosophical Context: Politics, Metaphysics, and Religion.* Archives Internationales d'Histoire des Idées. Dordrecht: Kluwer Academic Publishers, 1997.

Rogers, G. A. J. "Locke, Newton, and the Cambridge Platonists on Innate Ideas." *Journal of the History of Ideas* (1979): 191–205.

Röll, Johannes. "'Das Problem ist das vom Nachleben der Antike': Fritz Saxl 1890–1948." *Pegasus: Berliner Beiträge zum Nachleben der Antike* 1 (1999): 27–34.

Rommevaux, Sabine. *Clavius, un clé pour Euclide au XVIe siècle.* Paris: J. Vrin, 2005.

Rose, Paul L. "Humanist Culture and Renaissance Mathematics: The Italian Libraries of the Quattrocento." *Studies in the Renaissance* 20 (1973): 46–105.

———. *The Italian Renaissance of Mathematics: Studies on Humanists and Mathematicians from Petrarch to Galileo.* Travaux d'Humanisme et Renaissance. Genève: Droz, 1975.

Rossi, Paolo. *Francis Bacon: From Magic to Science.* London: Routledge and Kegan Paul, 1968.

Roth, Michael S. "Foucault's 'History of the Present.'" *History and Theory* (1981): 32–46.

Rothstein, Bret. "Making Trouble: Strange Wooden Objects and the Early Modern Pursuit of Difficulty." *Journal for Early Modern Cultural Studies* 13, no. 1 (2013): 96–129.

Rottman, Gerald. *The Geometry of Light: Galileo's Telescope, Kepler's Optics.* Baltimore: Gerald Rottman, 2008.

Rousseau, Jean-Jacques. *Oeuvres complètes.* L'Intégrale. 3 vols. Vol. 1. Paris: Éditions du Seuil, 1967.

Rowe, John Howland. "The Renaissance Foundations of Anthropology." *American Anthropologist* 67, no. 1 (1965): 1–20.

Rubiés, Joan-Pau. "Theology, Ethnography, and the Historicization of Idolatry." *Journal of the History of Ideas* 67, no. 4 (2006): 571–96.

Rummel, Erika. "Et cum theologo bella poeta gerit: The Conflict Between Humanists and Scholastics Revisited." *Sixteenth Century Journal* 23, no. 4 (1992): 713–26.

Runciman, David. *Political Hypocrisy: The Mask of Power, from Hobbes to Orwell and Beyond.* Princeton, N.J.: Princeton University Press, 2008.

Rupke, Nicolaas A. *Göttingen and the Development of the Natural Sciences.* Göttingen: Wallstein, 2002.

Rupprich, Hans. *Schriftlicher Nachlass.* Vol. 2. Berlin: Deutscher Verein für Kunstwissenschaft, 1956.

Ruysschaert, José. "Du Globe terrestre attribué à Giulio Romano aux globes et au planisphère oubliés de Nicolaus Germanus." *Bolletino dei monumenti, musei e gallerie pontifice* 6 (1985): 93–104.

Sacksteder, William. "Hobbes: The Art of the Geometricians." *Journal of the History of Philosophy* 18, no. 2 (1980): 131–46.
———. "Hobbes: Geometrical Objects." *Philosophy of Science* 48, no. 4 (1981): 573–90.
Sahlins, Marshall D. "The Apotheosis of Captain Cook." *Kroeber Anthropological Society Papers* 53/54 (1978): 1–31.
———. "Captain Cook at Hawaii." *Journal of the Polynesian Society* 98 (1989): 371–423.
———. *Culture and Practical Reason*. Chicago: University of Chicago Press, 1976.
Said, Edward W. *Orientalism*. New York: Vintage Books, 1979.
Saitta, Giuseppe. *Nicolò Cusano e l'umanesimo Italiano: Con altri saggi sul Rinascimento Italiano*. Bologna: Tamari, 1957.
Saliba, George. "Arabic Versus Greek Astronomy: A Debate over the Foundations of Science." *Perspectives on Science* 8, no. 4 (2000): 328–41.
———. "The First Non-Ptolemaic Astronomy at the Maraghah School." *Isis* 70, no. 4 (1979): 571–76.
———. *Islamic Science and the Making of the European Renaissance*. Transformations. Cambridge, Mass.: MIT Press, 2007.
Sambursky, Shmuel. *Physics of the Stoics*. London: Routledge and Kegan Paul, 1959.
Sarasohn, Lisa T. "Thomas Hobbes and the Duke of Newcastle: A Study in the Mutuality of Patronage Before the Establishment of the Royal Society." *Isis* (1999): 715–37.
Sartre, Jean-Paul. *Nausea*. A New Directions Paperbook. Norfolk: New Directions, 1964.
Sauter, Michael J. "Clock Watchers and Stargazers: Time Discipline in Early-Modern Berlin." *American Historical Review* 112, no. 3 (2007): 685–709.
Savage-Smith, Emilie. "Celestial Mapping." In *The History of Cartography: Cartography in Prehistoric, Ancient and Medieval Europe and the Mediterranean*, edited by J. B. Harley and David Woodward, 12–70. Chicago: University of Chicago Press, 1987.
Savage-Smith, Emilie, and Andrea P. A. Belloli. *Islamicate Celestial Globes, Their History, Construction, and Use*. Washington, D.C.: Smithsonian Institution Press, 1985.
Sbacchi, Michele. "Euclidism and Theory of Architecture." *Nexus Network Journal* 3, no. 2 (2001): 25–38.
Schaffer, Simon. "Instruments, Surveys and Maritime Empire." In *Empire, the Sea and Global History; Britain's Maritime World, c. 1760–c. 1840*, edited by David Cannadine, 83–104. Houndmills: Palgrave Macmillan, 2007.
Schechner, Sara J. *Comets, Popular Culture and the Birth of Modern Cosmology*. Princeton, N.J.: Princeton University Press, 1997.
———. "The Material Culture of Astronomy in Daily Life: Sundials, Science, and Social Change." *Journal for the History of Astronomy* 32 (2001): 189–222.
Scheler, Max. *Die Stellung des Menschen im Kosmos*. Darmstadt: O. Reichl, 1928.
———. "Die Stellung des Menschen im Kosmos." In *Philosophische Anthropologie*, edited by Werner Schüssler, 49–69. Alber-Texte Philosophie. München: Verlag Karl Alber Freiburg, 2000.
Schings, Hans-Jürgen. *Der ganze Mensch: Anthropologie und Literatur im 18. Jahrhundert: DFG-Symposion 1992*. Stuttgart: Metzler, 1994.
Schlachter, Alois, and Friedrich Gisinger. *Der Globus, seine Entstehung und Verwendung in der Antike nach den literarischen Quellen und den Darstellungen in der Kunst*. Leipzig: B. G. Teubner, 1927.
Schleiner, Winfried. *Melancholy, Genius, and Utopia in the Renaissance*. Wolfenbütteler Abhandlungen zur Renaissanceforschung. Wiesbaden: Harrassowitz, 1991.
Schmitt, Carl. *Der Leviathan in der Staatslehre des Thomas Hobbes: Sinn und Fehlschlag eines Politischen Symbols*. Hamburg: Hanseatische Verlagsanstalt, 1938.
Schmitt, Charles B. *Aristotle and the Renaissance*. Martin Classical Lectures. Cambridge: Published for Oberlin College by Harvard University Press, 1983.
Schmitt, Charles B., Quentin Skinner, and Eckhard Kessler, eds. *The Cambridge History of Renaissance Philosophy*. Cambridge: Cambridge University Press, 1988.

Schröder, Eberhard. *Dürer, Kunst und Geometrie: Dürers künstlerisches Schaffen aus der Sicht seiner "Underweysung."* Wissenschaft und Kultur. Basel: Birkhäuser Verlag, 1980.

Schuchard, Margret, ed. *Bernhard Varenius (1622–1650).* Brill's Studies in Intellectual History. Leiden: Brill, 2007.

Schuster, Peter-Klaus. *Melencolia I: Dürers Denkbild.* 2 vols. Berlin: Gebr. Mann Verlag, 1991.

Scott, John T. "The Theodicy of the Second Discourse: The 'Pure State of Nature' and Rousseau's Political Thought." *American Political Science Review* 86, no. 3 (1992): 696–711.

Sedley, David. "Philoponus' Conception of Space." In *Philoponus and the Rejection of Aristotelian Science,* edited by Richard Sorabji, 181–93. London: Duckworth, 1987.

Shakespeare, William. *The Complete Works.* The Oxford Shakespeare. Compact ed. Oxford: Clarendon Press 1988.

Shalev, Zur, and Charles Burnett, eds. *Ptolemy's Geography in the Renaissance.* Warburg Institute Colloquia, vol. 17. London: Warburg Institute, 2011.

Shapin, Steven. *The Scientific Revolution.* Chicago: University of Chicago Press, 1996.

Sheehan, Jonathan. "Enlightenment, Religion, and the Enigma of Secularization: A Review Essay." *American Historical Review* 108, no. 4 (2003): 1061–80.

Sheldon, Garrett Ward. *The History of Political Theory: Ancient Greece to Modern America.* American University Studies Series 10, Political Science. New York: Peter Lang, 1988.

Shirali, Shailesh A. "Marin Mersenne, 1588–1648." *Resonance* 18, no. 3 (2013): 226–40.

Simek, Rudolf. *Heaven and Earth in the Middle Ages: The Physical World Before Columbus.* Translated by Angela Hall. Woodbridge: Boydell Press, 1996.

Simon, Julia. "Natural Man and the Lessons of History: Rousseau's Chronotypes." *Clio* 26 (1997): 473–84.

Siorvanes, Lucas. *Proclus: Neo-Platonic Philosophy and Science.* Edinburgh: Edinburgh University Press, 1996.

Skinner, Quentin. *Visions of Politics.* 3 vols. Vol. 3. Cambridge: Cambridge University Press, 2002.

Smith, Adam. "To John Petty, 1st Earl of Shelburne, 4 April 1759." In *Electronic Enlightenment,* edited by Robert McNamee. Oxford: Oxford University Press, 2011. http://www.e-enlightenment.com.

Smith, David Eugene. "John Wallis as a Cryptographer." *Bulletin of the American Mathematical Society* 24, no. 2 (1917): 82–96.

Smith, Gregory B. *Nietzsche, Heidegger, and the Transition to Postmodernity.* Chicago: University of Chicago Press, 1996.

Solinas, Giovanni. "Newton and Buffon." *Vistas in Astronomy* 22, no. 4 (1979): 431–39.

Sorell, Tom. "Hobbes and Aristotle." In *Philosophy in the Sixteenth and Seventeenth Centuries: Conversations with Aristotle,* edited by Constance Blackwell and Sachiko Kusukawa, 364–79. Aldershot: Ashgate, 1999.

———. "Seventeenth-Century Materialism: Gassendi and Hobbes." In *The Renaissance and Seventeenth-Century Rationalism,* edited by G. H. R. Parkinson, 219–52. Routledge History of Philosophy. London: Routledge, 2003.

Spitz, Lewis William. *The Northern Renaissance.* Sources of Civilization in the West. Englewood Cliffs: Prentice-Hall, 1972.

———. *The Religious Renaissance of the German Humanists.* Cambridge, Mass.: Harvard University Press, 1963.

———. *The Renaissance and Reformation Movements.* Rand McNally History Series. Chicago: Rand McNally, 1971.

Springborg, Patricia. "The Enlightenment of Thomas Hobbes." *British Journal for the History of Philosophy* 12, no. 3 (2004): 513–34.

Stark, Werner. "Historical and Philological References on the Question of a Possible Hierarchy of Human 'Races,' 'Peoples,' or 'Populations' in Immanuel Kant—A Supplement." In *Reading Kant's Geography,* edited by Stuart Elden and Eduardo Mendieta, 87–102. SUNY Series in Contemporary Continental Philosophy. Albany: State University of New York Press, 2011.

Starobinski, Jean. *Histoire du traitement de la mélancolie des origines à 1900*. Acta Psychosomatica. Basel: Geigy, 1960.

Stephens, John N. *The Italian Renaissance: The Origins of Intellectual and Artistic Change Before the Reformation*. London: Longman, 1990.

Stevens, Henry N. *Ptolemy's Geography. A Brief Account of All the Printed Editions Down to 1730*. London: Henry Stevens, Son and Stiles, 1908.

Stevenson, Edward Luther. *Terrestrial and Celestial Globes: Their History and Construction, Including a Consideration of Their Value as Aids in the Study of Geography and Astronomy*. New Haven, Conn.: Published for the Hispanic Society of America by the Yale University Press, 1921.

Strauss, Leo. *The Political Philosophy of Hobbes, Its Basis and Its Genesis*. Chicago: University of Chicago Press, 1961.

Strauss, Leo, and Joseph Cropsey. *History of Political Philosophy*. 3rd ed. Chicago: University of Chicago Press, 1987.

Stromberg, Roland N. *European Intellectual History Since 1789*. 6th ed. Englewood Cliffs, N.J.: Prentice Hall, 1994.

Struik, Dirk J. *A Concise History of Mathematics*. 4th rev. ed. New York: Dover Publications, 1987.

Swerdlow, Noel M. "Essay Review: Ptolemy's Geography, an Annotated Translation of the Theoretical Chapters by J. Lennart Berggren and Alexander Jones." *Annals of Science* 60, no. 3 (2010): 313–20.

Sykes, James. "Der Erdglobus in Raphaels 'Die Schule Von Athen.'" *Der Globusfreund: Wissenschaftliche Zeitschrift für Globenkunde* 55/56 (2009): 53–73.

Sylla, Edith. "The Oxford Calculators." In *The Cambridge History of Later Medieval Philosophy from the Rediscovery of Aristotle to the Disintegration of Scholasticism, 1100–1600*, edited by Norman Kretzmann, Anthony Kenny and Jan Pinborg, 540–63. Cambridge: Cambridge University Press, 1982.

———. "The Oxford Calculators in Context." *Science in Context* 1, no. 2 (2008): 257–79.

Tabarroni, Giorgio. "Globi celesti e terrestri sulle monete romane." *Physis: rivista di storia della scienza* 8, no. 3 (1965): 318–53.

Tannery, Paul. *La Géométrie au XIe siècle*. Paris: Imprimé pour l'Auteur, 1897.

———. "La Géometrie Imaginaire et la notion d'espace." *Revue Philosophique de la France et de l'Étranger* 2 (1876): 433–51.

———."La Géometrie Imaginaire et la notion d'espace." *Revue Philosophique de la France et de l'Étranger* 3 (1877): 553–75

Teleki, József. "To Jean-Jacques Rousseau, 26 February 1778." In *Electronic Enlightenment*, edited by Robert McNamee. Oxford: Oxford University Press, 2011. http://www.e-enlightenment.com.

Terrall, Mary. *The Man Who Flattened the Earth: Maupertuis and the Sciences in the Enlightenment*. Chicago: University of Chicago Press, 2002.

Thomas, Keith. *Religion and the Decline of Magic: Studies in Popular Beliefs in Sixteenth and Seventeenth Century England*. 2nd ed. London: Weidenfeld and Nicolson, 1971.

Thorndike, Lynn. *The Sphere of Sacrobosco and Its Commentators*. Chicago: University of Chicago Press, 1949.

Tibbetts, Gerald R. "The Beginnings of a Cartographic Tradition." In *The History of Cartography: Cartography in Prehistoric, Ancient and Medieval Europe and the Mediterranean*, edited by J. B. Harley and David Woodward, 90–107. Chicago: University of Chicago Press, 1987.

Tinguely, Frédéric. "Le Vertige cosmographique à la Renaissance." *Archives Internationales d'Histoire des Sciences* 59, no. 163 (2009): 441–50.

Toews, John E. *Hegelianism: The Path Toward Dialectical Humanism, 1805–1841*. Cambridge: Cambridge University Press, 1980.

Toksvig, Signe. *Emmanuel Swedenborg: Scientist and Mystic*. New Haven, Conn.: Yale University Press, 1948.

Toomer, G. J. "Ptolemy." In *Dictionary of Scientific Biography*, edited by Charles C. Gillispie and Frederic L. Holmes, 186–206. New York: Scribner, 1981.

Trinkaus, Charles Edward. *In Our Image and Likeness: Humanity and Divinity in Italian Humanist Thought*. 2 vols. Notre Dame: University of Notre Dame Press, 1995.

Trismegistus, Hermes. *Liber viginti quattuor philosophorum*. Corpus Christianorum Continuatio Mediaevalis. Turnhout, Belgium: Brepols, 1997.

Tuck, Richard. *Philosophy and Government, 1572–1651*. Ideas in Context. Cambridge: Cambridge University Press, 1993.

Tully, James. *An Approach to Political Philosophy: Locke in Contexts*. Ideas in Context. Cambridge: Cambridge University Press, 1993.

Turner, Alice K. *The History of Hell*. New York: Harcourt Brace, 1993.

Vailati, Ezio. *Leibniz & Clarke: A Study of Their Correspondence*. New York: Oxford University Press, 1997.

Vanderjagt, Arie Johan, A. A. MacDonald, Z. R. W. M. von Martels, and Jan R. Veenstra, eds. *Christian Humanism: Essays in Honour of Arjo Vanderjagt*. Leiden: Brill, 2009.

Vansteenberghe, Edmond. *Le Cardinal Nicolás de Cues (1401–1464) l'action—la pensée*. Bibliothèque du XVᵉ siècle. Paris: H. Champion, 1920.

Véliz, Zahira. "Francisco Pacheco's Comments on Painting in Oil." *Studies in Conservation* 27, no. 2 (1982): 49–57.

Vermeulen, Han F. "Enlightenment Anthropology." In *Encyclopedia of Social and Cultural Anthropology*, edited by Alan Barnard and Jonathan Spencer, 279–82. London: Routledge, 1996.

Veselovsky, I. N. "Copernicus and Nasīr Al-Dīn Al-Tusī." *Journal for the History of Astronomy* 4, no. 2 (1973): 128–30.

Victor, Stephen K. *Practical Geometry in the High Middle Ages: Artis Cuiuslibet Consummatio and the Pratike de Geometrie*. Philadelphia: American Philosophical Society, 1979.

Vlastos, Gregory. "Minimal Parts in Epicurean Atomism." *Isis* (1965): 121–47.

———. "Plato's Supposed Theory of Irregular Atomic Figures." *Isis* (1967): 204–9.

Vyverberg, Henry. *Human Nature, Cultural Diversity, and the French Enlightenment*. New York: Oxford University Press, 1989.

Wagner, David L. "The Seven Liberal Arts and Classical Scholarship." In *The Seven Liberal Arts in the Middle Ages*, edited by David L. Wagner, 1–31. Bloomington: Indiana University Press, 1983.

———, ed. *The Seven Liberal Arts in the Middle Ages*. Bloomington: Indiana University Press, 1983.

Waldron, Jeremy. "John Locke: Social Contract versus Political Anthropology." *Review of Politics* 51, no. 1 (1989): 3–28.

Walker, Mack. *German Home Towns: Community, State, and General Estate, 1648–1871*. Ithaca, N.Y.: Cornell University Press, 1971.

Wallis, H. M. "The Molyneux Globes." *British Museum Quarterly* 16, no. 4 (1952): 89–90.

Walls, Laura Dassow. *The Passage to Cosmos: Alexander von Humboldt and the Shaping of America*. Chicago: University of Chicago Press, 2009.

Warburg, Aby Moritz. *Die Erneuerung der heidnischen Antike: Kulturwissenschaftliche Beiträge zur Geschichte der europäischen Renaissance; mit einem Anhang unveröffentlichter Zusätze*. Gesammelte Schriften. Leipzig: Teubner, 1932.

Warntz, William. "Newton, the Newtonians, and the Geographia Generalis Varenii." *Annals of the Association of American Geographers* 79, no. 2 (1989): 165–91.

Waszink, J. H. *Studien zum Timaioskommentar des Calcidius*. Philosophia Antiqua. Leiden: Brill, 1964.

Watts, Pauline Moffitt. "The European Religious Worldview and Its Influence on Mapping." In *The History of Cartography*, edited by J. B. Harley and David Woodward, 382–400. Chicago: University of Chicago Press, 1987.

———. *Nicolaus Cusanus, a Fifteenth-Century Vision of Man*. Studies in the History of Christian Thought. Leiden: Brill, 1982.

Weber, Max. *The Protestant Ethic and the Spirit of Capitalism*. Translated by Talcott Parsons. London: Routledge, 1992.

Weber, Paul. *Beiträge zu Dürers Weltanschauung: Eine Studie über die drei Stiche 'Ritter, Tod und Teufel', 'Melancholie', und 'Hieronymus im Gehäus'*. Studien Zur Deutschen Kunstgeschichte. Nendeln: Kraus, 1979.

Wehler, Hans Ulrich. *Die Herausforderung der Kulturgeschichte*. Beck'sche Reihe. Originalausg. ed. München: C. H. Beck, 1998.

Weidhaas, Peter. *A History of the Frankfurt Book Fair*. Translated by Carolyn Gossage and W. A. Wright. Toronto: Dundurn Press, 2007.

Weitzel, Hans. "Zum Polyeder auf A. Dürers Stich Melencolia I—Ein Nürnberger Skizzenblatt mit Darstellungen Archimedischer Körper." *Sudhoffs Archiv* 91, no. 2 (2007): 129–73.

Wessel, Günther. *Von Einem, der Daheim blieb, die Welt zu entdecken: Die Cosmographia des Sebastian Münster, oder, wie man sich vor 500 Jahren die Welt vorstellte*. Frankfurt am Main: Campus, 2004.

West, Hugh. "The Limits of Enlightenment Anthropology: Georg Forster and the Tahitians." *History of European Ideas* 10, no. 2 (1989): 147–60.

Westfall, Richard S. *Never at Rest: A Biography of Isaac Newton*. Cambridge: Cambridge University Press, 1983.

Westman, Robert S. *The Copernican Achievement*. Contributions of the UCLA Center for Medieval and Renaissance Studies. Vol. 7. Berkeley: University of California Press, 1975.

Wey Gómez, Nicolás. *The Tropics of Empire: Why Columbus Sailed South to the Indies*. Transformations. Cambridge, Mass.: MIT Press, 2008.

Wilson, Catherine. *Epicureanism at the Origins of Modernity*. Oxford: Oxford University Press, 2008.

Wilson, Gordon. "Henry of Ghent's 'Quodlibet I': Initial Departures from Thomas Aquinas." *History of Philosophy Quarterly* 16, no. 2 (1999): 167–80.

Windschuttle, Keith. "Foucault as Historian." *Critical Review of International Social and Political Philosophy* 1, no. 2 (1998): 5–35.

———. *The Killing of History: How Literary Critics and Social Theorists Are Murdering Our Past*. New York: Free Press, 1997.

Winkler, Friedrich. *Albrecht Dürer: Leben und Werk*. Berlin: Mann, 1957.

Wittkower, Rudolf. "Marvels of the East: A Study in the History of Monsters." *Journal of the Warburg and Courtauld Institutes* 5 (1942): 159–97.

Wolff, Larry. "Discovering Cultural Perspective: The Intellectual History of Anthropological Thought in the Age of the Enlightenment." In *The Anthropology of the Enlightenment*, edited by Larry Wolff and Marco Cipolloni, 3–32. Stanford, Calif.: Stanford University Press, 2007.

———. *Inventing Eastern Europe: The Map of Civilization on the Mind of the Enlightenment*. Stanford, Calif.: Stanford University Press, 1994.

Wolff, Larry, and Marco Cipolloni. *The Anthropology of the Enlightenment*. Stanford, Calif.: Stanford University Press, 2007.

Wölfflin, Heinrich. *Die Kunst Albrecht Dürers*. Pantheon-Colleg. 9., durchges. Aufl. ed. München: Bruckmann, 1984.

Wolfschmidt, Gudrun. *Nicolaus Copernicus (1473–1543): Revolutionär wider Willen*. Stuttgart: Verlag für Geschichte der Naturwissenschaften und der Technik, 1994.

Woodward, David. "The Image of the Spherical Earth." *Perspecta* 25 (1989): 3–15.

———. "Medieval *Mappaemundi*." In *The History of Cartography: Cartography in Prehistoric, Ancient and Medieval Europe and the Mediterranean*, edited by J. B. Harley and David Woodward, 286–370. Chicago: University of Chicago Press, 1987.

———. "Reality, Symbolism, Time, and Space in Medieval World Maps." *Annals of the Association of American Geographers* 75, no. 4 (1985): 510–21.

———. "Roger Bacon's Terrestrial Coordinate System." *Annals of the Association of American Geographers* 80, no. 1 (1990): 109–22.

Worth, Roland H. *Bible Translations: A History through Source Documents*. Jefferson, N.C.: McFarland and Company, 1992.

Yamaki, Kazuhiko, ed. *Nicholas of Cusa: A Medieval Thinker for the Modern Age*. Waseda/Curzon International Series. Richmond, Surrey: Curzon Press, 2002.

Yates, Frances A. *The Occult Philosophy in the Elizabethan Age*. London: Routledge and Kegan Paul, 1979.

Yolton, John W. *Locke and the Compass of Human Understanding: A Selective Commentary on the Essay.* Cambridge: Cambridge University Press, 1970.

Zagorin, Perez, ed. *Culture and Politics from Puritanism to the Enlightenment.* Berkeley: University of California Press, 1980.

Zagorin, Perez. *Francis Bacon.* Princeton, N.J.: Princeton University Press, 1998.

Zaitsev, Evgeny A. "The Meaning of Early Medieval Geometry: From Euclid and Surveyors' Manuals to Christian Philosophy." *Isis* (1999): 522–53.

Zambelli, Paola. "Scholastic and Humanist Views of Hermeticism and Witchcraft." In *Hermeticism and the Renaissance: Intellectual History and the Occult in Early Modern Europe,* edited by Ingrid Merkel and Allen George Debus. Folger Institute Symposia, 125–53. Washington, D.C.: Folger Books, 1988.

Zammito, John. "Policing Polygeneticism in Germany, 1775: (Kames,) Kant, and Blumenbach." In *The German Invention of Race,* edited by Sara Eigen and Mark Larrimore, 35–54. Albany: State University of New York Press, 2006.

Zammito, John H. *Kant, Herder, and the Birth of Anthropology.* Chicago: University of Chicago Press, 2002.

Zinner, Ernst. *Leben und Wirken des Joh. Müller von Königsberg genannt Regiomontanus.* Milliaria, 10,1. Edited by Helmut Rosenfeld and Otto Zeller. 2nd ed. Osnabrück: Otto Zeller, 1968.

Zuccato, Marco. "Gerbert of Aurillac and a Tenth-Century Jewish Channel for the Transmission of Arabic Science to the West." *Speculum* 80, no. 3 (2005): 742–63.

Index

Acknowledgments

An intellectual journey as long and complicated as this one should not be condensed into a simple list of thanks. Nevertheless, I must tread that reductive path here. Many people have either read parts of this book or listened patiently while I prated on about it. I would like to begin by thanking my colleagues in the División de Historia at the Centro de Investigación y Docencia Económicas (CIDE), including Catherine Andrews, Luis Barrón, Michael Bess, Clara García, Soledad Jiménez, José Juan López-Portillo, Luis Medina, Jean Meyer, Pablo Mijangos, David Miklos, Emma Nakatani, Camila Pastor, Andrew Paxman, Rafael Rojas, Catherine Vézina, and Marco Zuccato for their support of my efforts, even when these made no sense whatsoever. In addition, I would also like to recognize the quiet labors of my numerous research assistants, especially Alejandro Cheirif, whose remarkable research skills proved a great help in the writing's later stages.

Beyond CIDE, I would also like to note the support of mentors, friends, and colleagues who offered critiques, suggestions, or plain encouragement when I most needed the help. The list of names is long and includes Peter Reill, David Sabean, Hans-Erich Bödeker, John Zammito, Ian Hunter, Chris Laursen, Jason Coy, Jared Poley, Benjamin Marschke, Peter K. J. Park and Kimberly Garmoe. Other professionals also gave generously of their time, including Adam Mosley and Nicholas Jardine, who read early drafts of one chapter. Finally, as I look back on the personal history that brought me to this book's completion, I am aware of the array of unfailingly excellent teachers who have shaped my education. Included in the honor roll is one gentle librarian at Bertram F. Gibbs School in New Milford, New Jersey, who explained patiently to a fidgety boy what a terrestrial globe was. In doing so, she gave to this project a beginning.

As is always the case in academic research, many institutions have supported this research project generously. My home institution, CIDE, is at the top of this list, and I would like to express my thanks for the many ways it has scrounged up funds for me to waste. I would also like to thank the Deutscher Akademischer Austauschdienst (DAAD), which has awarded me a number of research grants; the Herzog-August-Bibliothek in Wolfenbüttel, whose support foisted me on their staff for three months in 2010; the Center for Seventeenth- and Eighteenth-Century Studies at the University of California, whose grant gave me access in 2009 to the collections at the Clark Library and the Charles E. Young Research Library; and the Max-Planck-Institut-für-Geschichte in Göttingen, which has hosted me more times than I can remember. Finally, I would like to thank the Zentrum für Geschichte der Naturwissenschaft und Technik at the University of Hamburg (and particularly its director, Gudrun Wolfschmidt) for hosting me while I was on sabbatical leave in 2013. Many of this text's wilder ideas were developed in the center's book-laden offices over pots of strong coffee and plates of delightful pastries.

Other institutions have aided my project in different but no less essential ways. Some parts of Chapter 3 and Chapter 4 have appeared as book chapters elsewhere. I would like to thank Tredition and Berghahn Books for permission to reuse those materials here. I would also like to thank the institutions that have specifically granted me permission to reproduce images from their collections. These include the Bayerische Staatsbibliothek München, the Bibliothèque nationale de France, the British Museum, the Charles E. Young Research Library at UCLA, the Dean and Chapter of Hereford Cathedral, the ETH-Bibliothek Zürich, the Germanisches Nationalmuseum Nürnberg, the Herzog-August-Bibliothek Wolfenbüttel, the Houghton Library at Harvard University, the Metropolitan Museum of Art in New York, the Museo di Capodimonte, the Museo Nacional del Prado, the Niedersächsische Staats- und Universitätsbibliothek Göttingen, the National Aeronautics and Space Administration, the SCALA Archives, the Staatsbibliothek zu Berlin, the Staatliche Bibliothek Regensburg, the Universiteitsmuseum Utrecht, the Warburg Institute, and the William Andrews Clark Memorial Library at UCLA. In this context, I would also like to thank especially Frans van den Hoven of Universiteitsmuseum Utrecht for the aid he lent me in arranging reproductions of Carl Bauer's anthropological globe—all on the basis of a few plaintive emails that I sent from Hamburg. It is always nice to work with a professional.

Having broached the theme of professionalism, I also feel compelled to thank everyone at the University of Pennsylvania Press who helped me to bring this project to its conclusion. First, thanks are due to Sophia Rosenfeld of Yale University, who serves as one of the academic editors for the series Intellectual History of the Modern Age, for having first suggested that I submit this text for review. Second, I owe thanks to my editor, Damon Linker, for having ushered this work through the review process. Third, I would like to thank the anonymous reviewers for not only reading the text carefully but also offering penetrating critiques. Finally, to Erica Ginsburg I owe the profoundest thanks for having overseen the hard work of copyediting my prose.

Most important, I wish to express my thanks to the members of my family who have put up with this project for far too long. To the jolly tag-team of Annabel and Gwendolyn, my own fidgety children, I offer my thanks for their occasionally allowing me to stare into my computer screen just a little longer. Above all others, however, I would like to express my thanks to Allyson Benton, my wife, for accompanying me steadfastly from this fatiguing work's beginnings to its final conclusion. I dedicate this book to her.